Compound-specific Stable Isotope Analysis

Compound-specific Stable Isotope Analysis

Maik A. Jochmann
Instrumental Analytical Chemistry, University of Duisburg-Essen, Essen, Germany
Email: maik.jochmann@uni-due.de

Torsten C. Schmidt
Instrumental Analytical Chemistry and Centre for Water and Environmental Research (ZWU), University of Duisburg-Essen, Essen, Germany
Email: torsten.schmidt@uni-due.de

RSCPublishing

ISBN: 978-1-84973-157-7

A catalogue record for this book is available from the British Library

© The Royal Society of Chemistry, 2012

All rights reserved

Apart from fair dealing for the purposes of research for non-commercial purposes or for private study, criticism or review, as permitted under the Copyright, Designs and Patents Act 1988 and the Copyright and Related Rights Regulations 2003, this publication may not be reproduced, stored or transmitted, in any form or by any means, without the prior permission in writing of The Royal Society of Chemistry or the copyright owner, or in the case of reproduction in accordance with the terms of licences issued by the Copyright Licensing Agency in the UK, or in accordance with the terms of the licences issued by the appropriate Reproduction Rights Organization outside the UK. Enquiries concerning reproduction outside the terms stated here should be sent to The Royal Society of Chemistry at the address printed on this page.

The RSC is not responsible for individual opinions expressed in this work.

Published by The Royal Society of Chemistry,
Thomas Graham House, Science Park, Milton Road,
Cambridge CB4 0WF, UK

Registered Charity Number 207890

Visit our website at www.rsc.org/books

Printed in the United Kingdom by CPI Group (UK) Ltd, Croydon, CR0 4YY, UK

We dedicate this book to our families who have suffered severely from our unavailability during the extended time of book writing. Ina, Ole, Katrin, David, Hannah, Birgit: we promise not to write another one in due time and hope you appreciate the result.

A good world needs knowledge, kindliness, and courage; it does not need a regretful hankering after the past or a fettering of the free intelligence by the words uttered long ago by ignorant men.
                                                                Bertrand Russell

Preface

There has been a rapid growth in the scientific field of stable isotope analysis in the last two decades, to a large extent stimulated by the combination of separation techniques such as gas chromatography (GC) and liquid chromatography (LC) with isotope ratio mass spectrometry (IRMS), which is nowadays commonly called compound-specific stable isotope analysis (CSIA).

Beyond its origin in geochemistry this technique has been established as one of the most powerful analytical tools in diverse branches of science such as geomicrobiology, food science, archaeology and environmental science. Findings largely based on results of CSIA measurements also affect common social issues, such as doping control in sports, food authenticity and forensics. Thus, more and more advanced students and researchers in these areas are confronted with carrying out measurements and interpreting data from stable isotope analysis. However, to date, this is hardly reflected in the curricula of analytical chemistry university courses. One of the pioneers of isotope ratio mass spectrometry, Karl E. Habfast, pointed out that 'isotope ratio mass spectrometers and its associated sample preparation and measuring methods are highly specialized and so demanding that special training and much experience is required from everybody working successfully in this fascinating field'.[1] Furthermore, the hyphenation of isotope ratio mass spectrometry with chromatographic separation requires a firm background in separation fundamentals and practice. Although stable isotope analysis in general has been the subject of several textbooks, in particular in its 'home discipline' geochemistry, CSIA has been treated only marginally in these books. We hope to fill this gap by focusing solely on stable isotope analysis of individual compounds in sometimes complex mixtures. Thereby, the book will fulfil the double function of a companion for a lecture on the subject and a consultant for advanced scientists working in this research area.

The book starts with a brief historical record of the field. In Chapter 2, stable isotopes are explained from scratch and the different ways to express isotope abundances are introduced together with isotope effects and isotopic fractionation as a basis for the use of stable isotope analysis. Chapter 3 gives a detailed account of the principles of technical equipment, starting from the mass spectrometer via the interface components to the gas chromatographic separation. In a separate section, liquid chromatography coupled to IRMS is discussed. Chapter 4 is devoted to the sample and necessary steps prior to the measurement by GC-IRMS or LC-IRMS to ensure accurate isotope analysis including sections on preconcentration and derivatization of analytes. Chapter 5 deals with the very important topics of referencing and calibration in CSIA. This is an issue that differs to a large extent from approaches used in quantitative analysis and thus often is difficult for the newcomer to comprehend. Chapter 6 gives examples of successful applications of CSIA in six different areas: food authenticity, forensics, archaeology, doping control, environmental science, and extraterrestrial materials. Along with the applications, isotope data treatment and presentation are discussed. Emphasis is given to general conclusions that can be drawn based on the presented use of CSIA. Chapter 7 briefly highlights further instrumental developments in the field, focusing on multidimensional chromatography coupled to IRMS, approaches for position-specific isotope analysis in individual compounds and recent developments in online chlorine and bromine isotope analysis in organic molecules.

As in all scientific areas, there is, nowadays, a multitude of information available on the internet. Therefore, we have picked a few interesting web links at the end of each section that can be used for further information or to give a deeper insight in specific topics.

Book writing is a bit like diving for wrecks of ships with valuable loads. You never know exactly what you will encounter but in the best case you surely find something worthy or even a treasure in your endeavours. We have certainly experienced that feeling during creation of this monograph. Despite our long experience in stable isotope analysis we certainly have benefited most ourselves by writing this book. We found basic principles revoked that may lead us to better acknowledge fundamental developments in the area and point us to new research directions. We saw exciting new applications of compound-specific stable isotope analysis that open new opportunities for further work. Our hope is that our still existing fascination and enthusiasm on CSIA is

Preface ix

conveyed to the readers of the book but that, at the same time, they realize with us the current limits of the method that need to be pushed further. Maybe some of the younger readers will pursue crossing these borders in their career and find themselves as much attracted by CSIA as we are.

Acknowledgements

We are indebted to Alfred V. Hirner (Environmental Analytical Chemistry, University of Duisburg-Essen) who provided Section 6.7 on extraterrestrial matter.

We thank Bernd Malonn (Thermo Fisher Scientific, Dreieich) for discussions about technical matters whenever needed.

The following colleagues reviewed sections of this textbook at various stages during its generation:

Chapter 1: Charles B. Douthitt (Thermo Fisher Scientific), Roland Diaz-Bone, Dorothea M. Kujawinski, Ursula Telgheder (all Instrumental Analytical Chemistry, University of Duisburg-Essen)

Chapter 3: Willi A. Brand (Max-Planck-Institute for Biogeochemistry, Jena), Charles B. Douthitt (Thermo Fisher Scientific), Armin Meyer (Isodetect GmbH, Munich), Wolfgang Schrader (Max-Planck-Institut für Kohleforschung, Mülheim a.d. Ruhr), Lijun Zhang (Instrumental Analytical Chemistry, University of Duisburg-Essen)

Chapter 4: Thomas B. Hofstetter (eawag, Dübendorf), Michaela Blessing (BRGM, Orléans), Wolfram Meier-Augenstein (Stable Isotope Laboratory, Scottish Crop Research Institute, Dundee), Thomas Piper (Swiss Laboratory for Doping Analysis, Lausanne)

Chapter 5: Arndt Schimmelmann (Department of Geological Sciences, Indiana University in Bloomington, Indiana), Willi A. Brand (Max-Planck-Institute for Biogeochemistry, Jena), Tyler B. Coplen (U.S. Geological Survey, Reston, Virginia)

Chapter 6: (Food, Forensics, Archaeology, Doping) Sabine Schneiders (Bundeskriminalamt, Wiesbaden) (Food) Dieter Juchelka (Thermo Fisher Scientific, Bremen), Markus Greule (Max-Planck-Institute for Chemistry, Mainz), (Forensics, Archaeology) Wolfram Meier-Augenstein, (Doping) Mario Thevis (German Sport University, Cologne), (Environment) Thomas B. Hofstetter, Michaela Blessing, (Extraterrestrial matter) Martin Sulkowksi (Environmental Analytical Chemistry, University of Duisburg-Essen).

We are very grateful for the time and effort they spent on commenting our drafts, thus substantially improving content and avoiding some errors and misunderstandings. Nevertheless, we as the authors are fully responsible for any remaining flaws of this work and welcome any future critical comments and suggestions. Finally, we want to thank our current students who contributed exemplary measurements, text drafts, numerous reviews and discussions. In particular, we acknowledge the input by Daniel Tabersky (now at Eidgenössische Technische Hochschule Zürich, Zurich), Andreas Kremser, Dorothea M. Kujawinski, Kamil Michalski, Marcel Schulte, Jens-Benjamin Wolbert, Oliver Würfel and Lijun Zhang (all Instrumental Analytical Chemistry, University of Duisburg-Essen).

REFERENCE

1. I. T. Platzner, K. Habfast, A. J. Walder and A. Goetz, *Modern Isotope Ratio Mass Spectrometry*, J. Wiley, Chichester, New York, 1997.

Contents

Chapter 1	**Introduction to Compound-specific Isotope Analysis**	1
	1.1 Development of Stable Isotope Analysis	1
	1.2 Instrumentation for SIA	2
	1.2.1 Mass Spectrometric Methods for SIA	2
	1.2.2 Classification of SIA Techniques	3
	1.2.3 Principle of BSIA and CSIA	5
	1.3 Historical Development of CSIA by GC- and LC-IRMS	7
	1.4 Spectroscopic Methods for SIA	9
	1.5 Web Findings	9
	References	10
Chapter 2	**Fundamental Aspects of Stable Isotopes and Isotopic Fractionation**	14
	2.1 The Atom Model and Nuclides	14
	2.2 Stable Isotopes	19
	2.2.1 Isotope Stability and Abundance	19
	2.2.2 Relative Atomic Mass	21
	2.2.3 Isotopologues, Isotopomers and Isotopocules	22
	2.3 Natural Abundance and How to Express It	24
	2.3.1 Isotope-amount Fraction	24
	2.3.2 Variations of Natural Isotopic Abundance	25
	2.3.3 Isotope Ratios	26
	2.3.4 The δ-Scale and its Limits	27

Compound-specific Stable Isotope Analysis
By Maik A. Jochmann and Torsten C. Schmidt
© The Royal Society of Chemistry 2012
Published by the Royal Society of Chemistry, www.rsc.org

2.4	Isotope Fractionation and How to Express It	30
	2.4.1 Thermodynamic and Kinetic Isotope Effects	30
	2.4.2 Isotopic Fractionation Factor, Isotopic Fractionation and Isotopic Difference	32
2.5	Isotope Mass Balances	35
2.6	Physical Kinetic Isotope Fractionation during Transport Processes	36
2.7	Isotopic Fractionation under Various System Conditions	39
	2.7.1 Reversible Reaction in a Closed System	40
	2.7.2 Irreversible Reaction in a Closed System	42
	2.7.3 Reversible Reaction in an Open System	45
	2.7.4 Irreversible Reaction in an Open System	45
2.8	Web Findings	46
References		47

Chapter 3 Instrumentation for Compound-specific Stable Isotope Analysis — 50

3.1	Isotope Ratio Mass Spectrometer	50
	3.1.1 Vacuum System	53
	3.1.2 Ion Source and Focusing Beam Optics	56
	3.1.3 Ion Separation in the Magnet Sector Field	63
	3.1.4 Ion Collection and Signal Pathway	70
	3.1.5 Ion Corrections	77
3.2	Inlet Systems	83
	3.2.1 Dual-inlet Systems	83
	3.2.2 Continuous-flow Inlet Systems	86
	3.2.3 Open-split Systems	86
	3.2.4 Reference Gas Inlet	89
3.3	Peripheral Devices for Continuous-flow IRMS	91
3.4	CSIA by GC-IRMS	93
	3.4.1 Carbon and Nitrogen Isotope Ratio Analysis	94
	3.4.2 Hydrogen Isotope Ratio Analysis	106
	3.4.3 Oxygen Isotope Ratio Measurement	115
3.5	Chromatographic Aspects of CSIA	115
	3.5.1 Separation Basics	115
	3.5.2 Peak Detection in CSIA	127
	3.5.3 Carrier Gas, Carrier Gas Flow and GC Columns for CSIA	129
	3.5.4 Multidetection	133
3.6	LC-IRMS	134
3.7	Web Findings	147
References		147

Contents   xiii

Chapter 4 Sample Preparation in Compound-specific Stable Isotope Analysis   **155**

    4.1 Scope   155
    4.2 Sampling, Sample Preservation and Storage   155
    4.3 Sample Processing   157
        4.3.1 General Remarks   157
        4.3.2 Sample Processing for Volatile Compounds   159
        4.3.3 Sample Processing for Semi-volatile Compounds   161
    4.4 Derivatization   168
        4.4.1 General Considerations and Corrections   168
        4.4.2 Overview of Derivatization Reactions   174
    References   180

Chapter 5 Referencing Strategies and Quality Assurance for Compound-specific Stable Isotope Analysis   **185**

    5.1 Accuracy, Uncertainty, Precision and Error   185
    5.2 International Primary Reference Materials and Certified Reference Materials   187
        5.2.1 Overview   187
        5.2.2 Hydrogen and Oxygen Calibration and Reference Materials   190
        5.2.3 Carbon Calibration and Reference Materials   191
        5.2.4 Nitrogen Calibration and Reference Materials   197
        5.2.5 Sulfur Calibration and Reference Materials   198
        5.2.6 Chlorine and Bromine Calibration and Reference Materials   199
    5.3 Normalization of Stable Isotope Data   200
    5.4 Referencing in CSIA   204
    5.5 Repeatability, Reproducibility, Linearity, Stability, Detection Limits and 'Total Uncertainty'   211
    5.6 Quality Assurance (QA) and Quality Control (QC) in CSIA   218
    5.7 Web Findings   221
    References   223

Chapter 6 Applications of Compound-specific Stable Isotope Analysis   **230**

    6.1 Scope   230
    6.2 Authenticity of Food and Related Commodities   234
        6.2.1 Scope   234
        6.2.2 Isotope Fractionation during Carbon Fixation in Photosynthesis   235

	6.2.3	Aromas, Flavours and Essential Oils	240
	6.2.4	Alcoholic Drinks	249
	6.2.5	Honey	253
	6.2.6	Juice	254
	6.2.7	Oils	256
	6.2.8	Natural Stimulants: Cocoa, Coffee, Tea	258
6.3	Forensics		259
6.4	Archaeology		264
6.5	Doping Control		273
	6.5.1	Analytical Procedure for Isotope Analysis of Steroids	277
	6.5.2	Further Developments	279
6.6	Environmental Science		280
	6.6.1	Scope	280
	6.6.2	Source Values of Organic Compounds	280
	6.6.3	Source Apportionment	283
	6.6.4	Use of Isotope Fractionation in Environmental Systems	295
	6.6.5	Further Reading	317
6.7	Extraterrestrial Matter		317
6.8	General Conclusions		326
6.9	Web Findings		329
	6.9.1	General Links	329
	6.9.2	European Isotope Application Projects	330
References			330

Chapter 7 Further Developments in Compound-Specific Isotope Analysis — 349

7.1	Scope	349
7.2	Multidimensional GC (GC-GC), Comprehensive GC (GC×GC)	349
7.3	CSIA by Multicollector Inductively Coupled Plasma Mass Spectrometry	353
7.4	CSIA of Chlorine- and Bromine-Containing Compounds	356
7.5	Position-specific Isotope Ratio Analysis	359
References		361

Appendix — 364

Prefixes	364
SI-units Used in this Book	364

Physical Constants 365
Useful Conversions in the Lab 365
Important ions 366
Reference 366

Subject Index **367**

CHAPTER 1
Introduction to Compound-specific Isotope Analysis

1.1 DEVELOPMENT OF STABLE ISOTOPE ANALYSIS

Since the 1940s,[1] stable isotope analysis (SIA) has found widespread application in various branches of science. First applications of SIA were mainly geochemical in nature and focused on fundamental aspects such as isotope variations caused by differences in properties of isotopes relating to (i) thermodynamics, (ii) chemical kinetics, (iii) their masses such as in diffusion processes, and (iv) the forces between atoms (thermal diffusion processes).[2] Also during that time, the theoretical underpinning of these processes was developed by Urey, Bigeleisen and others.[3-5] Furthermore, the precision of isotope abundance determinations was gradually improved[1] by changes to isotope ratio mass spectrometers (IRMS), including multiple Faraday cup collector systems and better amplifier electronics.[6,7] Groundbreaking applications of SIA during this time include a paleotemperature scale based on $^{18}O/^{16}O$ isotope ratio measurements of fossil carbonate.[1,8,9] The observation by Nier and Gulbransen in 1939 that carbon in nature varies in isotope composition, with living organisms and their remains such as coal, natural gas and petroleum containing less ^{13}C than carbonate of limestone, opened the way to using isotopes to study biogeochemical processes and interactions.[10,11] The isotope abundances of nitrogen, oxygen, hydrogen and sulfur caused by exchange processes gave fundamental insights into the evolution of the Earth's crust and atmosphere

---

Compound-specific Stable Isotope Analysis
By Maik A. Jochmann and Torsten C. Schmidt
© The Royal Society of Chemistry 2012
Published by the Royal Society of Chemistry, www.rsc.org

as well as the history and origin of life.[2] It was also recognized early on that variations in natural abundance of the stable isotopes caused by isotope fractionation processes could be utilized as nonradioactive natural or artificial tracers to follow complex geochemical and biological processes in geological cycles, ecosystems, organisms and chemical reactions.[1,12]

1.2 INSTRUMENTATION FOR SIA

1.2.1 Mass Spectrometric Methods for SIA

Nowadays, a variety of techniques and instrumentation using mass spectrometry or spectroscopic methods for the determination of isotope ratios are available. Mass spectrometry is still the most important technique, both for heavy and light elements.[13]

While in general all mass spectrometers are able to measure isotope abundances, dedicated mass spectrometers with precisions in the order of 10^{-4} to 10^{-6} are mandatory for isotope ratio determination at natural abundance level.[13,14] Such high precision can be obtained by magnetic sector field instruments with Faraday cups enabling simultaneous detection of ion currents from the different mass-to-charge ratios for isotope ratio determination. For heavier elements, thermal ionization[15] and, more recently, inductively coupled plasmas are used as ion sources (thermal ionization mass spectrometry, TIMS,[15] multicollector inductively coupled plasma mass spectrometry, MC-ICP-MS).[16–18] In this book, we mainly focus on the lighter elements carbon ($^{13}C/^{12}C$), nitrogen ($^{15}N/^{14}N$), oxygen ($^{18}O/^{16}O$), sulfur ($^{34}S/^{32}S$) and hydrogen ($^{2}H/^{1}H$), which represent the main elements in biological systems. These elements are typically introduced into an IRMS after conversion to low molecular weight gases such as CO_2, N_2, CO, SO_2 and H_2, which are ionized by electron impact in a tight gas source (see Chapter 3). This fundamental IRMS design was developed by Alfred Nier (see Figure 1.1) and co-workers in the 1940s and is, in principle, still the basis for all modern instruments.[6,19,20] An interesting biographical review about Alfred Nier and his mass spectrometer developments was written by De Laeter and Kunz.[21] A more detailed discussion of the mass spectrometer instrumentation is given in Chapter 3.

The conversion of analytes to gaseous form was traditionally carried out 'off-line'; for organic matter, this typically involved combustion or reduction in sealed quartz tubes (tube combustion), with the products cryogenically purified in vacuum lines and transferred to the IRMS via the 'dual viscous flow inlet system' or 'dual inlet'.[22] The dual-inlet was

Introduction to Compound-specific Isotope Analysis

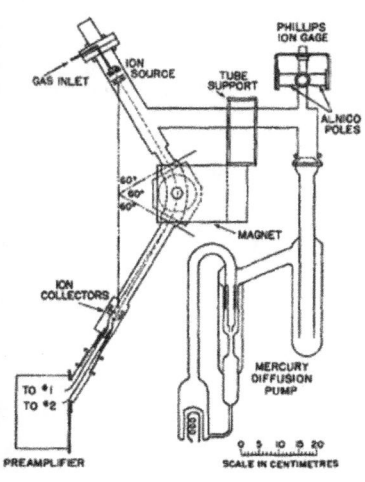

Figure 1.1 Nier's mass spectrometer.
The photograph on the left was taken in 1940 and shows Alfred O. Nier, *1911–†1994, with the glass mass spectrometer with which he conducted measurements relating to uranium fission. On the right side a schematic diagram of Nier's 60° magnetic sector field mass spectrometer is depicted.[6] Reprinted with permission of the American Institue of Physics. (Picture and graphic from De Laeter and Kurz reprinted with permission of Wiley InterScience.)

originally introduced by Murphey for thermal gas diffusion investigations.[23] Its incorporation in the IRMS by McKinney and co-workers[20] can be considered as the birth of high precision IRMS.[14] With modern dual-inlet systems, as they will be described in Chapter 3, relative ratios at highest precisions (<0.1‰) for the biologically relevant elements discussed here can be achieved.[14] Until the mid-1970s, isotope ratio analysis was carried out exclusively by gas IRMS using such dual-inlet systems.

Today one can distinguish between the dual-inlet or viscous flow inlet system in which the pure sample gas is introduced into the IRMS and 'continuous flow'-IRMS (CF-IRMS) in which the sample gas is introduced via an inert helium carrier gas stream into the IRMS ion source.

1.2.2 Classification of SIA Techniques

SIA techniques can be classified according to the kind of sample (bulk sample or individual compounds), of which the isotope ratio is analysed (see Figure 1.2).

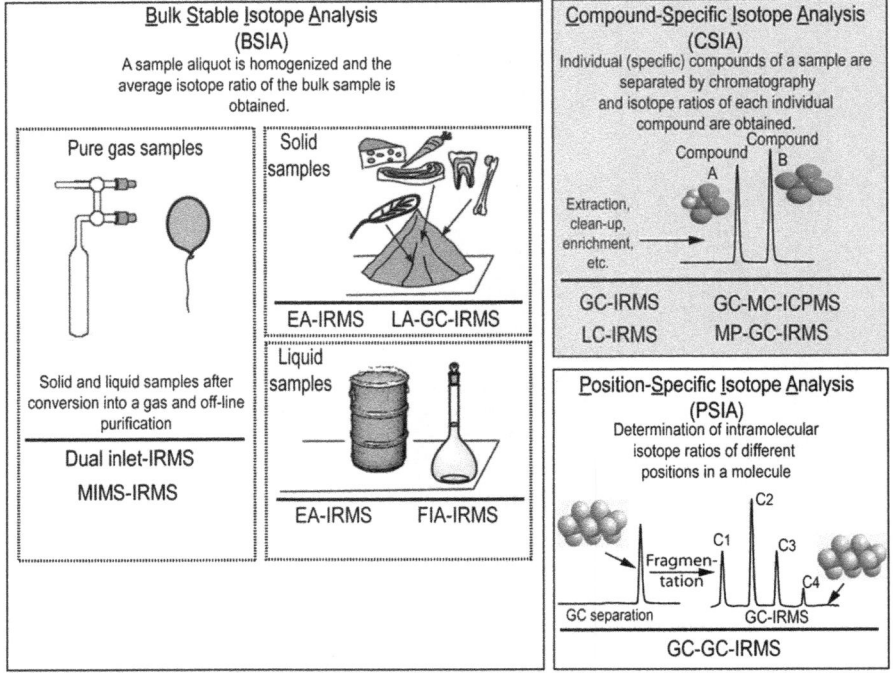

Figure 1.2 Differentiation between different techniques used for stable isotope ratio analysis by isotope ratio mass spectrometry. For a description see text.

In 'bulk stable isotope ratio analysis' (BSIA), the isotope ratio of the bulk sample is measured. If isotope ratios of individual compounds are analysed, the technique is referred to as 'compound-specific isotope analysis' (CSIA). Another interesting application of stable isotope analysis is the intramolecular measurement of isotope ratios of certain atoms within a molecule. Following the precedent set by CSIA, this technique is termed 'position-specific isotope analysis' (PSIA).[14]

It has to be mentioned that a variety of sample preparation devices for special purposes have been developed. Because the focus of this book is on gas chromatography (GC) and liquid chromatography (LC) in combination with IRMS we will mention only a few techniques and refer to the literature[24–26] for more detailed information on special devices, for example, for atmospheric gases, carbonates and water. The bulk analysis of water is carried out with equilibrium devices for controlled oxygen and hydrogen exchange[24] or with temperature conversion of water and organic materials on glassy carbon at temperatures $> 1400\,°C$.[27,28]

For the analysis of trace gases, pre-concentration devices[29] as well as membrane inlet devices (MIMS-IRMS)[30,31] and membrane permeation gas chromatography isotope ratio mass spectrometry (MP-GC-IRMS)[32] are used. The measurement of spatially narrow sample compartments, such as tree rings is nowadays carried out by laser ablation prior to GC-IRMS (LA-GC-IRMS).[33]

1.2.3 Principle of BSIA and CSIA

Continuous flow-IRMS can be hyphenated with gas chromatographs or, more recently, liquid chromatographs in order to analyse isotope ratios of individual compounds in a complex mixture. In analogy to 'CSIA', the term 'bulk stable isotope ratio analysis' (BSIA) was coined by K. Habfast.[24,34,35] In Figure 1.3, the fundamental difference between BSIA by EA-IRMS and CSIA by GC-IRMS is illustrated. In BSIA, the

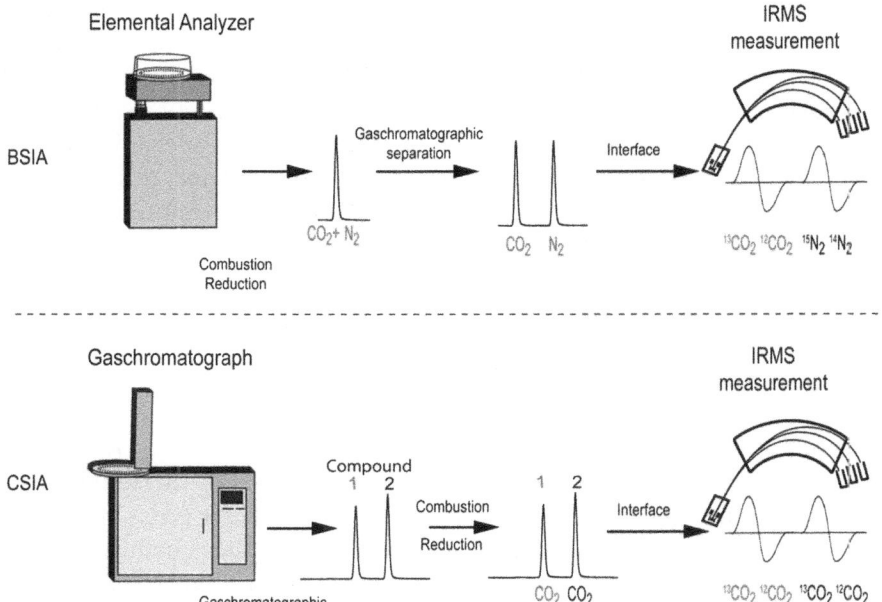

Figure 1.3 Comparison of bulk and compound-specific isotope ratio analysis. In BSIA (upper scheme) carbon and nitrogen of the bulk sample are first converted to CO_2 and N_2 via combustion and reduction in an elemental analyser (EA). Then, the developed gases are separated by GC and introduced into the IRMS via an interface or inlet system. In CSIA (lower scheme), complex mixtures of compounds are first separated via GC or LC. Combustion and reduction of the separated compounds is conducted on-line prior to introduction into an IRMS.

compounds are converted to low molecular weight gases (CO_2, N_2, CO, SO_2 and H_2) prior to separation by GC, thus allowing the analysis of isotope ratios of multiple elements, but only of the entire sample,[36] in contrast to CSIA, in which compounds in complex mixtures are first separated by GC or LC prior to conversion to low molecular weight gases and subsequent introduction into an IRMS. CSIA allows the differentiation of isotope ratios of individual compounds, but the isotope ratio of only one element at a time can be measured for the separated compounds.

An advantage of CSIA by GC- or LC-IRMS over BSIA is the ease and efficiency of on-line sample preparation and transfer, the ability to measure all compounds in a complex sample mixture in a single data acquisition run, and the significantly reduced sample size requirements.[37]

In the literature, various terms for hyphenation of GC or LC with IRMS are used, which can be confusing for users unfamiliar with this nomenclature. Here, reviews by Douthitt[36] and Sessions[22] provide insights into the nomenclature of CSIA instrumentation. Matthews and Hayes coined the term 'isotope ratio-monitoring GC-MS' (irm-GC-MS),[22,36] by analogy with the molecular GC-MS technique of selected-ion-monitoring, 'SIM'.[14] Although often used, we will not use this term in this book since it could easily be mistaken by non-specialists as just another detection mode of an organic mass spectrometer, thus clouding the fundamental differences with an IRMS and the requirements for high precision measurements at natural isotope abundance.

The term 'GC-IRMS' was introduced in the 1980s[22] and relates to the coupling of gas chromatography with an IRMS as detector. Other acronyms used today describe the nature of the post GC chemical conversion, including 'C', for combustion (GC-C-IRMS), 'P' or 'py' for pyrolysis (GC-P-IRMS, GC-py-IRMS) or, for direct introduction without modification, GP for 'general purpose' (GC-GP-IRMS). Thermo Fisher Scientific uses 'thermochemolysis' and 'thermal conversion' (TC) in the names of their commercial products (GC-TC-IRMS, TC/EA)[22] as a more accurate description of the chemistry often described as pyrolysis or (incorrectly) as 'chromium pyrolysis'. To prevent misunderstandings we will generally use GC- or LC-IRMS within this book and mention explicitly the elements measured.

The term 'compound-specific isotope analysis' (CSIA) was proposed by Martin Schoell in the title of a workshop at the 203rd ACS meeting in San Francisco in 1992.[38,39] However, Hayes *et al.* had already used this term in the title of the publication 'Compound-specific isotopic analyses: A novel tool for reconstruction of ancient biogeochemical processes' in 1990.[40] Although CSIA, in principle, relates to all methods for determination of isotope ratios of single compounds, including off-line isolation of compounds from mixtures[41] or bulk analysis of pure substances, it is most frequently restricted to isotope analysis of individual compounds separated by chromatographic methods coupled on-line to an IRMS. In this book we will follow that terminology and focus on CSIA in this sense.

1.3 HISTORICAL DEVELOPMENT OF CSIA BY GC- AND LC-IRMS

Stimulated by the advances of gas chromatography coupled to organic mass spectrometry (GC-MS) in the second half of the 20th century, attempts were made to combine separation techniques such as GC and LC with an IRMS.

In 1976, Sano et al. studied volatile metabolites after administration of 100 ng ^{13}C-labelled aspirin after a 24 h urine collection. To that end, the metabolites were separated by GC and combusted to CO_2 using an on-line combustion reactor. Isotope ratios of m/z $^{44}CO_2$ and $^{45}CO_2$ were measured by quadrupole-MS alternately in 0.5 s intervals.[42,43]

Two years later, Matthews and Hayes coupled a conventional single-collector sector field mass spectrometer with a GC. They used a copper-oxide-packed combustion furnace at 750 °C inserted between the gas chromatographic column outlet and a GC-MS interface attached to a computer-controlled beam-switching isotope ratio mass spectrometer.[44] By monitoring relevant ion currents of N_2 and CO_2 for continuous measurement of $^{15}N/^{14}N$ or $^{13}C/^{12}C$ ratios, directly comparable isotope ratios for all eluting compounds were obtained, regardless of composition and mass spectrometric behaviour of the parent compounds. Carbon and nitrogen isotope ratios could be measured with a precision of 5‰ or better with 20 nmol of CO_2 or 100 nmol of N_2. The obtained precisions approached the required precisions for natural abundance variation studies.

In 1984, Barrie from VG and Bricout and Koziet from the research laboratories of Pernod-Ricard in Paris coupled the first on-line combustion interface (a quartz furnace filled with Co_3O_4 at 700 °C) between a capillary GC and a dual collector isotope ratio mass spectrometer,[45] thus permitting continuous recording of isotope ratios by detecting two successive masses at the same time,[43,46] measuring carbon isotope ratios from 8 nmol CO_2 to precisions < 1‰. A vent valve at the end of the GC column acted as backflush to prevent solvent peaks from entering the combustion oven.[47] The aim of this coupling was to generate data useful for the authenticity control of flavour compounds and ethanol.[47]

GC-IRMS instrumentation has been commercially available since 1988, when devices for coupling GC with an IRMS were introduced at the 11th International Mass Spectrometry Conference in Bordeaux.[47]

Two systems that demonstrated GC-IRMS determination of nitrogen isotope ratios for derivatized amino acids were introduced in 1992.[22] Preston and Slater presented a system with a combustion furnace, liquid nitrogen cold trap for trapping interfering CO_2 and water, as well as a porous layer open tubular 'PLOT' column to resolve N_2 from any CO formed by poor conversion.[48] Merritt and Hayes presented a comparable system but with an additional reduction furnace, loaded with Cu wires maintained at 600 °C, to reduce N-oxides to N_2. They reported a precision of 0.2‰ for a sample size of 2 nmol of an amino acid.[49]

In 1994, Brand *et al.* introduced a GC-IRMS system for CSIA of oxygen by converting oxygen-containing organic compounds on-line to CO by a pyrolytic reaction (the so called 'Unterzaucher reaction') in a high temperature micro-furnace.[38] An interface for oxygen isotope ratio measurements with CSIA has been commercially available since 1996.

Compound-specific isotope ratio analysis of hydrogen by GC-IRMS posed a number of analytical challenges, discussed in detail in Chapter 3. Tobias and Brenna demonstrated CSIA for hydrogen by using an on-line combustion micro-reactor filled with CuO and held at 850 °C followed by a reduction reactor filled with nickel metal held at 950 °C.[50] Burgoyne and Hayes showed that quantitative pyrolysis can be achieved without metal reductants by using a carbon-lined non-porous alumina tube reactor heated to temperatures > 1440 °C,[51] a method that was commercialized in 1998.[22]

Apart from chlorine and bromine isotope ratio determination by GC-IRMS,[52–54] recently, GC-quadrupole-MS was used for the determination of chlorine isotope ratios at natural abundance level with acceptable levels of precision for environmental degradation studies.[55–58] Additionally, GC-MC-ICP-MS has been applied to determine chlorine[59] and bromine[60,61] isotope ratios of organic compounds. So far, no CSIA of sulfur by GC-IRMS is available, although ^{34}S-CSIA by GC-MC-ICP-MS has been reported.[62–64] A more detailed discussion can be found in Chapter 7.

Gas chromatography is restricted to GC-compatible compounds, which can be transferred into the gas phase without thermal degradation. For other compounds, derivatization into a GC-amenable form is necessary. For large molecules, or if isotope fractionation by kinetic isotope effects during derivatization is unavoidable, LC is the method of choice for separation. Several attempts were made over the last decades to combine LC with IRMS, including a chemical reaction interface[65] and a moving wire[66–69] to remove the water and/or organic molecules in the mobile phase, but these were not commercialized. The first commercially available instrument was introduced in 2004 and is based on wet chemical oxidation (peroxodisulfate and concentrated phosphoric acid) for the conversion of carbon in organic molecules to CO_2 after their elution from the LC column.[70] This method is restricted to an aqueous mobile phase without organic modifiers or solvents and can only be used for carbon isotope ratio measurements.[13,71]

The interface can also be used without a chromatographic column to measure isotopic signatures of water-soluble pure substances by flow injection analysis (FIA-IRMS).[70] A detailed discussion of LC-IRMS is given in Chapter 3.

1.4 SPECTROSCOPIC METHODS FOR SIA

A widely applied method to determine isotope ratios of pure liquid substances (for example, ethanol from wine samples) is site-specific natural isotope fractionation-nuclear magnetic resonance spectroscopy (SNIF-NMR).[72,73] So far, on-line coupling of SNIF-NMR with chromatographic techniques has not been realized and is beyond the scope of this book.

Recently, laser spectroscopy for isotope ratios is an emerging field of investigation and instrumental development. The near future will show which role these methods will play apart from already established isotope ratio measurements of water and carbon dioxide.[74–76] However, these spectroscopic methods are also beyond the scope of this book.

1.5 WEB FINDINGS

The Scripps Center for Metabolomics and Mass Spectrometry
The home page offers an interactive view of the history of mass spectrometry with links to key publications.
http://masspec.scripps.edu

National Academy of Sciences
Here one can find a detailed biography of Alfred O. C. Nier by John H. Reynolds.
http://www.nap.edu/readingroom.php?book=biomems&page=anier.html

The Official Web Site of the Nobel Prize
The site provides a biography of Harold C. Urey and a link to the Nobel Prize lecture of the year 1934.
 http://www.nobelprize.org/nobel_prizes/chemistry/laureates/1934/urey-bio.html

REFERENCES

1. H. C. Urey, *Science*, 1948, **108**, 489–496.
2. K. I. Mayne, *Rep. Prog. Phys.*, 1952, **15**, 24–48.
3. H. C. Urey, *J. Chem. Soc.*, 1947, 562–581.
4. J. Bigeleisen and M. G. Mayer, *J. Chem. Phys.*, 1947, **15**, 261–267.
5. J. Bigeleisen, *T. New York Acad. Sci.*, 1953, **16**, 823–828.
6. A. O. Nier, *Rev. Sci. Instrum.*, 1947, **18**, 398–411.
7. H. G. Thode and R. B. Shields, *Rep. Prog. Phys.*, 1948, **12**, 1–21.
8. H. C. Urey, S. Epstein, C. McKinney and J. McCrea, *Geol. Soc. Am. Bull.*, 1948, **59**, 1359–1360.
9. H. C. Urey, S. Epstein, H. A. Lowenstam and C. R. McKinney, *Science*, 1950, **111**, 462–463.
10. A. O. Nier and E. A. Gulbransen, *J. Am. Chem. Soc.*, 1939, **61**, 697–698.
11. B. F. Murphey and A. O. Nier, *Phys. Rev.*, 1941, **59**, 771–772.
12. J. Bigeleisen, *Science*, 1949, **110**, 14–16.
13. J.-P. Godin, L.-B. Fay and G. Hopfgartner, *Mass Spectrom. Rev.*, 2007, **26**, 751–774.
14. J. T. Brenna, T. N. Corso, H. J. Tobias and R. J. Caimi, *Mass Spectrom. Rev.*, 1997, **16**, 227–258.
15. T. Walczyk, *Anal. Bioanal. Chem.*, 2004, **378**, 229–231.
16. M. Moldovan, E. M. Krupp, A. E. Holliday and O. F. X. Donard, *J. Anal. At. Spectrom.*, 2004, **19**, 815–822.
17. M. E. Wieser and J. B. Schwieters, *Int. J. Mass Spectrom.*, 2005, **242**, 97–115.
18. F. Vanhaecke, L. Balcaen and D. Malinovsky, *J. Anal. At. Spectrom.*, 2009, **24**, 863–886.
19. A. O. Nier, *Rev. Sci. Instrum.*, 1940, **11**, 212–216.
20. C. R. McKinney, J. M. McCrea, S. Epstein, H. A. Allen and H. C. Urey, *Rev. Sci. Instrum.*, 1950, **21**, 724–730.
21. J. De Laeter and M. D. Kurz, *J. Mass Spectrom.*, 2006, **41**, 847–854.
22. A. L. Sessions, *J. Sep. Sci.*, 2006, **29**, 1946–1961.
23. B. F. Murphey, *Phys. Rev.*, 1947, **72**, 834–837.
24. I. T. Platzner, K. Habfast, A. J. Walder and A. Goetz, *Modern Isotope Ratio Mass Spectrometry*, J. Wiley, Chichester; New York, 1997.

25. P. A. de Groot, ed., *Handbook of Stable Isotope Analytical Techniques Vol. I: 1*, Elsevier Science, Amsterdam, 2004.
26. P. A. de Groot, ed., *Handbook of Stable Isotope Analytical Techniques Vol. II: 2*, Elsevier Science, Amsterdam, 2008.
27. M. Gehre, in *New Approaches for Stable Isotope Ratio Measurements, Proceedings of an Advisory Group meeting held in Vienna, 20–23 September 1999, IAEA-TECDOC-1247*, Vienna, 2001, pp. 33–38.
28. M. Gehre and G. Strauch, *Rapid Commun. Mass Spectrom.*, 2003, **17**, 1497–1503.
29. W. A. Brand, *Isot. Environ. Health Stud.*, 1994, **31**, 277–284.
30. L. K. Smith, M. A. Voytek, J. K. Böhlke and J. W. Harvey, *Ecol. Appl.*, 2006, **16**, 2191–2207.
31. W. Eschenbach and R. Well, *Rapid Commun. Mass Spectrom.*, 2011, **25**, 1993–2006.
32. P. Tremblay, M. M. Savard, A. Smirnoff and R. Paquin, *Rapid Commun. Mass Spectrom.*, 2009, **23**, 2213–2220.
33. B. Schulze, C. Wirth, P. Linke, W. A. Brand, I. Kuhlmann, V. Horna and E. D. Schulze, *Tree Physiol.*, 2004, **24**, 1193–1201.
34. R. A. Werner and W. A. Brand, *Rapid Commun. Mass Spectrom.*, 2001, **15**, 501–519.
35. W. A. Brand, *Personal Communication*, 2011.
36. C. B. Douthitt, *Analusis*, 1999, **27**, 197–199.
37. B. Sherwood-Lollar, S. K. Hirschorn, M. M. G. Chartrand and G. Lacrampe-Couloume, *Anal. Chem.*, 2007, **79**, 3469–3475.
38. W. A. Brand, A. R. Tegtmeyer and A. Hilkert, *Org. Geochem.*, 1994, **21**, 585–594.
39. M. Schoell and J. M. Hayes, *Org. Geochem.*, **21**, R5.
40. J. M. Hayes, K. H. Freeman, B. N. Popp and C. H. Hoham, *Org. Geochem.*, 1990, **16**, 1115–1128.
41. E. Lichtfouse, *Rapid Commun. Mass Spectrom.*, 2000, **14**, 1337–1344.
42. M. Sano, Y. Yotsui, H. Abe and S. Sasaki, *Biomed. Mass Spectrom.*, 1976, **3**, 1–3.
43. W. A. Brand, *J. Mass Spectrom.*, 1996, **31**, 225–235.
44. D. E. Matthews and J. M. Hayes, *Anal. Chem.*, 1978, **50**, 1465–1473.
45. A. Barrie, J. Bricout and J. Koziet, *Biomed. Mass Spectrom.*, 1984, **11**, 583–588.
46. W. Meier-Augenstein, *J. Chromatogr. A*, 1999, **842**, 351–371.
47. W. A. Brand, in *Adv. Mass Spectrom.*, ed. E. J. Karjalainen, A. E. Hesso, J. E. Jalonen and U. P. Karjalainen, Elsevier Science Publishers B. V., Amsterdam, 1998, **14**.
48. T. Preston and C. Slater, *P. Nutr. Soc.*, 1994, **53**, 363–372.

49. D. A. Merritt and J. M. Hayes, *J. Am. Soc. Mass Spectrom.*, 1994, **5**, 387–397.
50. H. J. Tobias and J. T. Brenna, *Anal. Chem.*, 1996, **68**, 3002–3007.
51. T. W. Burgoyne and J. M. Hayes, *Anal. Chem.*, 1998, **70**, 5136–5141.
52. O. Shouakar-Stash, R. J. Drimmie, M. Zhang and S. K. Frape, *Appl. Geochem.*, 2006, **21**, 766–781.
53. O. Shouakar-Stash, S. K. Frape, R. Aravena, A. Gargini, M. Pasini and R. J. Drimmie, *Environ. Forensics*, 2009, **10**, 299–306.
54. H. Holmstrand, M. Unger, D. Carrizo, P. Andersson and O. Gustafsson, *Rapid Commun. Mass Spectrom.*, 2010, **24**, 2135–2142.
55. K. Sakaguchi-Soder, J. Jager, H. Grund, F. Matthaus and C. Schuth, *Rapid Commun. Mass Spectrom.*, 2007, **21**, 3077–3084.
56. C. Aeppli, H. Holmstrand, P. Andersson and O. Gustafsson, *Anal. Chem.*, 2010, **82**, 420–426.
57. B. Jin, C. Laskov, M. Rolle and S. B. Haderlein, *Environ. Sci. Technol.*, 2011, **45**, 5279–5286.
58. A. Bernstein, O. Shouakar-Stash, K. Ebert, C. Laskov, D. Hunkeler, S. Jeannottat, K. Sakaguchi-Söder, J. Laaks, M. A. Jochmann, S. Cretnik, J. Jager, S. B. Haderlein, T. C. Schmidt, R. Aravena and M. Elsner, *Anal. Chem.*, 2011, **83**, 7624–7634.
59. M. Van Acker, A. Shahar, E. D. Young and M. L. Coleman, *Anal. Chem.*, 2006, **78**, 4663–4667.
60. S. P. Sylva, L. Ball, R. K. Nelson and C. M. Reddy, *Rapid Commun. Mass Spectrom.*, 2007, **21**, 3301–3305.
61. F. Gelman and L. Halicz, *Int. J. Mass Spectrom.*, 2010, **289**, 167–169.
62. A. Amrani, A. Sessions and J. Adkins, *Anal. Chem.*, 2009, **81**, 9027–9034.
63. R. Santamaria-Fernandez and R. Hearn, *Rapid Commun. Mass Spectrom.*, 2008, **22**, 401–408.
64. R. Santamaria-Fernandez, R. Hearn and J. C. Wolff, *J. Anal. At. Spectrom.*, 2008, **23**, 1294–1299.
65. F. P. Abramson, G. E. Black and P. Lecchi, *J. Chromatogr. A*, 2001, **913**, 269–273.
66. R. J. Caimi and J. T. Brenna, *Anal. Chem.*, 1993, **65**, 3497–3500.
67. R. J. Caimi and J. T. Brenna, *J. Mass Spectrom.*, 1995, **30**, 466–472.
68. W. A. Brand and P. Dobberstein, *Isot. Environ. Health Stud.*, 1996, **32**, 275–283.
69. A. L. Sessions, S. P. Sylva and J. M. Hayes, *Anal. Chem.*, 2005, **77**, 6519–6527.
70. M. Krummen, A. W. Hilkert, D. Juchelka, A. Duhr, H. J. Schluter and R. Pesch, *Rapid Commun. Mass Spectrom.*, 2004, **18**, 2260–2266.

71. J.-P. Godin and J. S. O. McCullagh, *Rapid Commun. Mass Spectrom.*, 2011, **25**, 3019–3028.
72. S. Akoka and G. Remaud, *Actualite Chimique*, 2003, 18–21.
73. N. Ogrinc, I. J. Kosir, J. E. Spangenberg and J. Kidric, *Anal. Bioanal. Chem.*, 2003, **376**, 424–430.
74. E. H. Wahl, B. Fidric, C. W. Rella, S. Koulikov, B. Kharlamov, S. Tan, A. A. Kachanov, B. A. Richman, E. R. Crosson, B. A. Paldus, S. Kalaskar and D. R. Bowling, *Isot. Environ. Health Stud.*, 2006, **42**, 21–35.
75. E. Kerstel and L. Gianfrani, *Appl. Phys. B*, 2008, **92**, 439–449.
76. R. N. Zare, D. S. Kuramoto, C. Haase, S. M. Tan, E. R. Crosson and N. M. R. Saad, *P. Natl. Acad. Sci. USA*, 2009, **106**, 10928–10932.

CHAPTER 2

Fundamental Aspects of Stable Isotopes and Isotopic Fractionation

2.1 THE ATOM MODEL AND NUCLIDES

The discovery of radioactivity by Henry Becquerel towards the end of the 19th century and the development of quantum mechanics in the first half of the 20th century led to the modern concept of the atom. This concept or atom model states that an atom consists of a core that is composed of nucleons or, more accurately, neutral neutrons (n^0) and positively charged protons (p^+). In order to ensure electric neutrality, the positively charged nucleus is surrounded by negatively charged electrons (e^-) in discrete energy levels. Electrons are fundamental particles that have a small rest mass of $m_e = 9.10938291(40) \times 10^{-31}$ kg and an electrical charge of $e^- = -1.602176565 \times 10^{-19}$ C.[1] Together with the knowledge of Planck's theory of black body radiation and Einstein's photoelectric effect, Bohr developed his theory of the hydrogen atom. However, Bohr's theory gave no exact values for the energy levels of atoms with more than one electron or even molecules.[2] With the rise of quantum mechanics in the mid-1920s, Werner Heisenberg, Erwin Schrödinger and others developed the quantum mechanical tools to describe the electronic structure of atoms and molecules more exactly. According to the quantum mechanical model of the atom, electrons occupy so-called orbitals. With the model for multi-electron systems the periodic behaviour of the elements as it is reflected by the periodic table of elements can be understood.[3] Due to differences in the energy

---

Compound-specific Stable Isotope Analysis
By Maik A. Jochmann and Torsten C. Schmidt
© The Royal Society of Chemistry 2012
Published by the Royal Society of Chemistry, www.rsc.org

states of the electrons or in other words their electron configuration, the variety of the elements' chemical behaviour can be explained.

Let us come back to the atomic nucleus for a closer examination. Ernest Rutherford showed, in scattering experiments with α-particles (helium cores, $^4_2$He), that atomic nuclei have a radius in the order of $r_n \sim 10^{-14}$–10^{-15} m, whereas the atomic radius is around 10^{-10} m.[4] Almost the whole mass (99.95–99.98%) of the entire atom is concentrated within the nucleus.[5] In more precise scattering experiments, i.e., the measurement of α-decay between mirror nuclei and electron scattering at nuclei by linear accelerators, it was found that nuclear radii are proportional to the cubic root of the mass number A: $r_n = (1.07 \pm 0.02)A^{1/3}$ fm.[6] The radii vary between ~1 fm for a proton (hydrogen core) and 10 fm for the heavy cores, and apart from the rare earths, the atomic cores are almost spherically shaped.[6]

Henry G. J. Moseley was able to show that there is a linear relation between the emitted K_α X-ray wavelengths of metals λ_{K_α} and the number of positive charges in the nucleus, when the cores are bombarded by high-energy electrons.[7]

$$\lambda_{K_\alpha} = \frac{4}{3\Re(Z-1)^2} \tag{2.1}$$

The physical coherence of Moseley's law is given in eqn (2.1), where Z is the atomic number and $\Re$ the Rydberg constant 10 973 731.568 539(55) m^{-1}.[1] By Moseley's results it was possible to sort and establish the atomic numbers of the elements and remaining gaps in the periodic table could be filled.[7] In 1919, Rutherford concluded from the results of scattering experiments with α-particles on simple gases such as nitrogen that particles identical to hydrogen cores were emitted by the nuclei. He suggested calling these particles protons from the Greek word 'protos', which means 'the first'.[8] These positively charged protons in the nuclei have a rest mass of $m_p = 1.672\,621\,777(74) \times 10^{-27}$ kg,[1] which is ~1836 higher than that of an electron.[2] It turned out that the presence of protons alone is not sufficient to explain the core mass. Thus, in 1920 William D. Harkins described nuclei built of protons and neutral particles named neutrons and Rutherford proposed a similar idea in the same year.[9] Walter Bothe and his student Herbert Becker reported that exposure of light elements, in particular beryllium, to α-rays leads to a highly penetrating radiation.[10] In 1931–1932 Frédéric and Irène Joliot-Curie reported that exposing hydrogen-containing material, particularly paraffin, to this new radiation leads to the ejection of high-velocity protons.[2] At the same time, James Chadwick interpreted both sets of results in terms of radiation consisting 'of particles of mass nearly equal to that of the proton and with no net charge', which he identified as the neutron species first postulated by Rutherford. The missing particle, the neutral neutron has a 0.14% greater mass ($m_n = 1.674\,927\,351(74) \times 10^{-27}$ kg)[1] than the proton.[2]

The number of protons, which is also called atomic number Z, defines the element E and thus its position in the periodic table, whereas the total number of nucleons, i.e., neutrons N and protons Z, is the integer mass number $A = N + Z$. Together with the atomic number it defines a nuclide, which can be expressed by the notation ^A_ZE. Additionally, a nuclide should have a nuclear energy state that provides a mean life time long enough to be observable.

There are more than 2500 known nuclides[11] of which only about 340 occur naturally[12] and only 265 are stable with respect to radioactive decay,[6] which means that they have a half-life greater than 1×10^{10} years. All isotopes of elements with atomic numbers greater than 83 have half-lives less than 10^{10} years, except for thorium-232.[13] A recent survey by the Institut d'Astrophysique Spatiale showed that bismuth-83 is a very long living isotope that decays by α-particle emission with a half-life time of $1.9 \pm 0.2 \times 10^{19}$ years.[14] Thorium-232 is an α-emitter with a half-life of $1.40 \pm 0.01 \times 10^{10}$ years, decaying through a branched series to lead-208 without long-living intermediate isotopes.[13]

Nuclides are tabulated in nuclide charts such as shown schematically in Figure 2.1, in which the atomic number is plotted against the number of neutrons. For each nuclide the chart provides physicochemical information and whether it is naturally occurring or artificially produced by nuclear reactions. In case of stable isotopes, for example, the atomic masses, their abundance as well as thermal neutron cross-sections are tabulated. For radioactive nuclides, the half-life time, the mode and energy of decay and other information are stated. For an interactive nuclide chart see Web findings at the end of this chapter.

In the lower right part of Figure 2.1 a magnified part of the nuclide chart up to the element sodium (atomic number $Z = 11$) is shown. It can be taken from this chart that more than one nuclide of the same element exists. Due to their variations in the A, N, Z values, nuclides can be categorized as isobars, isotones and isotopes.

As depicted in Figure 2.2, isobars are defined as nuclides having the same mass numbers A but different atomic numbers Z. Thus, isobars can be found in diagonal lines in the nuclide chart. Examples for isobars are $^{14}_{6}$C and $^{14}_{7}$N. In Chapter 3 we will face a closely related term, the

Fundamental Aspects of Stable Isotopes and Isotopic Fractionation 17

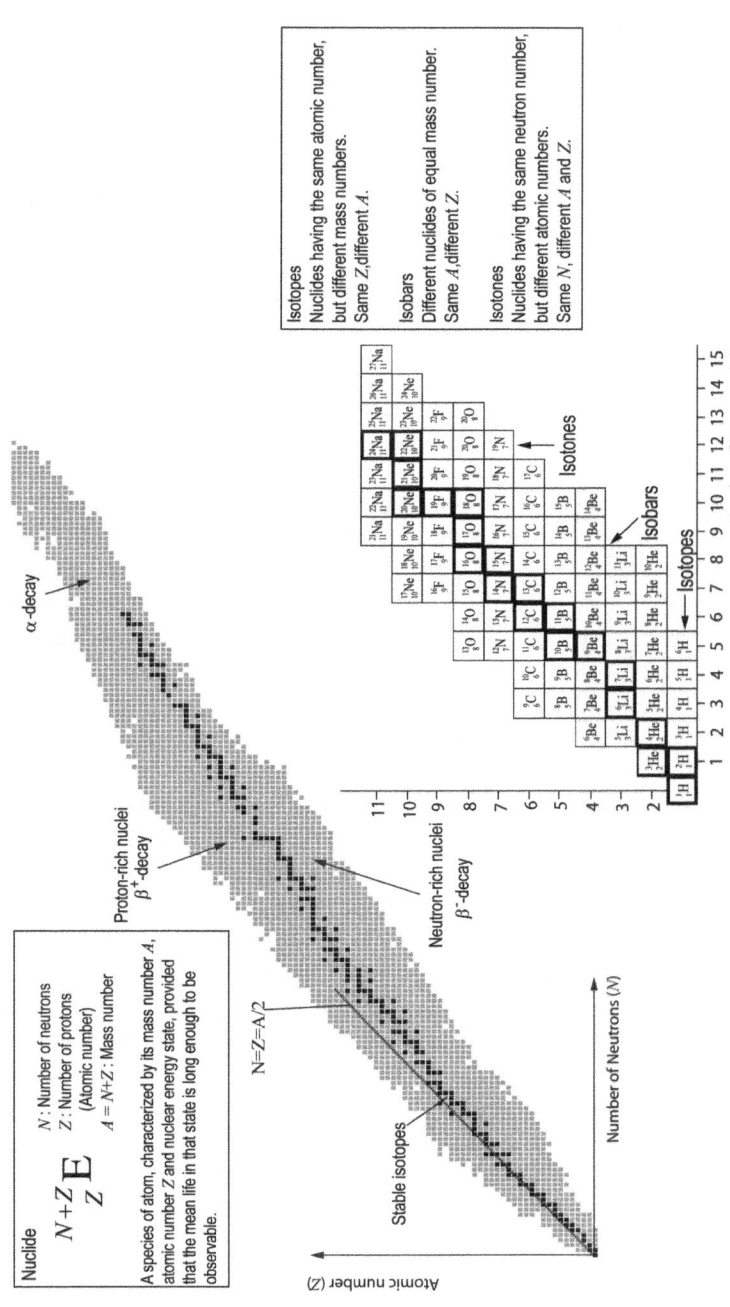

Figure 2.1 Schematic presentation of the chart of nuclides. Stable isotopes are represented by black squares. In the lower right part, a magnified detailed part of the first 11 elements is shown to explain the terms isotope, isobar and isotone. Stable isotopes are marked by thick black frames.

isobaric interferences, which may hamper (low-resolution) mass spectrometric differentiation of different species.

In contrast to isobars, isotones are defined as nuclides having the same neutron number N but different atomic numbers Z, or $N = A - Z = const$. Isotones can be found in the vertical columns of the nuclide chart. Some examples for isotones are $^{13}_{6}C$ and $^{14}_{7}N$, or $^{15}_{7}N$ and $^{16}_{8}O$.

In the context of this book, of course, the most relevant type of nuclides are the isotopes that are defined as nuclides of the same element (same number of protons Z) but different numbers of neutrons N. They are represented by horizontally neighbouring nuclides in the chart. In 1921, Frederick Soddy received the Nobel Prize for Chemistry for his investigations of radioactive substances and the theory of isotopes. This finding ended a controversial scientific debate that lasted for more than a hundred years. First indications for isotopes came from William Proud in 1815 and the so-called 'whole number rule' according to which masses of elements were multiples of the mass of hydrogen.[15] At that time, it was only possible to measure the relative atomic weight of an element by stoichiometric chemical methods and it turned out that the addition of hydrogen atoms alone was not able to explain the observed atomic weights completely.[15] According to Proud's model the atomic weights should have integer numbers. Neon and chlorine, in particular, violated this rule because they showed atomic weights of 20.20[16] and 35.46 in the oxygen scale, respectively.[15,17]

In 1912, Joseph J. Thomson found with a first type of mass spectrometer, two lines for neon at mass to charge ratios (m/z) of 20 (strong line) and 22 (weak line) and further lines at 10 and 11 for the double charged species.[15,18] He concluded that the weaker line could be due to NeH_2 but he favoured the idea that it corresponded to a new element.[15] In 1920, Aston wanted to show why the chemical atomic weight of neon violated the whole number rule. By exact mass determination of both the ^{20}Ne and the ^{22}Ne signals using a modified focusing mass-spectrograph with higher resolution, the values of 20.00 and 22.00 with an estimated ratio of the line intensities of 9:1 in accordance with the experimental average value of 20.20 were obtained.[15] A less intensive line, at a mass to charge ratio of 21 suggested the existence of a third isotope of neon.[19,20] These measurements provided the first exact proof of isotopes with a mass spectrometer for which Aston was awarded the Nobel Prize in Chemistry in 1922 (see Web findings at the end of this chapter).

Another strong indication for isotopes came from the radioactive decay series of uranium and thorium. Thorium-232 and uranium-238 decay by a certain number of α- and β-particle emissions into different stable lead isotopes of m/z 208 and 206, respectively.[13,15,21] As a result of the very careful stoichiometric atomic weight determinations of Theodore Richards it was found that samples of lead from different parts of the earth's crust had different atomic weights and, in particular, that samples of lead isolated from thorium minerals had atomic weights approaching 208, whereas those isolated from uranium minerals had atomic weights approaching 206.[22] Soddy coined the word 'isotope', to account for these chemically identical and, by chemical methods, non-separable elements, which occupy the same place in the periodic table of elements.[23] The name 'isotope' (Greek: *isos topos*: meaning 'in the same place') was suggested by Dr Margaret Todd a friend of the family during a discussion at a dinner party given by Soddy's father-in-law.[24]

2.2 STABLE ISOTOPES

2.2.1 Isotope Stability and Abundance

The nuclide chart in Figure 2.1 shows that in the case of light elements, stability is achieved when the number of neutrons equals the number of protons ($N/Z = 1$). This is indicated by the $N = Z$ stability line[6] which is a good approximation up to $Z = 10$.[12] With an increasing proton number the deviation from this line increases and the N/Z ratio of stable isotopes is always greater than unity. It has its maximum value at about 1.5, which indicates that a little neutron excess is necessary for stability.[7]

The black squares in Figure 2.1 indicate stable nuclides and the pale gray squares indicate the radioactive ones. The black squares can be imagined as the bottom of a valley of stability. The proton rich part above this valley contains nuclides that obey β^+-decay and the part underneath the valley follow a β^--decay. For elements with mass numbers higher than 208 mostly α-decay occurs.[6]

Of the chemical elements in the periodic table, 34 have no stable isotopes, e.g., radium, uranium, as well as the artificially produced plutonium.[25] The three elements thorium, protactinium and uranium have no stable isotopes but have characteristic isotopic compositions because of their long-living radioactive isotopes so that an atomic weight can be determined for them.[25] 61 elements of the periodic table have two or more stable isotopes. The elements with the highest numbers of stable isotopes are Xenon ($Z = 54$) with nine and tin ($Z = 50$) with ten.[26] In Table 2.1 the isotopic abundances of the isotopes considered in this book are tabulated.

The Commission on Isotopic Abundance and Atomic Weights (CIAAW) considers an element to be monoisotopic if it has one and

Table 2.1 Isotopic abundance of the elements covered in this book expressed in terms of mole fraction (from Berglund et al.).[26]

Isotope	Isotopic abundance (mole fraction)[26]
^{1}H	0.999 885(70)
^{2}H	0.000 115(70)
^{12}C	0.9893(8)
^{13}C	0.0107(8)
^{14}N	0.996 36(20)
^{15}N	0.003 64(20)
^{16}O	0.997 57(16)
^{17}O	0.000 38(1)
^{18}O	0.002 05(14)
^{32}S	0.9499(26)
^{33}S	0.0075(2)
^{34}S	0.0425(24)
^{36}S	0.0001(1)
^{35}Cl	0.7576(10)
^{37}Cl	0.2424(10)
^{79}Br	0.5069(7)
^{81}Br	0.4931(7)

only one isotope that is either stable or has a half-life greater than 1×10^{10} years.[13] There are 20 monoisotopic elements (Be, F, Na, Al, P, Sc, Mn, Co, As, Y, Nb, Rh, I, Cs, Pr, Tb, Ho, Tm, Au, Bi). In case of these elements the atomic masses can be stated up to eight decimal places because atomic masses are known to this accuracy.[25] Isotope abundance values that are free from all known sources of bias within stated uncertainties are referred to as 'absolute' isotopic abundances.[13,27] Absolute isotope abundance values can be determined by calibrating a mass spectrometer by means of gravimetrically prepared synthetic mixtures of materials in which an isotope is enriched (or depleted) by a known or measurable factor.[27]

Different empirical rules to grasp the nuclide stability in accordance with their abundance have been developed. It was recognized early on that nuclides with even proton and neutron numbers (so called gg-cores) are more abundant than cores with odd neutron and even proton or even neutron and odd proton numbers (ug-, or gu-cores) and those with odd neutron and proton numbers (uu-cores).[28] According to Harkin's rule, elements with even atomic numbers are more abundant in nature than those with odd ones.[29] Thus, 159 of the 265 stable nuclides possess an even proton number Z and neutron number N. Examples for this type of nuclides are $^4_2$He, $^{12}_6$C and $^{16}_8$O. 53 stable nuclides have even Z

and odd N such as $^3_2$He or $^{13}_6$C and 51 of them have an odd Z and an even N, e.g., $^{15}_7$N. Only four stable nuclides have odd Z and N. $^2_1$H and $^{14}_7$N belong to this type of nuclide. Aston's isotope rule states that elements with odd proton numbers have, at the most, two stable isotopes, whereas those with even proton numbers can have more.[28] In particular, for nuclides with so called 'magic numbers' (2, 8, 20, 28, 50, 82, 126) of neutrons and/or protons an above-average number of stable isotopes exist.

Comparable with the shell structure of the atom these numbers can be explained by the shell model of the atomic core. Nucleons pair preferentially with nucleons of the same type meaning that stable nuclides with even proton and even neutron numbers are much more abundant than those with odd proton and odd neutron numbers.[28]

Other models such as the 'nuclear drop model' can explain the binding energy of the nucleus.[6] At this point, however, we will leave the discussion about core stability and refer to nuclear physics monographs for a detailed in-depth treatment.

2.2.2 Relative Atomic Mass

As we saw from the historical developments that led to the discovery of isotopes, most of the elements are mixtures of two or more isotopes with differing natural abundances. This explains why relative atomic masses $A_r(E)$ generally show deviations from integer numbers.

A lot of confusion existed on the atomic weight scales up to the beginning of the 1960s. In the 19th century, two scales for atomic weights were used, which were based on hydrogen $A_r(H) = 1$ and oxygen $A_r(O) = 16$. In the first decades of the 20th century the oxygen scale was almost exclusively accepted, due to the fact that most of the stoichiometrically determined atomic weights were based on this scale. It turned out in 1929 that oxygen consists of three stable isotopes (^{16}O, ^{17}O, ^{18}O), after which chemists continued to use oxygen in its natural composition as $A_r(O) = 16$ as the scale whereas physicists adopted the $A_r(^{16}O) = 16$ scale. This led to the unpleasant situation whereby a conversion factor of 1.000 275 was necessary to change values from the physics scale into that used by the chemists. Later it turned out that the two scales were not even related by a fixed constant.[13]

This confusing situation motivated Alfred O. Nier and Arne Ölander to propose a new scale with $A_r(^{12}C) = 12$ as its basis.[13] In 1959 and 1960 the International Union of Pure and Applied Chemistry (IUPAC) and the International Union of Pure and Applied Physics (IUPAP) agreed to unify the atomic weight scale based on the mass of the isotope $^{12}C = 12$.[30] The term atomic weight was replaced by the term relative atomic mass $A_r(E)$. The SI-unit for the amount of a substance, the mole, was then derived from this definition such that one mole is defined as the amount of a substance that contains as many elemental entities as there are atoms in 0.012 kg of ^{12}C in its ground state. This number of entities is the so called Avogadro's constant $N_A = 6.022\,141\,29(27) \times 10^{23}\,\text{mol}^{-1}$. By using this definition the relative atomic mass of an element is given by:

$$A_r(E) = 12 \times m(E)/m(^{12}C) \qquad (2.2)$$

where $m(E)$ is the average mass of an atom of an element.

The relative atomic mass of a polyisotopic element can be determined from the isotopic abundances and the relative atomic masses of each of the k isotopes on the carbon-12 scale as given in the equation:

$$A_r(E) = \sum_{i=1}^{k} [x(^A E) A_r(^A E)]_i \qquad (2.3)$$

where $x(^A E)$ is the molar fraction of isotope $^A E$ in a mole of element E.[13,31]

2.2.3 Isotopologues, Isotopomers and Isotopocules

The term isotopologue is defined as a molecular entity that differs only in its isotopic composition or, in other words, in the number of isotopic substitutions.[32] In Figure 2.2 the five isotopologues of methane substituted by a varying number of deuterium (^2H) atoms are shown. Another example for isotopologues of a molecule is perchloroethylene substituted with different numbers of ^{37}Cl.

Other species of isotopically substituted molecules are related to the position that they occupy within the molecule. Therefore, the term used for these species is a contraction of 'isotopic isomer' and these species are called 'isotopomers'.[27] Isotopomers are isomers of a molecule having the same number of isotopic substitutes but these differ in their positions in the molecule.[32] If the isotopic substitutes differ only in their position it is a constitutional isomer[32] such as the acetic acid in the lower left part of Figure 2.2. In cases with a symmetry centre within the molecule we have isotopic stereoisomers such as deuterated ethanol or deuterated propene shown in the bottom right of Figure 2.2.

Isotopocules is a recently introduced definition that comprises both isotopologues and isotopomers, i.e., all molecular species that differ in either the number or positions of isotopic substitutions. The term is a

Fundamental Aspects of Stable Isotopes and Isotopic Fractionation 23

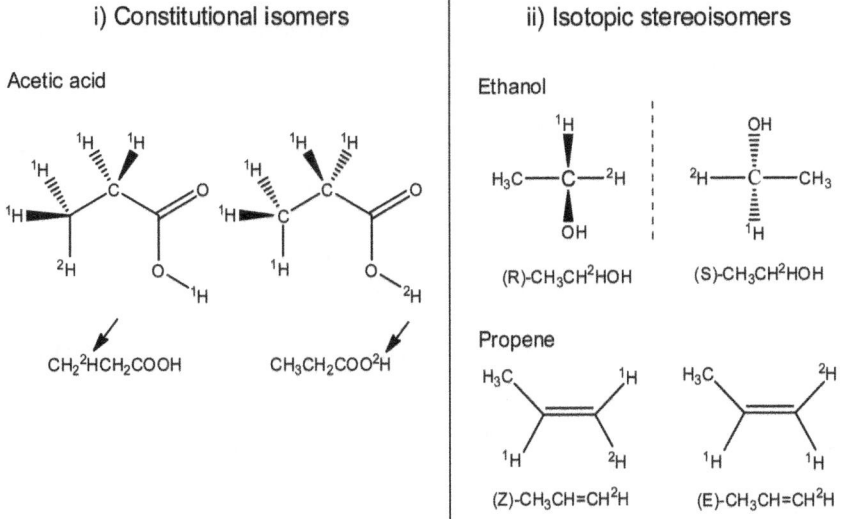

Figure 2.2 Isotopologues and isotopomers. Top: isotopologues of methane with respect to hydrogen and perchloroethylene with respect to chlorine. Bottom left: two constitutional isomers of acetic acid. Bottom right: isotopic stereoisomers of deuterated ethanol and propene.

contraction of 'isotopically substituted molecules'.[27] Examples of isotopocules are $^{15}N_2^{16}O$, $^{14}N^{15}N^{16}O$, $^{15}N^{14}N^{16}O$.

These species are not all isotopologues because the latter two are not different in their isotopic composition, nor are they all isotopomers (only the latter two are).[27]

2.3 NATURAL ABUNDANCE AND HOW TO EXPRESS IT

2.3.1 Isotope-amount Fraction

According to the IUPAC definition, the natural isotopic abundance of a specified isotope of an element is the isotopic abundance in the element as found in nature.[32] The isotopic abundances of a certain compound can be expressed by the 'atom fraction or isotope-amount fraction' $x(^{A}E)_c$, which is defined by the amount of an atom (isotope) of a chemical element $n(^{A}E)_c$ divided by the total entity or number of atoms of the element within the compound:[27]

$$x(^{A}E)_c = \frac{n(^{A}E)_c}{\sum_{i=1}^{k} n(E)_{c_i}} = \frac{N(^{A}E)_c/N_A}{\sum_{i=1}^{k} N(E)_{c_i}/N_A} \quad (2.4)$$

where $N(^{A}E)_c$ is the number of entities and N_A the Avogadro constant. The isotope-amount fraction is a dimensionless quantity. For elements with two isotopes the equation is reduced to:

$$x_c(^{h}E) = \frac{n(^{h}E)_c}{n(^{l}E)_c + n(^{h}E)_c} \quad (2.5)$$

Here, the superscripts h and l denote a heavier (higher atomic mass) and a lighter (lower atomic mass) isotope, respectively.

In Table 2.2 the obtained variation ranges of natural isotopic abundance for the biologically relevant elements are tabulated in terms of

Table 2.2 Observed ranges of natural abundance variations on earth.[26]

Element	^{A}E	Observed ranges of natural variations in terms of isotope-amount fraction $x(^{A}E)$
Hydrogen	^{1}H	0.999 816–0.999 974
	^{2}H	0.000 026–0.000 184
Carbon	^{12}C	0.988 53–0.990 37
	^{13}C	0.009 63–0.011 47
Nitrogen	^{14}N	0.995 79–0.996 54
	^{15}N	0.003 46–0.004 21
Oxygen	^{16}O	0.997 38–0.997 76
	^{18}O	0.001 88–0.002 22
Sulfur	^{32}S	0.944 54–0.952 81
	^{34}S	0.039 76–0.047 34
Chlorine	^{35}Cl	0.756 44–0.759 23
	^{37}Cl	0.240 77–0.243 56

isotope-amount fraction.[26] In literature, the term fractional abundance ^hF_c is used as a pseudonym for isotope-amount fraction.[33] In the case of carbon the expression for fractional abundance is:

$$^{13}F_c = x(^{13}C)_c = \left(\frac{n(^{13}C)_c}{n(^{12}C)_c + n(^{13}C)_c}\right) \qquad (2.6)$$

Although the term fractional abundance is not officially recommended anymore, it is still utilized in Chapter 3 when we deal with ion corrections since it is often easier and less confusing.

The term isotope-amount fraction is also identical with the widely used, but not IUPAC conforming, atom percent (atom-%, AP):

$$\text{atom-\%} \, (^hE)_c = x(^hE)_c \times 100 \equiv {}^hF_c \times 100 \qquad (2.7)$$

In particular, for metabolic investigations (tracer studies) and mixing calculations by mass balance equations it is convenient to use the isotope-amount fraction because often enriched tracers, with abundances very different from the natural abundance range, are used (see Section 2.3.4).

2.3.2 Variations of Natural Isotopic Abundance

The average isotopic abundance of the terrestrial elements was fixed around the time when the earth was formed.[34] However, in contrast to what we learn in high school or basic university chemistry courses, the natural isotope abundances of elements are only stable in nature to a first approximation. There are several reasons for these isotopic variations, such as:

(i) decay of naturally occurring and long-lived radionuclides;
(ii) interaction of terrestrial matter with cosmic rays;
(iii) artificial enrichment of isotopes;
(iv) mass-dependent and mass-independent isotopic fractionation.

In this book we will limit the discussion mainly to mass-dependent isotopic fractionation. Isotopic fractionation is thereby a measurable quantity that can occur during transport, phase transfer and chemical transformation processes and has its origin in isotope effects, which are caused by the slight differences in mass trough variable isotopic substitution of a molecule. To investigate the small variations in isotopic abundance, highly precise abundance measurements in the order of 10^{-4}

to 10^{-6} are necessary. In Table 2.2 the obtained variation ranges of natural isotopic abundance for the biologically relevant elements are tabulated in terms of isotope-amount fractions.[26] A more detailed compilation of isotopic abundances variations of different elements can be found in Coplen et al.[31]

The isotopic variations between different terrestrial compartments are nowadays widely used to trace and understand environmental processes. Most prominent processes that have been elucidated by SIA variations are the tracing of the hydrological cycle,[7] the differentiation between carbon fixation mechanisms in plants,[35] palaeoclimate reconstruction[36,37] and carbon dioxide impact on global climate change.[38] Nowadays the measurement of these variations has been established in a variety of branches of science. A more detailed treatise of application fields can be found in Chapter 6 of this book.

2.3.3 Isotope Ratios

Apart from isotope-amount fraction, fractional abundance and atom percent, isotope ratios are usually used to express isotope abundances. The isotope ratio $R(^hE/^lE)_c$ is defined by eqn (2.8),

$$R(^hE/^lE)_c = \frac{N(^hE)_c}{N(^lE)_c} \tag{2.8}$$

where N_c are the numbers of entities of a certain compound. The indices h and l denote the heavy and the light isotope, respectively, and for the elements discussed here (H, C, N, O, S, Cl, Br) the heavier isotope is also the less abundant one and appears in the denominator, whereas the more abundant lighter isotope is written in the numerator. Because of this practice, the expression $R(^hE/^lE)_c$, which is useful for polyisotopic elements, is preferred but can be reduced to the abbreviated form ^hR_c.

When elements with only two stable nuclides such as carbon, hydrogen and nitrogen are used, the relationship between isotope-amount fraction $x(^hE)_c$ and the isotope ratio is given by eqn (2.9).[33]

$$x(^hE)_c = \frac{R(^hE/^lE)_c}{1 + R(^hE/^lE)_c} \tag{2.9}$$

Vice versa, to calculate the isotope ratio from the isotope-amount fraction eqn (2.10) can be used.

$$R(^hE/^lE)_c = \frac{x(^hE)_c}{1 - x(^hE)_c} \tag{2.10}$$

2.3.4 The δ-Scale and its Limits

Another commonly used expression is the so-called 'delta-' or 'δ-scale', which is widely applied in almost all disciplines in which stable isotope ratio measurements are carried out. This scale was introduced in the Chicago laboratories of Harold Urey in the late 1940s and occurred for the first time in a publication by McKinney.[39] The δ-value of a compound $\delta^h E_c$ is defined by eqn (2.11) as the relative difference between the isotope ratio (e.g., $^2H/^1H$ or $^{13}C/^{12}C$) of a sample or compound $R(^hE/^lE)_c$ and the isotope ratio defining an international reference scale $R(^hE/^lE)_{ref}$.

$$\delta^h E_{c,ref} = \frac{R(^hE/^lE)_c - R(^hE/^lE)_{ref}}{R(^hE/^lE)_{ref}}$$

$$= \frac{(N(^hE)_c/N(^lE)_c) - (N(^hE)_{ref}/N(^lE)_{ref})}{N(^hE)_{ref}/N(^lE)_{ref}} \quad (2.11)$$

$$= \frac{N(^hE)_c/N(^lE)_c}{N(^hE)_{ref}/N(^lE)_{ref}} - 1 = \frac{R(^hE/^lE)_c}{R(^hE/^lE)_{ref}} - 1$$

For example, for carbon isotope ratios the internationally accepted reference material is Vienna Pee Dee Belemnite (VPDB), a marine carbonate, which has an accepted $^{13}C/^{12}C$ isotope ratio of 0.0111802 and a defined corresponding δ-value of zero.[40] The internationally accepted reference materials that define the international scales for the light elements are listed in Table 2.3 and will be discussed in detail in Chapter 5.

The δ-notation is not an SI-unit and in the literature it is frequently defined as

$$\delta^h E_{c,ref} = \left(\frac{R(^hE/^lE)_c}{R(^hE/^lE)_{ref}} - 1\right) \cdot 10^3 \quad (2.12)$$

The values are than expressed in parts per thousand or per mil with the symbol ‰. For example, the δ-value of benzene is $\delta^{13}C_{benzene} = -0.0284$ according to eqn (2.11) and will be expressed as $\delta^{13}C_{benzene} = -28.4‰$ when using eqn (2.12). According to eqn (2.11) a positive δ-value means that the ratio of the heavy to light isotope is higher in the sample than it is in the standard, and vice versa for a negative δ-value. A $\delta^{13}C$-value of $+5‰$ corresponds to a sample with an isotope ratio that is 0.5% higher than that of the international

Table 2.3 Compilation of the international stable isotope ratio scales. Data are adapted from Werner et al.[40]

Element	Ratio	International scale	Kind of international reference material	Accepted isotope ratio ($\times 10^{-6}$)
Hydrogen	$^2H/^1H$	Vienna Standard Mean Ocean Water (VSMOW)	water	(155.75 ± 0.08)[41]
Carbon	$^{13}C/^{12}C$	Vienna Pee Dee Belemnite (VPDB)	carbonate	$(11\,180.2 \pm 2.8)$[42]
Nitrogen	$^{15}N/^{14}N$	AIR-N_2[a]	nitrogen gas	(3678.2 ± 1.5)[43]
Oxygen	$^{18}O/^{16}O$	Vienna Standard Mean Ocean Water (VSMOW)	water	(2005.2 ± 0.45)[44]
		Vienna Pee Dee Belemnite (VPDB)	carbonate	2067.2
Sulfur	$^{34}S/^{32}S$	Vienna Cañon Diabolo Troilite (V-CDT)	troilite	44159.9 ± 11.7[45]

[a]It has to be noted that atmospheric nitrogen is not distributed as a calibration material. It is actually hard and labour intensive to produce nitrogen from air without fractionation and argon contamination. For this purpose the IAEA and NIST distribute ammonium salts instead that need to be combusted to generate nitrogen gas.

standard. The $\delta^{13}C$-value of $+0.005$ or $+5‰$ for the sample then corresponds to a $^{13}C/^{12}C$-ratio of 0.011 2361 compared to the reference material value of 0.0111 802, which demonstrates the very subtle changes that need to be measured and the advantage of the δ-scale to deal with more manageable numbers (see Figure 2.3). An apparently large difference of 100‰ is still only a 10% difference in the isotope ratio. Note that we have omitted the common multiplier 1000 in all equations in this book in order to avoid confusion and be consistent with the recommendations given recently by Coplen et al.[27]

The δ-scale was introduced for several reasons. The first is a rather technical aspect: mass spectrometers typically used for measuring isotope abundances in natural materials are not really suited to obtain absolute isotope ratios or mole fractions of isotopes. Thus, rather than absolute values, the differences in relative ratios are reported to allow a correction for mass-discriminating effects in a single instrument and to facilitate the comparison of published data. At the time when Urey and his coworkers started to measure oxygen isotope ratios for paleoclimate investigations it was not possible to measure accurately absolute ^{18}O abundances,[46] this was only done much later by Baertschi

Fundamental Aspects of Stable Isotopes and Isotopic Fractionation

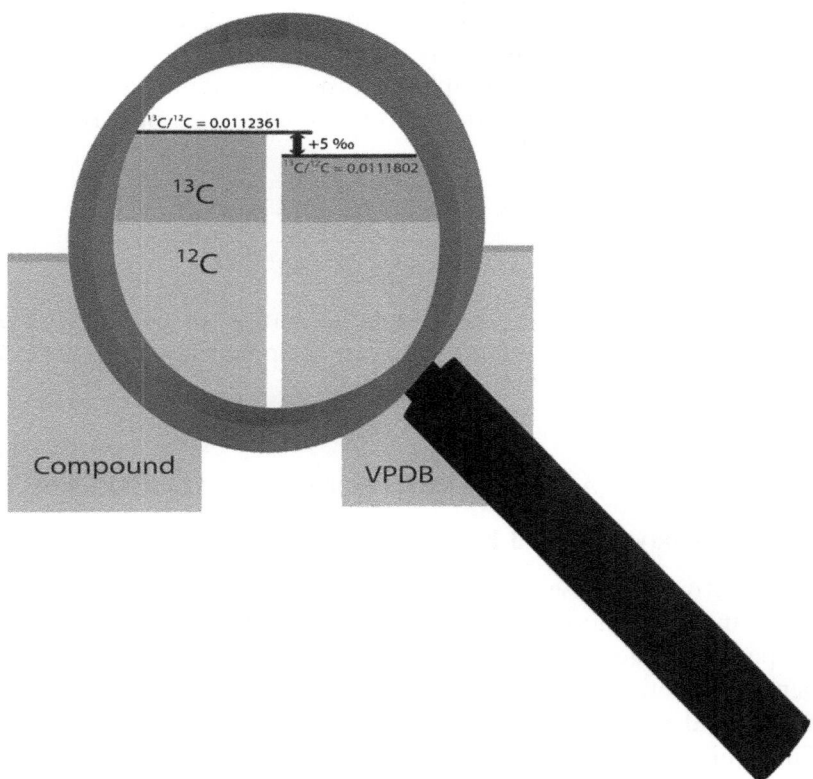

Figure 2.3 Magnifying effect of the δ^{13}C-scale.

in 1976.[44] A second reason is that abundances of the heavy and the light isotope of an element are different by orders of magnitude (see Table 2.1), which results in very small isotope ratios.[33] A third is that natural isotope variations due to isotope fractionation processes are, except for hydrogen, rather small. Recently, Brand and Coplen proposed to introduce the unit Urey to get rid of the per mil sign '‰'. A sub multiple of 1 defined the unit Urey 'Ur' and a milliurey (symbol 'mUr') would be equivalent to the per mil.[25] For the above example of benzene with a $\delta^{13}C_{benzene} = -0.0284$ or $\delta^{13}C_{benzene} = -28.4‰$ this would equal $\delta^{13}C_{benzene} = -28.4$ mUr.

The validity of the δ-scale is restricted to the natural abundance range. In the case of tracer studies with artificially enriched samples or when perfectly accurate mass balance calculations are needed, the isotope-amount fraction $x(^hE)_c$ has to be used. Due to the huge natural abundance range of hydrogen, for that element the δ-scale can also lead to unacceptable errors. In Figure 2.4 the relation between the isotope-amount fraction and the δ-scale is shown for carbon.

It can be taken from these graphs that there is a nonlinear relationship between isotope-amount fraction and the δ-value as explicitly shown in eqn (2.13).

$$x(^hE)_c = \frac{1}{1 + \frac{1}{(\delta^h E_{c,ref} + 1) \times R(^hE/^lE)_{ref}}} \quad (2.13)$$

If we zoom in to the natural abundance range for carbon in Figure 2.4c, it is evident that the relationship can be approximated as linear and thus be used to express isotope ratios.

2.4 ISOTOPE FRACTIONATION AND HOW TO EXPRESS IT

2.4.1 Thermodynamic and Kinetic Isotope Effects

Mass-sensitive processes that result in isotope fractionation can be divided into thermodynamical or equilibrium isotope effects (*EIE*) (isotope exchange reactions, partitioning) and unidirectional, rate-dependent kinetic isotope effects (*KIE*) (effusion, diffusion, evaporation and chemical reaction rates).

The equilibrium stable isotope fractionation is a quantum mechanical phenomenon, which is mainly driven by the differences in the vibrational energies of molecules of different masses.[47] Considering a diatomic molecule AB, the isotopologue that contains the heavier isotope will vibrate slower (or with a lower frequency v), than that containing the lighter isotope. Correspondingly, their dissociation energies are different. The energy to break a bond of the isotopologue containing the light isotope is lower. At equilibrium, the heavy isotopes of an element will tend to be concentrated in compounds where the element forms the stiffest (strongest) bonds.[47,48] The magnitude of isotopic fractionation will be roughly proportional to the difference in bond stiffness between the equilibrated substances. Bond stiffness is greater for short, strong chemical bonds, which correlates with:

(i) high oxidation state in the element of interest;
(ii) in the case of anions such as Cl^- and O^{2-} a high oxidation state in the atoms to which the element of interest is bound;
(iii) bonds involving elements near the top of the periodic table;
(iv) the presence of highly covalent bonds between atoms with similar electronegativity;

Fundamental Aspects of Stable Isotopes and Isotopic Fractionation

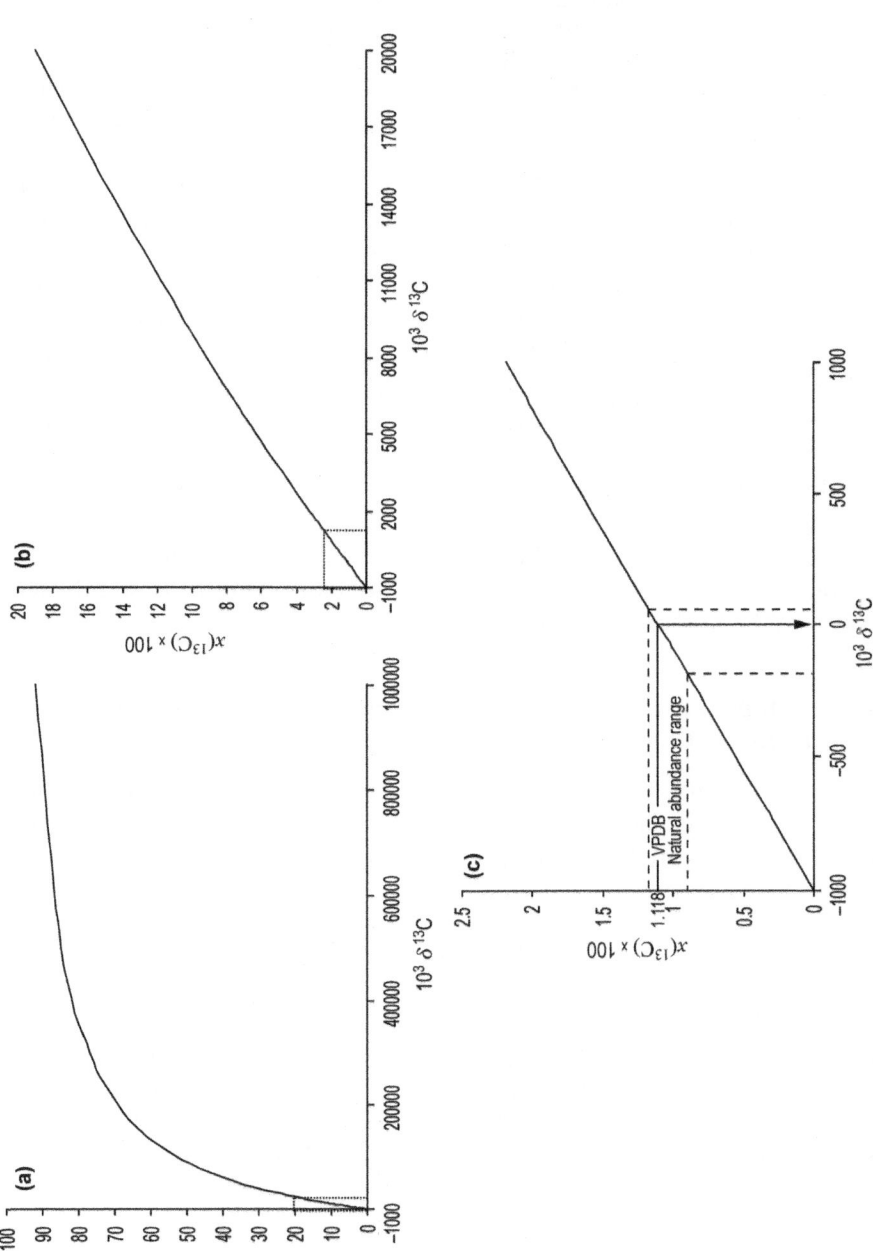

Figure 2.4 Relation of isotope-amount fraction and the δ-scale. (a) to (c) successive zooms on the more restricted δ-ranges in the boxed areas.

(v) low-spin electronic configurations for transition elements, also d^3 or d^8 electronic structure for octahedrally coordinated atoms;
(vi) low coordination number.

In general, equilibrium isotopic fractionation decreases as temperature increases. This increase is roughly proportional to $1/T^2$ for most compounds.[47]

EIEs can be understood by statistical thermodynamics. In a landmark paper in 1947, Urey developed the basis for the thermodynamic understanding of equilibrium isotopic fractionation.[49]

A *KIE* has its origin in the energy differences for isotopologues during the formation of an active complex. In the case of *KIEs* there is a distinction between primary *KIEs* and secondary isotope effects. Primary *KIEs* in which the isotope effect occurs directly in the reacting bond are much bigger than secondary isotope effects in which the element is not directly involved.[27]

A thorough introduction to equilibrium and kinetic stable isotope fractionation by isotope effects is beyond the scope of this book and we refer to articles and monographs such as those by Schauble,[47] Criss,[7] Melander and Saunders,[50] Kohen and Limbach[51] and an in-depth discussion by Wolfsberg, van Hook and Paneth.[52]

2.4.2 Isotopic Fractionation Factor, Isotopic Fractionation and Isotopic Difference

The most common and exact term to express isotopic fractionation of two isotopologues between two substances A and B is the isotopic fractionation factor. It is defined by

$$\alpha^h E_{A/B} = \frac{R(^hE/^lE)_A}{R(^hE/^lE)_B} \qquad (2.14)$$

where, $R(^hE/^lE)_A$ and $R(^hE/^lE)_B$ are the isotope ratios of two substances. The isotopic fractionation factor is dimensionless and a relationship between the fractionation factor and the δ-scale as expressed in eqn (2.15) can be easily derived by rearranging eqn (2.11) for the δ-scale to the corresponding isotope ratios for A and B and subsequent substitution into eqn (2.14).

$$\alpha^h E_{A/B} = \frac{R(^hE/^lE)_A}{R(^hE/^lE)_B} = \frac{(\delta^h E_A + 1) \times R(^hE/^lE)_{ref}}{(\delta^h E_B + 1) \times R(^hE/^lE)_{ref}} = \frac{\delta^h E_A + 1}{\delta^h E_B + 1} \qquad (2.15)$$

A distinction can be made between equilibrium and kinetic isotopic fractionation factors, which can be indicated by the subscripts *eq* and *kin*, respectively. If we assume an isotopic exchange reaction under equilibrium conditions in which isotopes are exchanged between the two substances AX and BX and related stoichiometric coefficients a and b:

$$aA^lX_b + bB^hX_a \rightleftharpoons aA^hX_b + bB^lX_a \qquad (2.16)$$

then the equilibrium isotopic fractionation factor $\alpha_{eq}\,^hX_{AX/BX}$ is exactly the inverse of the equilibrium isotope effect when the reaction is written in the same direction.[27]

The equilibrium fractionation factor of a chemical reaction is directly related to the reaction's equilibrium constant K_{eq} but the conversion between these two is often complicated because the equilibrium constant is always related to a specific chemical reaction, therefore it is often necessary to account for stoichiometry and symmetry of the involved molecules. The equilibrium constant for the reaction is the ratio of the product and reactant concentrations, given by

$$K_{eq} = \frac{[A^hX_b]^a[B^lX_a]^b}{[A^lX_b]^a[B^hX_a]^b} = \frac{\left(\frac{[A^hX_b]}{[A^lX_b]}\right)^a}{\left(\frac{[B^hX_a]}{[B^lX_a]}\right)^b} = \frac{(R(^hX/^lX)_{AX})^a}{(R(^hX/^lX)_{BX})^b} \qquad (2.17)$$

In the easiest case when a and b are equal, then

$$K_{eq} = \left(\frac{R(^hX/^lX)_{AX}}{R(^hX/^lX)_{BX}}\right)^a \qquad (2.18)$$

and for the isotopic fractionation factor eqn (2.19) is valid.

$$\alpha_{eq}\,^hX_{AX/BX} = K_{eq}^{1/a} \qquad (2.19)$$

When the stoichiometric coefficients a and b are not equal, the equilibrium isotopic fractionation factor can be described by the following expression,

$$\alpha_{eq}(^hX)_{AX/BX} = \left(\frac{S_{A^hX}}{S_{A^lX}}\right)^{\frac{1}{a}} \left(\frac{S_{B^hX}}{S_{B^lX}}\right)^{\frac{1}{b}} (K_{eq})^{\frac{1}{ab}} \qquad (2.20)$$

in which $S_{A^hX}, S_{A^lX}, S_{B^hX}, S_{B^lX}$ are the molecular symmetry numbers for the reactant and product molecules.[47]

In the case of a kinetic isotopic fractionation factor in which a product P is formed irreversibly from reactant Q, the isotopic fractionation factor can be written as:

$$\alpha_{kin}{}^h E_{P/Q} = \frac{R(^hE/^lE)_{P_{ins}}}{R(^hE/^lE)_Q} = \frac{R(^hE/^lE)_{Q_{rea}}}{R(^hE/^lE)_P} \quad (2.21)$$

where $R(^hE/^lE)_{P_{ins}}$ is the corresponding isotope ratio of the weighted average of all reaction products that are being formed in an infinitesimally short time. $R(^hE/^lE)_{Q_{rea}}$ is the isotope ratio of reactant molecules that are consumed through reaction in this short time, and $R(^hE/^lE)_Q$ is the isotope ratio of the remaining reactant and $R(^hE/^lE)_P$ that of the product. Only for small molecules in which all isotopes are located in the same reactive position, the isotopic fractionation factor is directly related to the *KIE* by:

$$\alpha_{kin}{}^h E_c = \frac{^l k}{^h k} = \frac{1}{KIE} \quad (2.22)$$

where $^l k$ is the rate constant of the light isotope and $^h k$ that of the heavy, respectively. In other cases the kinetic isotopic fractionation factor differs from a *KIE* in that *KIEs* are position-specific, whereas isotopic fractionation factors are derived from isotope ratios of reacting molecules and express average effects. Taking into account non-reacting positions and intramolecular competition, it is possible to estimate *KIEs* from kinetic isotopic fractionation factors as further described in Chapter 6.[53]

If $^l k/^h k > 1$ we talk about a normal isotope effect, and in the case where $^l k/^h k < 1$ the isotope effect is inverse. A short discussion about the estimation of the magnitude of isotope effects can be found in the environmental part of Chapter 6.

Another way to express isotopic fractionation is to use the isotopic enrichment factor as defined in eqn (2.23).

$$\varepsilon^h E_{P/R} = \alpha^h E_{P/R} - 1 \quad (2.23)$$

As with the isotopic fractionation factor α, the isotopic enrichment factor ε is a dimensionless quantity.

The 'isotopic difference' or 'separation' denoted by Δ is the difference between the δ-values of two compounds P and R:

$$\Delta^h E_{P,R,ref} = \delta^h E_P - \delta^h E_R \quad (2.24)$$

It is important that the reference used is clearly specified. The isotopic difference is related to the isotopic enrichment factor ε, the isotopic fractionation factor α and the δ-value via the following relations:

$$\Delta^h E_{P,R,ref} \approx \varepsilon^h E_{P/R} = \alpha^h E_{P/R} - 1 = \frac{\delta^h E_{P,ref} + 1}{\delta^h E_{R,ref} + 1} \qquad (2.25)$$

Additionally, the isotopic difference can be approximated by the natural logarithm of the isotopic fractionation factor:

$$\Delta^h E_{P,R,ref} \approx \ln \alpha^h E_{P/R} \qquad (2.26)$$

These approximations provide convenient but inaccurate estimates of the isotope fractionation [Ref. 7 p. 31].

In tracer studies with enriched samples, it is more convenient to express isotope data in terms of atom percent excess APE.[33] By definition, APE indicates the excess isotopic tracer in a compound in atom-% $(^hE)_c$ relative to a background (reference) value atom-% $(^hE)_{ref}$:

$$APE = \text{atom-\%}\ (^hE)_c - \text{atom-\%}\ (^hE)_{ref} \qquad (2.27)$$

The alternative formulation, APE' is also widely used but the obtained quantities are not equivalent. However, at natural abundance level similar values are obtained:

$$APE' = \frac{R - R_{ref}}{1 + (R - R_{ref})} \times 100 \qquad (2.28)$$

It was pointed out earlier that atom percent AP is not IUPAC conforming. For this reason, Coplen *et al.* recommend using the excess isotope-amount fraction,

$$x(^hE)_{c,ex} = x(^hE)_c - x(^hE)_{ref} \qquad (2.29)$$

where the subscript *ex* indicates an excess quantity.[27]

2.5 ISOTOPE MASS BALANCES

Mass balance equations are enormously important in stable isotope studies. They can be used for (i) isotope dilution analysis, (ii) blank correction, (iii) derivatization reactions, and (iv) determination of relative contributions of compounds to a pool. In Section 2.7 the importance of mass balance equations for the calculation of isotopic compositions of reactants and products in different chemical systems will be discussed. The following general equation can be used for all possible cases in which accurate calculations are necessary:

$$n_{tot}x(^hE)_{c,tot} = \sum_{i=1}^{k} n_i x_c(^hE)_i = n_1 x_c(^hE)_1 + \cdots + n_k x_c(^hE)_k \quad (2.30)$$

where n_i represents the amounts of a compound and $x_c(^hE)_i$ are the related isotope-amount fractions of a compound. As an approximation the isotope-amount fraction can be replaced by $\delta^h E_{c,i}$:

$$n_{tot}\delta^h E_{c,tot} = \sum_{i=1}^{k} n_i \delta^h E_{c,i} = n_1 \delta^h E_{c,1} + \cdots + n_k \delta^h E_{c,k} \quad (2.31)$$

The magnitude of error caused by this approximation depends on the element and whether the compounds are enriched or at natural abundance. In case of carbon at natural abundance the error induced is negligible.[33] The application of mass balances in case of compound derivatization is discussed in Chapter 4. For an in-depth discussion of mass balance applications for blank correction we refer to Brenna[33] as well as Hayes.[54] Hayes demonstrates an error propagation for the blank correction via a mass balance (see also Web findings).[54] Finally, Brand and Coplen discuss mass balances and Keeling plots.[25]

2.6 PHYSICAL KINETIC ISOTOPE FRACTIONATION DURING TRANSPORT PROCESSES

Kinetically driven transport processes such as effusion and diffusion can cause isotope fractionation that depends on molecular translation velocities and thus on isotope masses as can be deduced from the average translational kinetic energy of a molecule, expressed by:

$$\langle E_{kin} \rangle = \frac{3}{2} k_B T = \frac{1}{2} m \langle v^2 \rangle \quad (2.32)$$

where m is the mass of the molecule in kg, $\langle v \rangle$ its mean velocity in m s^{-1}, k_B the Boltzmann constant and T the temperature. At a given temperature all gases have the same average kinetic energy so that:

$$\frac{1}{2} m_{Gas\,A} \langle v^2 \rangle_{Gas\,A} = \frac{1}{2} m_{Gas\,B} \langle v^2 \rangle_{Gas\,B} \quad (2.33)$$

Accordingly, molecules containing different isotopes will have different average velocities due to their mass differences.

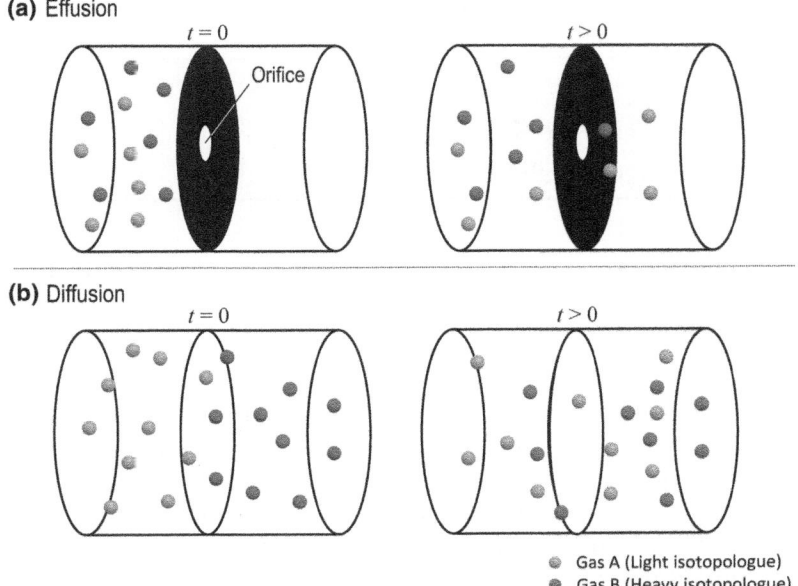

Figure 2.5 Effusion and diffusion. (a) Effusion of two gases through an orifice into a vacuum. (b) Diffusion of a *Gas A* into a *Gas B*.

However, effusion and diffusion differ from each other as will be discussed in the following. In Figure 2.5a, the process of effusion of a Gas A through an orifice into a vacuum is illustrated. The effusion rate φ of a gas through a hole is given by:

$$\varphi_{Gas} = \frac{pA_h N_A}{\sqrt{2\pi MRT}} \qquad (2.34)$$

where p is the pressure, A_h the area of the orifice through which the gas effuses, N_A Avogadro's constant, M the molar mass, R the universal gas constant and T the temperature. The effusion rate of two gases through an orifice is proportional to the mean velocities of the gas molecules and, according to Graham's law of effusion, the following relation is valid:

$$\frac{\varphi_{Gas\,A}}{\varphi_{Gas\,B}} \propto \frac{\langle v \rangle_{Gas\,A}}{\langle v \rangle_{Gas\,B}} = \sqrt{\frac{M_{Gas\,B}}{M_{Gas\,A}}} = \alpha_{Gas\,A/Gas\,B} \qquad (2.35)$$

If we assume that *Gas B* is the isotopologue containing the heavy isotope, for example, $M_{^{13}CO_2} = 45$ g mol^{-1} and *Gas A* the isotopologue with

Table 2.4 Effusion related fractionation factors.

Fractionation factor α
$\alpha_{^{12}CO_2/^{13}CO_2} = 1.0113$
$\alpha_{^{1}H_2/^{2}H_2} = 1.4142$
$\alpha_{^{14}N_2/^{15}N_2} = 1.0177$

the light isotope $M_{^{12}CO_2} = 44\,\mathrm{g\,mol^{-1}}$, then the fractionation factor is:

$$\alpha_{^{12}CO_2/^{13}CO_2} = \sqrt{\frac{45}{44}} = 1.0113 \qquad (2.36)$$

The light isotopologue $^{12}CO_2$ will leave the compartment 1.13% faster by effusion through an orifice than the heavier isotopologue. Table 2.4 gives examples of a few effusion related fractionation factors.

The driving force of diffusive transport is a gradient of the chemical potential of a compound μ_c which results in a flux of matter in the direction of the decreasing chemical potential. From the spatial derivation of the chemical potential $d\mu_c(x)/d(x)$, the drift velocity v_c as well as the flux density j_c of a compound can be derived. According to Fick's first law the net flux density of a gas, through a unit surface area is

$$j_c = -D_c \frac{dc_c}{dx} \qquad (2.37)$$

where dc_c/d_x is the concentration gradient of the compound in the direction of diffusion and D_c is the diffusion coefficient in $\mathrm{m^2\,s^{-1}}$.

In Figure 2.5b a diffusion process for a *Gas A* that diffuses into a *Gas B* is shown. The following proportionality for the diffusion coefficient is valid:

$$D_c \propto \sqrt{\frac{kT}{m}} \propto \sqrt{\frac{kT}{\mu}} \qquad (2.38)$$

This proportionality stems from the fact that all molecules in a gas mixture at a given temperature possess an equal average kinetic energy as given by eqn (2.32). If the diffusion process involves the movement of *Gas A* into *Gas B*, however, m has to be replaced by the reduced mass μ defined as:

$$\mu = \frac{m_{Gas\,A} m_{Gas\,B}}{m_{Gas\,A} + m_{Gas\,B}} \qquad (2.39)$$

The isotope fractionation expressed by the fractionation factor α can be expressed by the ratio of the diffusion coefficients of the heavy

Fundamental Aspects of Stable Isotopes and Isotopic Fractionation

Table 2.5 Diffusion-related isotopic fractionation factors of different gases.

Isotopic fractionation factor α
$\alpha^{13}C_{^{13}CO_2-He} = 0.9991$
$\alpha^{13}C_{^{13}CO_2-air} = 0.9956$
$\alpha^{18}O_{^{12}C^{18}O^{16}O-He} = 0.9982$
$\alpha^{2}H_{^{2}H_2-He} = 0.7746$
$\alpha^{2}H_{^{2}H_2-air} = 0.7188$

An average molar mass of air 29 g mol^{-1} was used.

isotopologue or isotope species ^hD and the light one ^lD. The molecular masses can be replaced by the molar weights M in numerator and denominator so that

$$\alpha\,^h E_{Gas\,A-Gas\,B} = \frac{^hD_{Gas\,A}}{^lD_{Gas\,A}} = \sqrt{\frac{^l\mu}{^h\mu}} = \sqrt{\frac{^hM_{Gas\,A} + M_{Gas\,B}}{^hM_{Gas\,A}M_{Gas\,B}} \times \frac{^lM_{Gas\,A}M_{Gas\,B}}{^lM_{Gas\,A} + M_{Gas\,B}}} \quad (2.40)$$

is valid.

An example in nature is the diffusion of CO_2 through air, an example important in instrumental measurements of the diffusion of CO_2 or other used conversion gases in helium. The isotopic fractionation factor of a heavy and a light CO_2 isotopologue ($^{13}CO_2$ and $^{12}CO_2$) in helium can be calculated according to eqn (2.40) by:

$$\alpha^{13}C_{^{13}CO_2}-He = \sqrt{\frac{M_{^{13}CO_2} + M_{He}}{M_{^{13}CO_2}M_{He}} \times \frac{M_{^{12}CO_2}M_{He}}{M_{^{12}CO_2} + M_{He}}} \quad (2.41)$$

With the molar masses $M_{^{13}CO_2} = 45$ g mol^{-1}, $M_{^{12}CO_2} = 44$ g mol^{-1} and $M_{He} = 4$ g mol^{-1} an isotopic fractionation factor of $\alpha^{13}C_{^{13}CO_2}-He = 0.9991$ can be calculated. A compilation of diffusion-related isotopic fractionation factors of different gases is given in Table 2.5.

2.7 ISOTOPIC FRACTIONATION UNDER VARIOUS SYSTEM CONDITIONS

In the following, we describe isotopic fractionation processes and how the isotopic compositions of reactants and products under different system and reaction conditions can be treated. As for all chemical systems we can distinguish between open and closed systems. Open systems can exchange matter and energy with their surroundings, whereas in closed systems only energy can be exchanged across the system boundaries. Another important factor is the reversibility or irreversibility of the reaction.

2.7.1 Reversible Reaction in a Closed System

The first isotopic fractionation process we discuss is a reversible reaction in a closed system. To that end, we look at a reversible reaction such as the nucleophilic aliphatic substitution S_N2 leading to a halogen exchange in halomethanes (see eqn (2.42)) in a closed vessel.

$$^{13}CH_3I + CH_3F \rightleftharpoons CH_3I + {}^{13}CH_3F \qquad (2.42)$$

The isotopic difference between products and educts will be controlled by the fractionation factor $\alpha_{eq}{}^{13}C_{CH_3I/CH_3F} = 1.0306$ given for 25 °C. If we work at natural abundance level we can form the mass balance equation:

$$\delta^{13}C_{tot} = f_{CH_3F} \times \delta^{13}C_{CH_3F} + (1 - f_{CH_3F}) \times \delta^{13}C_{CH_3I} \qquad (2.43)$$

where $\delta^{13}C_{tot}$ refers to the weighted average isotopic composition of all material involved in the equilibrium, f_{CH_3F} is the fraction of the product methyl fluoride and $1 - f_{CH_3F}$ is the fraction of the reactant methyl iodide. By using eqn (2.15) we can rearrange eqn (2.43) to:

$$\delta^{13}C_{CH_3I} = \alpha_{eq}{}^{13}C_{CH_3I/CH_3F} \times \delta^{13}C_{CH_3F} + (\alpha_{eq}{}^{13}C_{CH_3I/CH_3F} - 1) \qquad (2.44)$$

The second term on the right side of the equation is equivalent to the enrichment factor as it is given by eqn (2.23) and eqn (2.44) then becomes:

$$\delta^{13}C_{CH_3I} = \alpha_{eq}{}^{13}C_{CH_3I/CH_3F} \times \delta^{13}C_{CH_3F} + \varepsilon^{13}C_{CH_3I/CH_3F} \qquad (2.45)$$

A substitution of $\delta^{13}C_{CH_3I}$ in eqn (2.43) leads to

$$\begin{aligned}\delta^{13}C_{tot} = & f_{CH_3F} \times \delta^{13}C_{CH_3F} + (1 - f_{CH_3F}) \\ & \cdot (\alpha_{eq}{}^{13}C_{CH_3I/CH_3F} \times \delta^{13}C_{CH_3F} + \varepsilon^{13}C_{CH_3I/CH_3F})\end{aligned} \qquad (2.46)$$

Solving this equation for $\delta^{13}C_{CH_3F}$, we get an expression which allows the calculation of $\delta^{13}C_{CH_3F}$ as a function of the fractionation factor and the average isotopic composition of all material involved:

$$\delta^{13}C_{CH_3F} = \frac{\delta^{13}C_{tot} - (1 - f_{CH_3F}) \times \varepsilon^{13}C_{CH_3I/CH_3F}}{\alpha_{eq}^{13}C_{CH_3I/CH_3F} \times (1 - f_{CH_3F}) + f_{CH_3F}} \quad (2.47)$$

If we substitute eqn (2.47) in eqn (2.45) we obtain an expression for the calculation of $\delta^{13}C_{CH_3I}$:

$$\delta^{13}C_{CH_3I} = \frac{\alpha_{eq}^{13}C_{CH_3I/CH_3F} \times \delta^{13}C_{tot} + f_{CH_3F} \times \varepsilon^{13}C_{CH_3I/CH_3F}}{\alpha_{eq}^{13}C_{CH_3I/CH_3F} \times (1 - f_{CH_3F}) + f_{CH_3F}} \quad (2.48)$$

However, eqn (2.47) and eqn (2.48) for the isotopic composition of the product and reactant are rarely employed. Because the isotopic fractionation factor is close to unity, the approximation $\alpha \approx 1$ leads to the far simpler expressions:

$$\delta^{13}C_{CH_3F} \approx \delta^{13}C_{tot} - (1 - f_{CH_3F}) \times \varepsilon^{13}C_{CH_3I/CH_3F} \quad (2.49)$$

and

$$\delta^{13}C_{CH_3I} \approx \delta^{13}C_{tot} + f_{CH_3F} \times \varepsilon^{13}C_{CH_3I/CH_3F} \quad (2.50)$$

There are some circumstances in which these approximations should not be used. One is when highly precise calculations are necessary, the other is when hydrogen isotopic fractionations are investigated. In such cases the fractionation factor often differs by more than 10% from 1, which leads to strong deviations from the accurate value. In Figure 2.6 accurate and approximate calculations of isotopic composition of reactants and products of the reversible $S_N 2$ reaction given by eqn (2.42) are presented. The general construction of such a graph is easy, for $f_{CH_3F} = 1$ we have $\delta^{13}C_{CH_3F} = \delta^{13}C_{tot}$ and for $f_{CH_3F} = 0$ we have $\delta^{13}C_{CH_3I} = \delta^{13}C_{tot}$. In this particular case the initial δ-value of the reactants is assumed to be $\delta^{13}C_{CH_3I} = -25‰$, and because the fractionation factor is bigger than 1 we get a positive enrichment factor. Thus, CH_3I will be enriched in ^{13}C relative to CH_3F, which can be seen by the downward slope with decreasing f_{CH_3F}. The intercepts at the opposite ends of the curves are given by $\delta^{13}C_{tot} \pm \varepsilon^{13}C_{CH_3I/CH_3F}$.

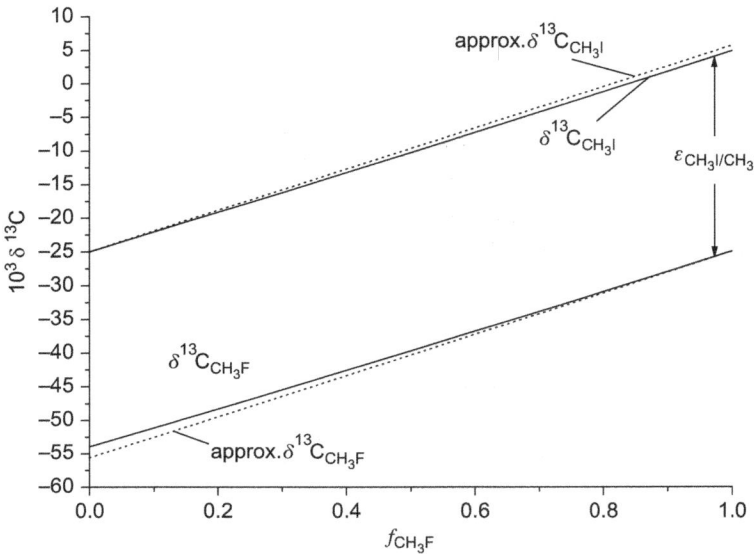

Figure 2.6 The graph shows results of both accurate and approximate calculations of isotopic composition of reactants and products of the reversible S_N2 reaction given by eqn (2.42). The solid lines are based on the accurate eqn (2.47) and eqn (2.48), respectively. The difference between these two lines is the enrichment factor of the reaction. The dotted lines are calculated according to the approximations given by eqn (2.49) and eqn (2.50).

2.7.2 Irreversible Reaction in a Closed System

If we assume an irreversible reaction of the type

$$Q \rightarrow P \qquad (2.51)$$

where Q is the reactant and P the resulting product of the reaction, then:

$$\alpha_{kin}^{h}E_{P/Q} = \frac{R(^hE/^lE)_P}{R(^hE/^lE)_Q} = \frac{dN(^hE)_P/dN(^lE)_P}{N(^hE)_Q/N(^lE)_Q} \qquad (2.52)$$

According to a mass balance eqns (2.53) are valid.

$$\begin{aligned} dN(^hE)_P &= -dN(^hE)_Q \\ dN(^lE)_P &= -dN(^lE)_Q \end{aligned} \qquad (2.53)$$

Combination of eqn (2.52) and eqns (2.53) and separation of the variables leads to

$$\frac{dN(^hE)_Q}{N(^hE)_Q} = \alpha_{kin}^h E_{P/Q} \cdot \frac{dN(^lE)_Q}{N(^lE)_Q} \qquad (2.54)$$

Integration of eqn (2.54) from $N(^hE)_{Q_0}$ to $N(^hE)_{Q_t}$ and $N(^lE)_{Q_0}$ to $N(^lE)_{Q_t}$ at the start of the reaction $t=0$ and time t,

$$\int_{N(^hE)_{Q_0}}^{N(^hE)_{Q_t}} \frac{1}{N(^hE)_Q} dN(^hE)_Q = \alpha_{kin}^h E_{P/Q} \cdot \int_{N(^lE)_{Q_0}}^{N(^lE)_{Q_t}} \frac{1}{N(^lE)_Q} dN(^lE)_Q \qquad (2.55)$$

leads to:

$$\ln \frac{N(^hE)_{Q_t}}{N(^hE)_{Q_0}} = \alpha_{kin}^h E_{P/Q} \cdot \ln \frac{N(^lE)_{Q_t}}{N(^lE)_{Q_0}} \qquad (2.56)$$

This equation can be rearranged to

$$\frac{N(^hE)_{Q_t}}{N(^hE)_{Q_0}} = \left(\frac{N(^lE)_{Q_t}}{N(^lE)_{Q_0}}\right)^{\alpha_{kin}^h E_{P/Q}} \qquad (2.57)$$

and division of both sides with $N(^lE)_{Q_t}/N(^lE)_{Q_0}$ leads to the following expression,

$$\frac{R(^hE/^lE)_{Q_t}}{R(^hE/^lE)_{Q_0}} = \left(\frac{N(^lE)_{Q_t}}{N(^lE)_{Q_0}}\right)^{(\alpha_{kin}^h E_{P/Q} - 1)} \qquad (2.58)$$

The fraction of the reactant can be written by:

$$f = \frac{c}{c_0} = \frac{N(^lE)_{Q_t} + N(^hE)_{Q_t}}{N(^lE)_{Q_0} + N(^hE)_{Q_0}} = \frac{N(^lE)_{Q_t} \cdot (1 + R(^hE/^lE)_{Q_t})}{N(^lE)_{Q_0} \cdot (1 + R_0(^hE/^lE)_{Q_0})} \qquad (2.59)$$

so that

$$\frac{R(^hE/^lE)_{Q_t}}{R(^hE/^lE)_{Q_0}} = f^{(\alpha_{kin}^h E_{P/Q} - 1)} \qquad (2.60)$$

where $R(^hE/^lE)_{Q_t}$ and $R(^hE/^lE)_{Q_0}$ are the ratios of the heavy isotope to the light isotope in the reactant at times $t=0$ and t, respectively, f is the remaining fraction of the reactant at time t $(=c_t/c_0)$, and α is the fractionation factor.

This expression is the so-called Rayleigh equation. An equation for the accumulated product can be derived by using an isotope mass balance equation that links the isotope ratio of reactant and accumulated product, and assuming that all atoms of an element are transferred to the product (see eqn (2.61)).

$$\delta^{13}C_P = \delta^{13}C_0 - \varepsilon \frac{f \cdot \ln f}{1-f} \qquad (2.61)$$

Figure 2.7 shows that the greater the amount of a compound that is degraded the larger is the isotope shift in the residual fraction. In order to determine the fractionation factor α or enrichment factor ε from such data, the Rayleigh equation is linearized (isotope ratios in δ-scale):

$$\ln\left(\frac{\delta^h E_{Q_t}+1}{\delta^h E_{Q_0}+1}\right) = \ln\left(\frac{\delta^h E_{Q_0}+\Delta\delta^h E_{Q_t}+1}{\delta^h E_{Q_0}+1}\right) = \varepsilon \ln f = (\alpha_{kin}^h E_{P/Q} - 1) \ln f \qquad (2.62)$$

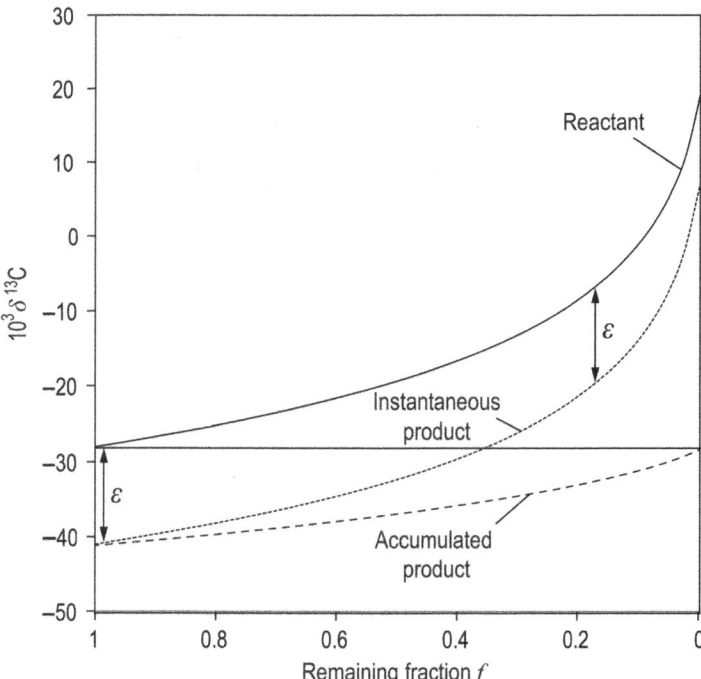

Figure 2.7 Representation of an irreversible isotopic fractionation in a closed system. The δ-value is plotted against the remaining fraction. The solid curve represents the reactant and the fine dashed line the instantaneously formed product. The lower dashed line describes the accumulated product.

Isotope data can then be presented in a double logarithmic plot of $\ln\left(\frac{\delta^h E_{Q_t}+1}{\delta^h E_{Q_0}+1}\right)$ vs. $\ln f$ where $\alpha - 1$ or ε are equivalent to the slope. It is not recommended to force the linear regression line through the origin.[53] Scott et al. have thoroughly compared data analysis methods for Rayleigh-type isotope data. They conclude that the best generally applicable method involves the use of Pitman estimators to combine datasets, in particular if the level of error varies among datasets.[55] Rayleigh fractionation in the context of environmental applications is discussed in more detail in Chapter 6.

2.7.3 Reversible Reaction in an Open System

A reversible reaction in an open system, in which the product is lost can be described by Rayleigh fractionation in the same way as for an irreversible process in a closed system. The most prominent example is the condensation of water in clouds. The heavy isotopes accumulate in the condensed phase and precipitate as rain or snow. The remaining water vapour fraction within the cloud contains more light water isotopologues. The shape of the involved curves is therefore the inverse of Figure 2.7.

2.7.4 Irreversible Reaction in an Open System

Finally, we will discuss an irreversible reaction in an open system in which an infinitely available reactant Q is transformed into an accumulating product P_{ac}. A prominent example for such a system is the accumulation of biomass in plants by assimilation of CO_2 and H_2O from the atmosphere.[54] For the relationship between the δ-value of the reactant $\delta^h E_Q$ and the accumulated product $\delta^h E_{P_{ac}}$, eqn (2.15) can be used to describe the isotopic fractionation factor of the system so that eqn (2.63) results:

$$\alpha^h E_{P_{ac}/Q} = \frac{\delta^h E_{P_{ac}} + 1}{\delta^h E_Q + 1} \qquad (2.63)$$

By rearranging eqn (2.63) to eqn (2.64), the δ-value of the accumulated product $\delta^h E_{P_{ac}}$ can be derived.

$$\delta^h E_{P_{ac}} = \alpha^h E_{P_{ac}/Q} \cdot \delta^h E_Q + \varepsilon^h E_{P_{ac}/Q} \qquad (2.64)$$

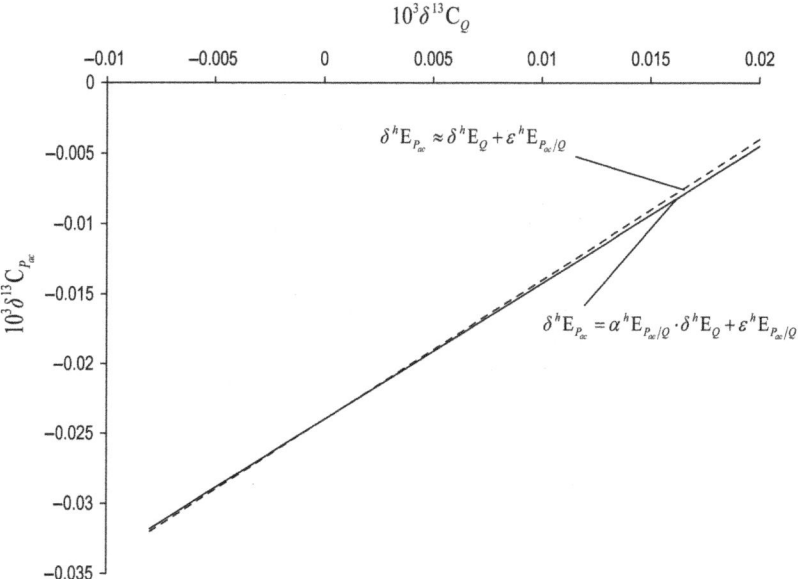

Figure 2.8 Irreversible reaction in an open system in which an infinitely available reactant Q is transformed into an accumulated product P_{ac}. The solid line shows the exact calculation including the isotopic fractionation factor (see eqn (2.64)). The dotted line represents the approximation according to eqn (2.65).

Eqn (2.64) can be simplified by the approximation that the isotopic fractionation factor is close to unity, so that

$$\delta^h E_{P_{ac}} \approx \delta^h E_Q + \varepsilon^h E_{P_{ac}/Q} \qquad (2.65)$$

results. It has to be emphasized that this approximation is valid for carbon at natural abundance, but leads to huge errors when hydrogen is involved. An example for such an isotopic fractionation is shown in Figure 2.8. For the calculation of the δ-value of the accumulated product an isotopic fractionation factor of $\alpha^{13}C_{P_{ac}/Q} = 0.976$ and thus $\varepsilon^{13}C_{P_{ac}/Q} = -0.024$ has been used.

2.8 WEB FINDINGS

Interactive Nuclide Chart
Provided by National Nuclear Data Center, Brookhaven National Laboratory. http://www.nndc.bnl.gov/chart/

The Official Web Site of the Nobel Prize
Francis W. Aston and a link to the Nobel Prize lecture of the year 1922: http://www.nobelprize.org/nobel_prizes/chemistry/laureates/1922/aston-bio.html
Joseph John Thomson and a link to the Nobel Prize lecture of the year 1906: http://www.nobelprize.org/nobel_prizes/physics/laureates/1906/thomson.html

An Introduction to Isotopic Calculations
John M. Hayes provides a detailed overview of isotopic calculations.
http://www.whoi.edu/nosams/page.do?pid=50996&tid=282&cid=74886
Practice and Principles of Isotopic Measurements in Organic Geochemistry
John M. Hayes, edited by Alex Sessions, Revision 2, August 2002. The text provides a more detailed introduction to isotopic calculations. Different systems in which isotopic fractionation occurs are treated in this text.
http://www.whoi.edu/nosams/page.do?pid=50996&tid=282&cid=74886

REFERENCES

1. *Physical Measurement Laboratory of NIST*; http://physics.nist.gov/cuu/index.html
2. L. Pauling, *General Chemistry*, 3rd edn., Dover Pubn. Inc., New York, 1988
3. P. W. Atkins, *Physical Chemistry*, 5th edn., Oxford University Press, Oxford, 1994.
4. H. Haken and H. C. Wolf, *The Physics of Atoms and Quanta: introduction to experiments and theory*, 7th rev. and enl. edn., Springer, Berlin, 2005.
5. E. Wiberg, N. Wiberg and A. F. L. d. a. C. Holleman, *Inorganic Chemistry*, 1st English edn./[edited] by Nils Wiberg, Academic Press, San Diego, London, 2001.
6. P. A. Tipler and R. A. Llewellyn, *Modern Physics*, 5th edn., W. H. Freeman, New York, 2008.
7. R. E. Criss, *Principles of Stable Isotope Distribution*, Oxford University Press, New York, 1999.
8. B. M. Peake, *J. Chem. Educ.*, 1989, **66**, 738–738.
9. H. A. Shadduck, *J. Chem. Educ.*, 1936, **13**, 303–308.

10. E. Segrè, *From X-rays to Quarks: Modern Physicists and their Discoveries*, W. H. Freeman, San Francisco, 1980.
11. D. R. Lide, *CRC Handbook of Chemistry and Physics: a Ready-Reference Book of Chemical and Physical Data*, 90th edn./editor-in-chief, David R. Lide, CRC, Boca Raton, FL; London, 2009.
12. G. Faure and T. M. Mensing, *Isotopes: Principles and Applications*, 3rd edn., Wiley, Hoboken, NJ, 2005.
13. J. R. De Laeter, J. K. Böhlke, P. De Bièvre, H. Hidaka, H. S. Peiser, K. J. R. Rosman and P. D. P. Taylor, *Pure Appl. Chem.*, 2003, **75**, 683–800.
14. P. de Marcillac, N. Coron, G. Dambier, J. Leblanc and J.-P. Moalic, *Nature*, 2003, **422**, 876–878.
15. H. Budzikiewicz and R. D. Grigsby, *Mass Spectrom. Rev.*, 2006, **25**, 146–157.
16. H. E. Watson, *J. Chem. Soc.*, 1910, **97**, 810–833.
17. J. R. De Laeter, *Mass Spectrom. Rev.*, 2009, **28**, 2–19.
18. J. J. Thomson, *Proc. R. Soc. London, Ser. A*, 1914, **89**, 1–20.
19. F. W. Aston, *Philos. Mag.*, 1920, **39**, 449–455.
20. F. W. Aston, *Nature*, 1935, **135**, 686–687.
21. H. C. Urey, *Science*, 1948, **108**, 489–496.
22. O. W. Richards and M. E. Lembert, *J. Am. Chem. Soc.*, 1914, **36**, 1329–1344.
23. F. Soddy, *Radioactivity. Annual Reports on the Progress of Chemistry for 1913*, The Chemical Society, London, 1914.
24. M. C. Nagel, *J. Chem. Educ.*, 1982, **59**, 739–740.
25. W. A. Brand and T. B. Coplen, *Isot. Environ. Health Stud.*, 2012, **48**, 393–409.
26. M. Berglund and M. E. Wieser, *Pure Appl. Chem.*, 2011, **83**, 397–410.
27. T. B. Coplen, *Rapid Commun. Mass Spectrom.*, 2011, **25**, 2538–2560.
28. G. M. Barrow, *Physikalische Chemie*, 6 edn., Bohmann-Verlag, Wien, 1984.
29. H. E. Suess and H. C. Urey, *Rev. Mod. Phys.*, 1956, **28**, 53–74.
30. H. E. Duckworth and A. O. Nier, *Int. J. Mass Spectrom. Ion Processes*, 1988, **86**, 1–19.
31. T. B. Coplen, J. K. Böhlke, P. De Bièvre, T. Ding, N. E. Holden, J. A. Hopple, H. R. Krouse, A. Lamberty, H. S. Peiser, K. Révész, S. E. Rieder, K. J. R. Rosman, E. Roth, P. D. P. Taylor, R. D. Vocke Jr and Y. K. Xiao, *Pure Appl. Chem.*, 2002, **74**, 1987–2017.
32. *IUPAC Gold Book*, http://goldbook.iupac.org/
33. J. T. Brenna, T. N. Corso, H. J. Tobias and R. J. Caimi, *Mass Spectrom. Rev.*, 1997, **16**, 227–258.

34. Z. Muccio and G. P. Jackson, *Analyst*, 2009, **134**, 213–222.
35. M. H. O'Leary, S. Madhavan and P. Paneth, *Plant. Cell Environ.*, 1992, **15**, 1099–1104.
36. H. C. Urey, S. Epstein, C. McKinney and J. McCrea, *Geol. Soc. Am. Bull.*, 1948, **59**, 1359–1360.
37. H. C. Urey, S. Epstein, H. A. Lowenstam and C. R. McKinney, *Science*, 1950, **111**, 462–463.
38. P. Ghosh and W. A. Brand, *Int. J. Mass Spectrom.*, 2003, **228**, 1–33.
39. C. R. McKinney, J. M. McCrea, S. Epstein, H. A. Allen and H. C. Urey, *Rev. Sci. Instrum.*, 1950, **21**, 724–730.
40. R. A. Werner and W. A. Brand, *Rapid Commun. Mass Spectrom.*, 2001, **15**, 501–519.
41. J. C. De Wit, C. M. Van der Straaten and W. G. Mook, *Geostand. Newslett.*, 1980, **4**, 33–36.
42. Q. L. Zhang and T. Ding, *Chinese Sci. Bull.*, 1989, **34**, 1086–1089.
43. P. De Bièvre, S. Valkiers, H. S. Peiser, P. D. P. Taylor and P. Hansen, *Metrologia*, 1996, **33**, 447–455.
44. P. Baertschi, *Earth Planet. Sci. Lett.*, 1976, **31**, 341–344.
45. T. P. Ding, R. M. Bai, Y. H. Li and D. F. Wan, *Sci. China Ser. D-Earth Sci.*, 1999, **42**, 45–51.
46. P. D. P. Taylor, P. De Bièvre and S. Valkiers, in *Handbook of Stable Isotope Analytical Techniques*, ed. P. A. De Groot, Elsevier, Amsterdam, 2004, vol. 1, pp. 907–927.
47. E. A. Schauble, *Reviews in Mineralogy & Geochemistry*, 2004, **55**, 65–111.
48. J. Bigeleisen, *J. Chem. Phys.*, 1949, **17**, 675–678.
49. H. C. Urey, *J. Chem. Soc.*, 1947, 562–581.
50. L. C. S. Melander and W. H. Saunders, *Reaction Rates of Isotopic Molecules*, Wiley, New York, 1980.
51. A. Kohen and H.-H. Limbach, *Isotope Effects in Chemistry and Biology*, Taylor & Francis, Boca Raton, 2006.
52. M. Wolfsberg, W. A. van Hook and P. Paneth, *Isotope Effects in the Chemical, Geological, and Bio Sciences*, Springer, Dordrecht, 2010.
53. M. Elsner, L. Zwank, D. Hunkeler and R. P. Schwarzenbach, *Environ. Sci. Technol.*, 2005, **39**, 6896–6916.
54. J. M. Hayes, *An Introduction to Isotopic Calculations, Teaching notes, Woods Hole Oceanographic Institution*, http://www.whoi.edu/nosams/page.do?pid=50996&tid=282&cid=74886.
55. K. M. Scott, X. Liu, C. M. Cavanaugh and J. S. Liu, *Geochim. Cosmochim. Acta*, 2004, **68**, 433–442.

CHAPTER 3
Instrumentation for Compound-specific Stable Isotope Analysis

3.1 ISOTOPE RATIO MASS SPECTROMETER

Mass spectrometry is currently the standard technique to measure isotope ratios of compounds at natural abundance.[1] In principle, all types of mass spectrometers can be used to measure isotope abundances but mostly not with the necessary precision.[2] For the light elements (C, H, N and O), the most commonly used mass spectrometers are isotope ratio mass spectrometers (IRMS) and scanning mass spectrometers (for example, quadrupole: Q-MS; ion trap: IT-MS; and time-of-flight instruments: TOF-MS).[1] In Figure 3.1, precision of these instruments for measurement of isotope ratios is compared. Measurement of isotope ratios at natural abundance level requires precision on the order of 10^{-2}–10^{-4}%.[1,2] For the light elements such precision can only be achieved by gas source magnetic sector field instruments which employ simultaneous measurement of ionized gaseous species in arrays of fixed Faraday collectors. For isotope ratio measurements by gas source instruments, the isotopes of interest in the molecules being analysed must be converted into specific analytes, including CO_2, N_2, CO or H_2 for carbon, nitrogen, oxygen and hydrogen measurements, respectively. In scanning mass spectrometry, isotope ratios of ionized organic molecules are measured sequentially on the ionized molecules or fragments thereof, with no intermediate conversion to a common analyte. Generally, precisions for molecular MS are in the range of 0.05% and 2%

Compound-specific Stable Isotope Analysis
By Maik A. Jochmann and Torsten C. Schmidt
© The Royal Society of Chemistry 2012
Published by the Royal Society of Chemistry, www.rsc.org

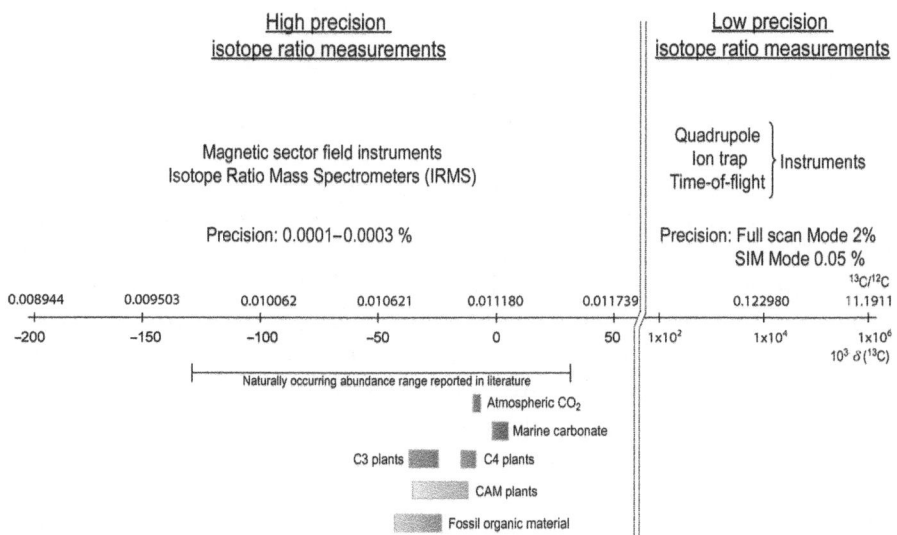

Figure 3.1 Mass spectrometric precision and employed instruments. Mass spectrometric instruments for the measurement of carbon isotope ratios ranging from natural abundance level up to ^{13}C enriched samples. $\delta^{13}C$ scale as well as isotope ratios are shown. According to the achievable precision the scheme classifies instruments in two categories: (i) IRMS instruments that obtain high precision isotope ratio measurements (between 0.0001% and 0.0003%) and (ii) molecular MS instruments (quadrupole, ion trap, time-of-flight) that reach precisions in the 0.05–2% range.
The figure was redrawn based on Godin et al.[1]

when using single ion monitoring (SIM) or full scan (total ion current: TIC) mode, respectively.[1] For the light elements, this precision of molecular MS is sufficient only for determination of isotope ratios of compounds artificially enriched in one isotope.

For chlorine and bromine isotope ratios at natural abundance level it has been shown that Q-MS can be used to determine isotope ratios with a precision sufficient for environmental applications.[3,4] The on-line CSIA of sulfur-containing compounds was achieved by coupling a gas chromatograph with multicollector inductively coupled plasma mass spectrometry (MC-ICP-MS) by Amrani and co-workers.[5] For a detailed discussion of these two techniques see Chapter 7.

In this chapter, the instrumentation and measurement fundamentals of high precision gas source IRMSs will be discussed in detail.

The basic design for an IRMS was developed by Alfred Nier in the 1940s and is the basis for all modern instruments.[6–9] All gas source mass spectrometers for isotope ratio determination comprise a vacuum system, an ion source for ion generation, ion beam focusing and acceleration, a flight tube with a magnetic field separator for mass analysis, in which the ions are separated according to their mass-to-charge ratio (m_{ion}/q_{ion}) as well as an ion detection system of two or more Faraday cups.

Isotope ratio mass spectrometers for continuous-flow applications required special modifications to analyse transient signals arising from the introduction of compounds in a carrier gas from, for example, a gas chromatograph into the ion source of the mass spectrometer.[10] These modifications include an open-split to ensure constant flow of helium into the ion source, differential pumping systems to manage high gas flows into the ion source, and voltage-to-frequency converters that are designed to deal with large dynamic ranges of ion currents.[10,11] In particular, this set-up requires a high ratio linearity L_R (see eqn (5.11)) as well as a high stability (no instrumental drift; see Chapter 5).[11,12] In the following the main components of isotope ratio mass spectrometers (see Figure 3.2) will be discussed with a focus on continuous-flow inlet systems.

Figure 3.2 Typical stable isotope ratio mass spectrometer. A stable isotope ratio mass spectrometer (MAT 253™, Thermo Fisher Scientific, Bremen, Germany), with a) ion source, b) high vacuum differential pumping system, c) flight tube, d) electromagnet, e) and f) amplifier housings.

3.1.1 Vacuum System

All gas source isotope ratio mass spectrometer systems possess a vacuum system for evacuation of the ion source, the analyser and (if installed) the amplifier housing.

The source vacuum is generally around 10^{-6} to 10^{-7} mbar[13] and around 10^{-9} mbar or better[14] with and without sample inflow, respectively. The operation of the mass spectrometer under vacuum is important mainly for the following reasons:

(i) Ion collisions with residual gas molecules will alter ion trajectories between the ion source and the collectors, resulting in peak broadening or even detection in the wrong cups.
(ii) Formation of other charged species during collision can lead to isobaric interferences.
(iii) Filament lifetime is reduced at higher gas pressures.
(iv) In continuous-flow, the system has to deal with high amounts of helium, which has to be handled by the use of differential pumping systems.[12]

Depending on the pressure, vacuums can be classified into three categories.[15] These categories are: rough vacuum between 1 bar and 10^{-3} mbar; high vacuum between 10^{-3} and 10^{-8} mbar; and ultra-high vacuum $<10^{-8}$ mbar. An important characteristic of a gas that can be used to define the various pressure categories is the mean free path length λ in metres (m). It determines the mean distance a molecule or atom can move before it collides with another molecule and is given by:

$$\lambda = k_B T / \sqrt{2}\, p\pi(2r_0)^2 \tag{3.1}$$

where k_B is the Boltzmann constant (1.38×10^{-23} J K^{-1}), T the thermodynamic temperature of the gas in kelvin (K), r_0 the atom or molecule radius in metres and p the pressure in pascal (Pa). According to this equation, the mean free path length in air at 298 K and a pressure of 1 bar is in the order of 10^{-7} m, whereas between 10^{-4} and 10^{-6} mbar it is in the order of ~ 1 to ~ 100 m, respectively. According to the mean free path length, gas flow through a tube or an orifice of diameter d can be divided into three gas flow regimes characterized by the Knudsen number.[15] First, if the mean free path length of a gas molecule is short compared to the dimension of the tube or orifice ($\lambda \ll d \Rightarrow Kn \ll 1$), the flow is *viscous*. The flow in this region is mainly determined by collisions between the gas molecules. This is the case at pressures greater than

~10^{-2} mbar, and during pumping, gas molecules move like a fluid. Second, if the mean free path length is long compared to the system dimensions ($\lambda \gg d \Rightarrow Kn \gg 1$), a *molecular* flow regime prevails. The gas molecules are so far apart from each other that they no longer exert any interdependence, and their motion is strictly random: they only collide with the walls. This flow regime prevails in the high vacuum region of the mass spectrometer. Because of the randomly moving molecules, tubes with relatively large diameter are required for evacuation of gases from a high vacuum vessel. A third region of gas flow is the *Knudsen* or *transition* flow ($\lambda \approx d \Rightarrow Kn \approx 1$), which is often used to describe the region in between the *viscous* and *molecular* flow regime. In Section 3.2 the viscous and molecular flow regimes will be discussed in the context of dual-inlet and continuous-flow inlet systems.

The most commonly used rotary vane oil pumps produce rough or fore-line vacuum conditions. For oil-free vacuum, also membrane pumps or scroll pumps can be used. A rotary vane pump consists of a housing with an intake port and an exhaust port that is equipped with a discharge valve. In the housing, spring loaded rotor vanes compress the gas and transport it out of the exhaust port. The housing is immersed in an oil bath and most pumps possess two stages for better vacuum production. The oil seals the pump against the atmospheric pressure but also cools and lubricates the pump. The achievable pressure of 10^{-2}–10^{-3} mbar is partly determined by the vapour pressure of the vacuum pump oil. With time, the oil becomes loaded with water or other impurities, which can reduce the pump performance. Therefore, the usually installed gas ballast valve should be opened periodically (see Chapter 5) to remove gaseous impurities. At low pressures, rotary vane pumps tend to backstream oil vapour into the rough vacuum line. A fore-line trap filled with a molecular sieve can be used to prevent such contaminations. The pump exhaust gas is generally guided via a hose into a lab exhaust and sometimes an oil mist filter is used.

For generating high vacuum under molecular flow conditions, turbomolecular pumps are used. A turbomolecular pump consists of a number of rotor-stator pairs with multiple blades arranged at an angle of, usually, 45°. The rotor-stator pairs are mounted in series, and the rotor blades rotate at 10 000 to 100 000 revolutions per minute so that they possess the mean thermal velocity of the gas molecules.[15] The pumping is based on a momentum transmission from the rotor blades to the atoms and molecules. A fore-line vacuum is necessary to prevent the pump overheating by friction, thus the pump only operates if the rotary vane pump is working. Additionally, a self-regulation mechanism accelerates the turbo pump when the pressure drops. In case of a system shut down, when the turbomolecular pump is turned off while still connected to the high vacuum region of the mass spectrometer, hydrocarbons will diffuse through the rotor-stator arrangement of the turbomolecular pump into the high vacuum region of the pump. Diffusion into the high vacuum region of the mass spectrometer can be kept at a minimum by opening a vent valve that admits vent gas to the turbomolecular pump when the pump speed drops below 50% of the full operating rotational speed (consult manufacturer manuals). The inflowing air can be noticed by a sibilant. For pump maintenance the lubricant brushes (oil wicks) should be replaced regularly (for example, once a year, see also Chapter 5).

Nowadays, most IRMS systems are equipped with differential pumping units. Such systems possess an additional turbomolecular pump that improves the vacuum in the analyser section of the IRMS, so that the analyser section is virtually separated from the ion source section.[12] In particular, in continuous-flow applications such as GC- or EA-IRMS with high helium carrier gas loading in the ion source, this can be advantageous, resulting in higher abundance sensitivities and a lower background.[12]

The fore-line vacuum of IRMS systems is monitored by a Pirani vacuum gauge located in the inlet system. In contrast, the high vacuum is monitored by an ionization vacuum gauge (Penning, Bayard-Alpert).

Pirani gauges are employed at pressures between 1 bar and 10^{-3} mbar. A Pirani gauge consists of a measuring tube and a compensator chamber. The measuring tube contains a heated filament, while the compensator chamber accommodates a resistor, which is held at constant temperature and pressure. Both the filament in the sampling chamber and the resistor in the compensator chamber are part of a balanced Wheatstone bridge circuit. In such a balanced bridge circuit, the current flow is equal and an amperemeter placed at the centre reads zero. The filament resistance depends on the heat conductance of the gas, which changes with the amount of gas and thus with pressure.[15] The resulting current is calibrated for air pressure.

Penning vacuum gauges are ionization gauges, in which electrons are emitted from a cathode and are accelerated to an anode by an electric potential of 3 kV within a magnetic field. As described in more detail in Section 3.1.2, the magnet forces the electrons on spiral paths, prolonging their travelling distance to increase the number of collisions with residual gas molecules and thus the ionization probability. Penning vacuum gauges are also called 'active inverse magnetron gauges'. The positively charged ions are collected at the anode and the resulting current is a function of the pressure within the high vacuum region of the mass spectrometer. This correlation is valid if the mean free path length of the electrons is large compared with the distance between the filament and the anode. Under such circumstances the number of formed ions is proportional to the number of gas molecules within the gauge.[15] The measurement range of the Penning gauges is between 10^{-2} and 10^{-9} mbar.

3.1.2 Ion Source and Focusing Beam Optics

Gas source isotope ratio mass spectrometers usually are equipped with electron-impact (EI) ion sources. The first EI source was introduced by Dempster in the 1920s and, to date, all commercially available sources are similar to the 'Nier type' ion source design.[16] In the upper part of Figure 3.3, the scheme of a typical EI ion source with electron optics for ion beam focusing is depicted. IRMS ion sources are used to provide high sensitivity, maximum ionization probability, high stability and low energy spread of the ion beam.[2,10,14]

Contrary to organic mass spectrometers for analysis of organic molecules, 'tight' ion sources are used in IRMS.[2] In molecular mass spectrometry it is typically undesirable for organic molecules to collide with each other and form cluster ions with masses greater than the analyte mass. The formation of such ions complicates the EI mass spectrum evaluation and, therefore, the analyte stream generally passes the ion beam only once, so that the ions are either ionized and accelerated into the mass analyser or pumped out of the 'open' source.[2,17] IRMS ion sources are tailored to provide a maximum ionization probability. Therefore, the source is designed to be gas-tight. The gas molecules still undergo (undesired) collisions with the walls of the ion box with some enhancement of the probability that they will collide with other molecules as well. Because of this, EI sources of an IRMS have only the entrance and exit holes of the electrons plus the gas inlet and ion exit slit.

The conductance of the ion box openings is much lower than the pumping rate of the vacuum pumps, which results in a roughly 100 times higher pressure within the ion box compared to the outside.[12] Ceramic spacers that insulate the different parts in the ion source also provide its tightness. For ionization, electrons are emitted from a resistively heated tungsten filament ($\sim 2300\,°C$) or a thoriated iridium filament ($\sim 1800\,°C$). Tungsten is mainly used because of its high temperature stability as well as the high current densities produced due to its rather low work function of ~ 4.5 eV. The emitted electrons are accelerated by an electrostatic potential between the filament and the ionization box as well as the electron trap (see Figure 3.3). Electrons that hit the ion box

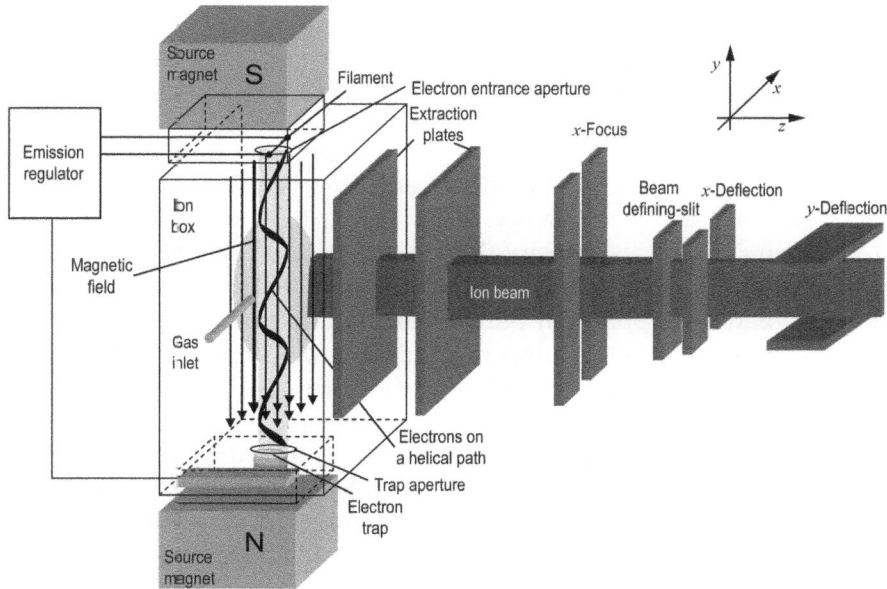

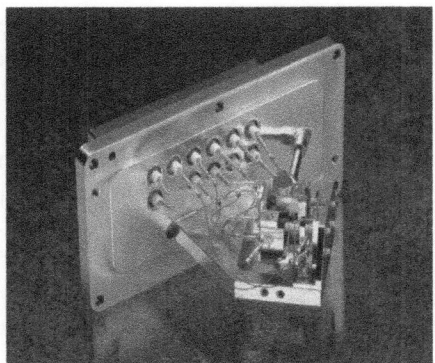

Figure 3.3 Ion source design scheme of an IRMS. Top: a generic scheme of an EI ion source. The ion source has a closed or tight design. Ceramic insulating spacers that also provide an enclosure for the whole source are omitted for clarity. Bottom left: the ion source of a MAT 253™ from Thermo Fisher Scientific, Bremen (picture by Jens Laaks). Bottom right: ion source of an Isoprime Prism™ mass spectrometer.
(Reproduced by permission of Elementar, Hanau, Germany).

and electrons reaching the electron trap cause currents; box current and trap current, respectively. The electrons are accelerated by the first electrostatic potential to energies between 50 and 150 eV before they enter the ionization box via the electron entrance aperture. The electron emission current is typically around 1 mA.[12] The velocity v_e of such an electron can be calculated by:

$$v_e = \sqrt{2E_e/m_e} = \sqrt{2eU/m_e} \qquad (3.2)$$

where e is the elementary charge, and m_e the rest mass of the electron (see Appendix), U is the accelerating potential in volts. Thus, electrons with a typical energy E_e of 150 eV have a velocity of about 7×10^8 cm s^{-1}. At this velocity they traverse the ~1 cm long ionization region in about 2 ns. A magnetic field with field strengths between 100 and 500 gauss is applied in parallel to the electron beam by two permanent source magnets. In this field the electrons describe helical paths, with radii <1 mm.[12] By extending the electron path length, the probability of collision and thus the ionization efficiency is slightly increased (less than 10%), when passing through the ionization box. Mostly (at least for Thermo instruments) the trap current plus the box current are kept constant. This results in better stability and avoids errors in focusing.[18] The trap current is measured and held constant by an emission regulator circuit that controls the emission from the filament via the filament current that heats the cathode.

Gases such as CO_2, H_2, N_2, CO, etc. that enter the ion box volume via the gas inlet are ionized by inelastic collision with some of the thermionic electrons. Compared with the electrons, the gases are apparently motionless with thermal velocities in the order of 10^{-4} cm s^{-1}.[12] In EI sources, typical ionization reactions can be observed. For atoms, the ionization reactions by electron impact follow the general relationship:

$$A + e^- \rightarrow A^{+\bullet} + 2e^- \qquad (3.3)$$

A similar relationship can be found for molecules:

$$AB + e^- \rightarrow AB^{+\bullet} + 2e^- \qquad (3.4)$$

and, in addition, dissociation can occur:

$$AB^{+\bullet} + e^- \rightarrow A^+ + B^\bullet + 2e^- \qquad (3.5)$$

Protonation is an important (unwanted) reaction, which often interferes with the isotope measurement (see isobaric interferences below).

Examples of multiply charged ions are the well-known triplet m/z 22, 22.5 and 23, of doubly charged CO_2 ions. Another is doubly charged He, which interferes with the H_2 measurement. This is the reason why the electron energy must be low for on-line D/H analysis

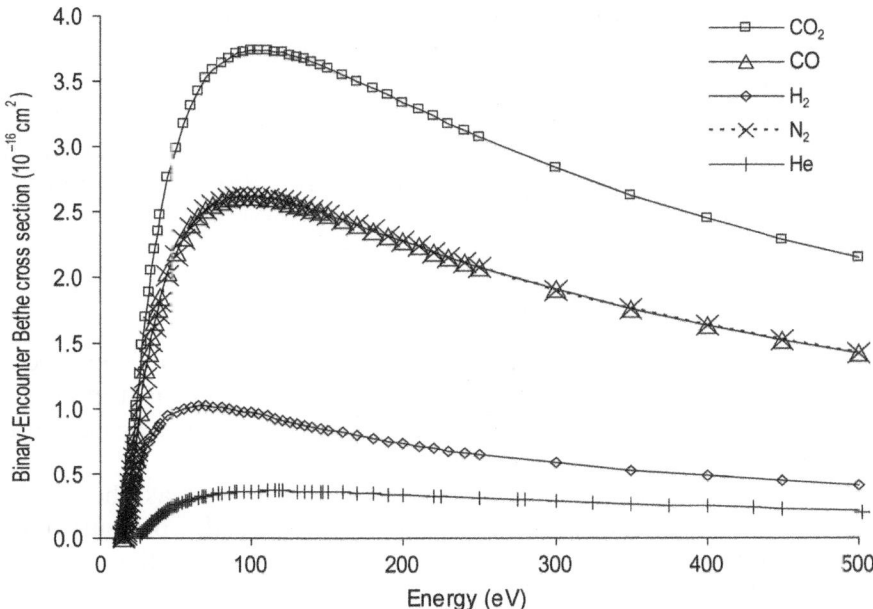

Figure 3.4 Electron-impact ionization cross-sections for CO_2, H_2, N_2, He based on the BEB model show maxima at electron energies between 70 and 100 eV. Data are obtained from NIST.[19]

(see Section 3.4.2). The ionization efficiency of the collision of an electron with a gas molecule depends on the electron energy and the ionization energy of the gas. Thus, the ionization efficiency differs for gases such as CO_2, CO, H_2, N_2 and He and can be described by their electron-impact ionization cross-section. In Figure 3.4 ionization cross-sections based on the Binary-Encounter Bethe (BEB) model show maxima in the region between 70 and 100 eV. These energies correspond to the electron energies obtained by the acceleration of the electrons into the ion box. Data on ionization cross-sections are available from NIST.[19] As can be seen in Figure 3.4, except for helium and hydrogen, typical ionization cross-sections are $\sim 3 \times 10^{-16}\,cm^2$. However, the discussion of the models describing ionization cross-sections are beyond the scope of this book and we refer to specialized literature.[20]

The resulting ion current in the source is given quantitatively by:[14]

$$i = E_Q p_Q \quad (3.6)$$

where E_Q is the source pressure sensitivity in $A\,mbar^{-1}$ and p_Q is the gas pressure in the ion box in mbar. However, the ion source pressure is not precisely known and therefore the system pressure sensitivity E_P and the measured residual gas pressure in the system near the vacuum pump p_P are used instead.[14]

The isotope ratio $R(^hE/^lE)$ of two isotopologues is given by the amounts of these isotopes $N(^hE)$, $N(^lE)$, which can be expressed in terms of the ratio of gas pressures $p(^hE)$, $p(^lE)$ and the ratio of ion currents ^hi, ^li, respectively.

$$R(^hE/^lE) = \frac{N(^hE)}{N(^lE)} = \frac{p(^hE)}{p(^lE)} = \frac{^hf^hi}{^lf^li} \quad (3.7)$$

In principle, the sensitivity factors ^hf and ^lf have to be incorporated in eqn (3.7), when the ionization probabilities for different isotopologues are different. However, in electron ionization, this is not the case. Thus, the ratio ^hf to ^lf equals 1.[21]

Because the source pressure sensitivity E_Q as well as the system pressure sensitivity E_P depend on the location of the pressure measurement, it is hardly possible to compare sensitivities of different IRMS instruments.[14] Therefore, generally the absolute sensitivity E_{abs}

$$E_{abs} = \Delta n_{mol}/n_{ion} \quad (3.8)$$

in molecules per ion is more convenient. Δn_{mol} is the number of consumed sample molecules and n_{ion} the number of detected ions at the collector cup. The sensitivities typically specified by the manufacturers are between 1000 to 2000 molecules ion^{-1}.[12] For a rough estimation of the sensitivity provided by a GC- or LC-IRMS system the following equation can be used:

$$E_{abs} \approx \frac{n_c N_{E_c} N_A R_{amp} e}{M_c s_r s_{os} A_c} \quad (3.9)$$

where n_c is the mass of a compound injected in g, s_r the split ratio during injection, s_{os} the open-split ratio, M_c the molar mass of the compound in g mol^{-1}, N_A Avogadro's constant, N_{E_c} the number of atoms of the measured element in the compound, A_c the peak area in Vs, R_{amp} the feedback resistance of the ion current amplifier in ohms (Ω) and e the elementary charge.

As an example that we have carried out ourselves, 1 µL of a 880 ng µL^{-1} solution of benzene in n-octane is injected with an injection split ratio s_r of 20:1. With a column outlet flow of 2 mL min^{-1}, the split ratio at the open-split s_{os} is around 5:1. Under these conditions, a peak area A_c of 14 Vs for the mass-to-charge ratio m/z 44 ion beam is obtained. The feedback resistance for this mass-to-charge ratio is 300 MΩ, resulting in a molar sensitivity E_{abs} of around 1400 molecules ion^{-1}, which agrees very well with manufacturer specifications. One can and should check E_{abs} on each instrument once in a while with injections of real target compound solutions because the whole measurement procedure is thus taken into account. If the obtained values deviate strongly from manufacturer specifications, this is a clear indication of poor instrument performance that should be dealt with prior to further measurements, particularly when maximum sensitivity is required.

The term isotope ratio linearity as used in isotope ratio analysis indicates that (within an acceptable range) the obtained isotope ratio is independent of the amount of gas present in the ion source and thus the amount of compound injected (see eqn (5.11)). In the case of isotope ratio measurements, linearity is of the utmost importance. Deviations from the linear behaviour described by the above equations can occur through the presence of:

(i) ion–molecule reactions with hydrogen-bearing molecules (in particular hydrocarbons or water) to form protonated species that interfere with the mass-to-charge ratio of the target ions (isobaric interferences such as, for example, $^{12}C^{16}O_2H^+$), which can be minimized by rapid extraction and rigorous exclusion of water;[13]

(ii) the permanent collimating magnetic field in the source that leads to mass-dependent dispersion of the flight path of nascent ions, when they are pulled out of the ion box;[2,12]

(iii) auto-interference by space charge formation in the ion source. The ion production itself causes a space charge in the ionization volume, which is large when the positively charged ions are slow. The electrons are much faster and cannot compensate the charge. The positive charge can partly compensate the draw out field and thus ions are hindered in leaving the source.[14] The consequence is that the amount of pre-dispersion can vary with signal height.[10] However, the effect is much lower when helium is used as sample carrier gas, as is used in continuous-flow IRMS instruments, because the relative change is lower in the presence of a huge but constant amount of He^+.[10]

As mentioned, the source pressure is important for the determination of isotope ratios. In case of carrier gas absence such as in dual-inlet systems, the source pressure has to be balanced and thus the ion current. This is carried out by matching the gas flow rate and pressure of the ion source when either sample or reference gas enters the ion source.

The balancing can be achieved by adjustable welded-metal bellows (see Section 3.2.1). In continuous-flow applications the helium carrier gas can eliminate the influence of source pressure essentially by buffering analyte gas concentration changes.[2]

Following ionization, the ions are drawn out of the ion box by a lateral potential, which is established either by an extraction lens outside (see Figure 3.3) or a repeller plate inside the ionization box. As mentioned under point (i) of the linearity impairing effects, high extraction potentials are applied, which result in high linearity but also a higher energy dispersion of the ion beam. Hence, ion–molecule reactions are diminished because collisions between the gaseous species are decreased.[2,10,11,14]

In Figure 3.5, the relation between sensitivity focusing and linearity focusing is illustrated.

After extraction, the ions are further accelerated by an electrical potential between 3 and 10 kV, depending on the instrument. Generally, larger instruments with higher accelerating voltage can provide higher sensitivities, resolutions and improved peak shapes.[12] The ion acceleration between the place of ion production and the beam defining slit (often also named source, object, entrance or alpha slit) can be described by Liouville's theorem of phase space preservation.[14] The ion beam is electrically focused in the x- and y-directions, for example, by Einzel lenses, before it enters the ion separation unit by passing the beam defining slit (~ 0.2 mm width) and the aperture of the mass spectrometer. The aperture has the function to limit the width of the ion beam so that it enters the magnetic field accurately (collisions with the flight tube should be avoided at all cost). By passing the acceleration potential U_{acc}, the electrical energy E_{el} is transformed into kinetic energy E_{kin} according to:

$$E_{el} = q_{ion} U_{acc} = \frac{1}{2}(mu) v_{ion}^2 = E_{kin} \quad (3.10)$$

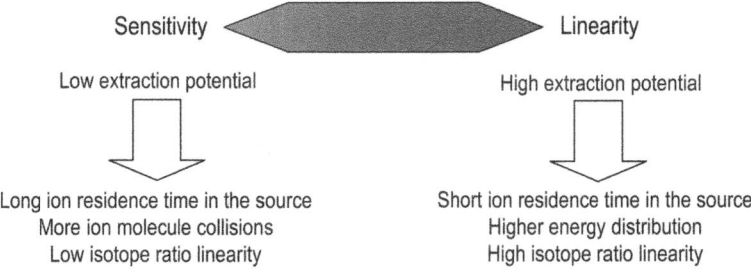

Figure 3.5 Relation between sensitivity focusing and linearity focusing.

The ion charge in Coloumb (C) is given by $q_{ion} = ze$ where z is the number of charges of the ion (mostly one), m is the molecular mass of the ion in dalton (Da), u is the atomic mass unit (see Appendix) and v_{ion} is the resulting velocity of the ion. By rearranging eqn (3.10) and extraction of the constants ($\sqrt{2e/u}$), the attained ion velocity in cm s^{-1} can be obtained:

$$v_{ion} = \sqrt{2q_{ion}U_{acc}/(mu)} = 1.39 \times 10^6 \times \sqrt{zU_{acc}/m_{ion}} \qquad (3.11)$$

In accordance with this equation, an $^{12}C^{16}O_2^{+\bullet}$ ion with m/z 44 and an accelerating voltage of 3 kV has a velocity of 1.15×10^7 cm s^{-1}.

At this point we need to mention that there is some confusion in the mass spectrometry literature on how to express the abscissa in mass spectra. Although from physical principles it is clear that a mass-to-charge ratio is governing the separation of ions or fragments in mass spectrometry, the dimensionless entity m/z is more frequently used than m_{ion}/q_{ion} since the latter has awkward numbers and units of kg C^{-1} associated with it. Current IUPAC nomenclature recommends the use of m/z 'to denote the dimensionless quantity formed by dividing the mass number of an ion by its charge number' although it is acknowledged that m/z does not represent a mass-to-charge ratio. However, a revision of IUPAC-endorsed mass spectrometry terms is in progress. A new unit 'Thomson' that has been suggested by some researchers to replace m/z has not found widespread acceptance. Within this book, we use m_{ion}/q_{ion} in all mass spectrometric equations and define $m_{ion} = mu$ and $q_{ion} = ze$ to derive m/z by extraction of the physical constants e and u as done above. For denoting mass-to-charge ratios in the following we will then stick to m/z to avoid confusing the reader. As already done above, the $^{12}C^{16}O_2^{+\bullet}$ ion thus represents m/z 44.[22]

3.1.3 Ion Separation in the Magnet Sector Field

For high precision isotope ratio analysis some prerequisites have to be fulfilled to obtain the required precision. One is that the lateral mass dispersion has to be large enough to place several Faraday collector cups for the simultaneous detection of ion beams with a minimum mass difference of 1 Da in the focal plane of the mass spectrometer.[14] An additional prerequisite is that the optical transmission should be stable, mass-independent and close to 100%.[12] In modern magnetic sector fields these conditions are largely fulfilled and will be discussed in the following section.

In current instruments, electromagnets are employed exclusively for spatial separation of ions according to their mass-to-charge ratio.[12] A magnetic sector field analyser consists of a magnetic north and south pole separated by a narrow distance of around 1 cm between the pole shoes. It is important that the magnetic field between the pole shoes is homogeneous to produce defined ion beams towards the collector cups (see Figure 3.6a).[2] Most commercially available IRMS instruments have deflection angles ϕ_m of 90° or 120°.

As shown in the diagram in Figure 3.6b a magnetic sector field analyser can be divided into five areas. The first area denoted by the roman number I is the field-free entrance space and the numbers II and IV indicate the entrance and exit fringe fields of the electromagnet, respectively. Area III is the main magnetic field and V the field-free exit space. A detailed discussion of ion trajectories in the magnetic sector field is beyond the scope of this book and not necessary for understanding the fundamental principles required here. Thus, we refer the interested reader to more specialized literature.[23–25] When positively charged ions enter a homogeneous magnetic field they will be deflected perpendicular to the flight path by the Lorentz force $\vec{F}_L$. The vector of the Lorentz force is given by the vector product of the magnetic flux density vector $\vec{B}$ and the scalar product of the ion velocity $\vec{v}_{ion}$ vector and the ion charge q_{ion}.

$$\vec{F}_L = \vec{B} \times q_{ion}\vec{v}_{ion} \qquad (3.12)$$

By applying the right hand rule for positively charged ions ($\vec{v}$ thumb, $\vec{B}$ index finger, $\vec{F}_L$ middle finger), which is shown in Figure 3.6a, it is possible to determine the direction of the Lorentz force $\vec{F}_L$. A positively charged ion with a certain mass-to-charge ratio that travels with a velocity v perpendicular to the homogeneous magnetic field will follow a circular path of radius r_m because of a compensation of the Lorentz force by the centripetal force F_c as described by eqn (3.13):

$$F_L = q_{ion}v_{ion}B = \frac{(mu)v_{ion}^2}{r_m} = F_c \qquad (3.13)$$

A rearrangement of eqn (3.13) leads to the circular path radius r_m and shows that the ions are separated according to their momentum:

$$r_m = \frac{(mu)v_{ion}}{q_{ion}B} \qquad (3.14)$$

When the ion velocity is substituted by eqn (3.11) one obtains eqn (3.15) in which the constants u and e can be combined so that r_m is given in cm.

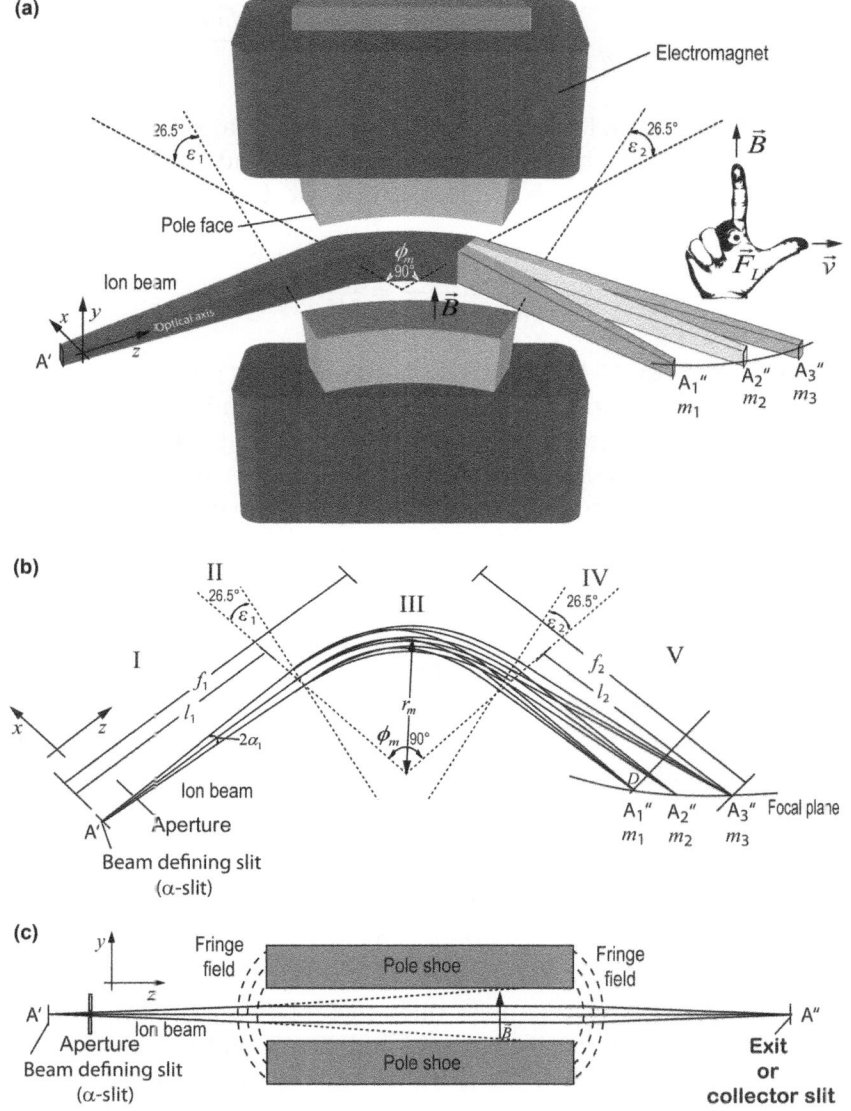

Figure 3.6 Diagram of a 90° magnetic sector field analyser, with mass dispersion of the ion beams. a) A 90° magnetic sector field analyser, with mass dispersion of the ion beams. On the right side, the right hand rule (for positively charged particles) is illustrated (see text). b) Diagram of the different areas (I)–(V) an ion beam has to traverse in the magnetic field as well as a construction of the focal and dispersive properties of a magnetic sector ion optical system. Details of the parameters shown are discussed in the text. c) Difference between ion beam focusing in the hypothetical case of entrance angle and exit angles $\varepsilon = 0°$ (sketched lines indicate ion beam spread in the y-direction) and the realized entrance angle and exit angles $\varepsilon = 26.5°$ for axial focusing or 'stigmatic focusing' in the y-direction.

$$r_m = \frac{1}{B}\sqrt{\frac{2(mu)U_{acc}}{q_{ion}}} = 1.44 \times 10^{-2} \times \frac{1}{B}\sqrt{\frac{mU_{acc}}{z}} \qquad (3.15)$$

A more widely used expression which gives a relation to the mass-to-charge ratio m_{ion}/q_{ion} by rearranging eqn (3.14) and substitution of v_{ion} leads to eqn (3.16), which is often called the basic equation of mass spectrometry:

$$\frac{m_{ion}}{q_{ion}} = \frac{(mu)}{(ze)} = \frac{r_m^2 B^2}{2U_{acc}} \qquad (3.16)$$

As is seen in eqn (3.15), the circle radius depends on the ion mass, the ion charge as well as the acceleration potential and the magnetic flux density. Giving an example, with an accelerating voltage of 3 kV a $^{12}C^{16}O_2^{+\bullet}$ ion possesses an ion velocity of 1.15×10^7 cm s^{-1} and will describe a radius of around 7 cm through a homogeneous magnetic field of 0.75 Tesla. It is also evident from eqn (3.15) that singly charged monoenergetic ions with larger molecular masses describe a larger circle than those with lower molecular masses, as shown in Figure 3.6b for $m_3 > m_2 > m_1$. In principle, the magnetic field can be compared with a prism in optics. In the same way that light with different wavelengths is dispersed by passing through a glass prism, the magnetic field disperses ions with different m_{ion}/q_{ion} in the homogeneous magnetic field so that they hit different collector cups that are arranged along the focal plane of the mass spectrometer. The lateral mass dispersion D of the sector field mass spectrometer determines the separation between adjacent mass peaks and is given by:

$$D = \frac{\Delta m}{2m} k_m \qquad (3.17)$$

Here, $\Delta m/m$ is the relative mass difference of two adjacent ion beams and k_m the dispersion coefficient, which depends on the deflection angle ϕ_m, the radius r_m as well as the entrance and the exit angles, the distance between the beam defining slit and the magnetic sector field l_1 and the distance between the magnetic sector field and the exit slit (also image slit, collector, or detector slit) l_2. In the case of a symmetrical analyser system and normal beam entrance angle $\varepsilon_1 = 0°$ the dispersion coefficient reduces to $k_m = 2r_m$ and $D = r_m \Delta m/m$.

Furthermore, the magnetic sector field acts like a convex optical lens in which ions that possess the same m_{ion}/q_{ion}, a given lateral spread and an object distance between the beam defining slit and the magnetic sector field entrance of l_1, will be refocused at the exit slit after leaving the magnetic field at a radial image distance l_2. For a symmetrical system in which $l_1 = l_2 = l_m$ the distance of the focal point l_m behind the magnetic field is given by:

$$l_m = r_m \cot \phi_m \tag{3.18}$$

So far we have discussed the divergence of the ion beam in space and its focusing by the lens effect only in the x-direction. In Figure 3.6c the ion beam spread in the y-direction across the flight path is shown by the broken line where the ion beam hits the upper walls of the flight tube, which must not happen. An accumulation of residues on the flight tube walls can lead to static charge formation on the surface, which can deteriorate peak shapes of the ion beams and scattering of ions into the wrong detector cups.[12] Changing the pole face angle ε will improve the focusing capabilities of a mass spectrometer in the y-direction, which is called 'stigmatic focusing'. This angle is noted with a prime subscript ε_1 to indicate the entrance of the magnet and ε_2 to indicate the exit of the magnet. The pole face angle ε is positive if the normal to the pole face is farther from the origin of r_m than the optical axis; ε is negative if it is closer to the origin of r_m. A positive entrance angle exhibits a focusing action in the y-direction. All current IRMS instruments use some form of stigmatic focusing. In 90° deflection systems, optimal focusing conditions in the y-direction are obtained when the ion beam enters the magnetic field boundary or fringe at an angle of $\varepsilon_1 = 26.5°$, traverses the 90° magnetic sector field and exits the magnetic field boundary also at an angle of $\varepsilon_2 = 26.5°$.[12,26] A major advantage of the focusing in the y-direction is a maximized ion optical transmission.[14] A positive side effect of a symmetric sector field instrument with an oblique entrance angle is that the dispersion coefficient is doubled to $k_m = 4r_m$ so that the system will provide the same mass resolving power as a system with $\varepsilon = 0°$ but with half the radius[14] thus requiring a smaller footprint.

The peak shape observed in IRMS instruments is likely the most prominent difference between an IRMS and a high resolution organic mass spectrometer. The peak shape is defined by the source and the exit slit widths, denoted by s_s and s_a, respectively. It has already been mentioned that the mass analyser is constructed such that the image of the beam defining slit is focused at the exit slit after mass separation. In organic mass spectrometers that should provide high resolution between the analysed mass-to-charge ratios, the exit slit s_a is much narrower than the beam defining slit s_s and, therefore, than the beam width b, so that $b \gg s_a$. In this case, the width of the exit slit determines the minimum peak width and a 'Gaussian', or sometimes called 'triangular', peak shape is observed when the ion beam image of the beam defining slit passes the exit slit, as shown in the right part of Figure 3.7a. In contrast, if the beam defining slit s_s and therefore the beam width b is much narrower than the exit slit s_a ($b \ll s_a$), a flat-topped peak shape is observed. This arrangement is used in the case of IRMSs as explained below. In such instruments the exit slit s_a is chosen to be at least two times the width of the beam defining slit s_s so that $s_a \geq 2b$ prevails.[14]

As an example, in the case of a Delta V™ IRMS instrument (Thermo Fisher Scientific, Bremen) used for carbon isotope ratio determination, the beam defining slit width is 0.2 mm whereas the width of the exit slit, when using a universal triple collector (see below) is between 1.2 mm and 3 mm. Under these circumstances, the beam defining slit determines the width of a flat-topped peak. In IRMS it is important that adjacent ion beams with a mass difference of 1 Da can be separated and detected. As described before, the lateral dispersion of two adjacent ion beams depends on the relative mass difference, which is the mass resolution of the mass spectrometer:

$$R = m/\Delta m \tag{3.19}$$

As shown in Figure 3.7b, two neighbouring mass-to-charge ratios are assumed to be sufficiently separated when the valley separating their peak maxima has decreased to 10% of their intensity (10% valley definition of resolution). For symmetrical peaks equal in size, the 10% valley condition is fulfilled at the individual peak width at 5% intensity.[27] The influence of the exit slit width s_a on resolution and intensity is shown in Figure 3.7c. The slit width is decreasing from the left to the right signal, so that in the left graph ($s_a \cong 2b$), and for the middle as well as the graph on the right side ($s_a < b$) is valid. With decreasing exit slit width an increased resolution but lowered intensity is obtained. Since in IRMS instruments, flat-topped peaks are aimed for, as shown in the left graph, their mass resolution with $m/\Delta m \approx 100\text{--}200$ is rather low.[28] However, mass resolution is of minor importance in IRMS compared with signal stability and high transmission.[28] In the case of flat-topped peaks these are maximized because the wide slit ensures that the entire ion image produced in the source is depicted on the detector, and small shifts in the position of the ion image within the focal plane due to fluctuations of the acceleration voltage or the magnetic field do not result in variations in measured ion current intensities.[2] In practice, though, the corners of the flat-topped peak are rounded, which is caused by some scattering of the ions by the residual gas molecules within the mass spectrometer. The recording of the slope of intensity at half-peak height can be used as a simple tool to examine the stability of the accelerating potential and the magnetic field and thus the quality of the

Instrumentation for Compound-specific Stable Isotope Analysis 69

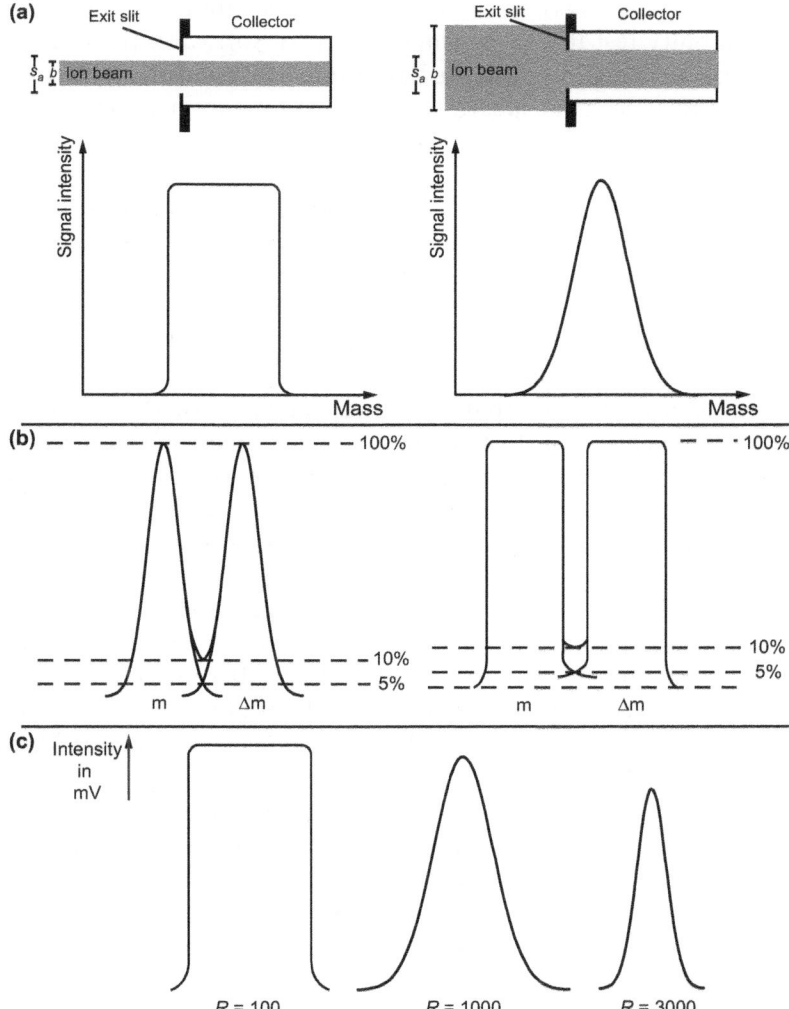

Figure 3.7 Peak shape in dependence of the exit slit s_a (peaks are drawn artificially). a) Right: the exit slit s_a is much narrower than the beam defining slit s_s and, therefore, than the beam width b so that ($b \gg s_a$) and a 'Gaussian' or 'triangular' peak shape results. Left: the beam defining slit s_s and, therefore, the beam width b is much narrower than the exit slit s_a ($b \ll s_a$) and a flat-topped peak shape is observed. b) Illustration of the 10% valley definition of resolution. c) The influence of the exit slit width s_a on mass resolution and intensity. With decreasing slit width (from the left to the right signal), the mass resolution increases but intensity decreases. In IRMS flat-topped peaks are used due to the signal stability on the flat-topped region of the peak. Note that the beams in actual measurements cannot rise as sharply as shown here schematically.

performance of a mass spectrometer. This is often incorporated as a system stability test in the instrument control software. In contrast, the stability of the signal intensity is determined by measuring the intensity on the top of a peak for a certain period of time (consult manufacturer manuals and see Chapter 5).

Different types of scans can be carried out with the sector field instrument. The selection of m/z for measuring an element isotope ratio is achieved by adjusting the magnetic field strength. For the isotope ratio measurement, the high voltage and the magnetic field are held constant so that the ion beams of interest simultaneously hit the relevant collector cups, which are arranged along the focal plane. However, the high voltage can alter over time.

A peak centre ensures that a safe flat position of the ratio trace is found even when the peak shape has changed over time. It also ensures that the high voltage is reset after a possible sparking. It is designed to find the centre of a flat portion of the two overlapping ion beams. Another high voltage scan is the so-called 'peak shape' scan, which is shown in Figure 3.8a. The high voltage scan at a constant magnetic field depends on the cup geometry and arrangement when measuring the different gases. During such a scan, the ion beams are moved over the respective cups by changing the high voltage. It represents the optical image as it occurs in the respective cups. In Figure 3.8a, an exemplary peak shape scan of a universal triple collector such as used in a Delta V™ (Thermo Fisher Scientific, Bremen, Germany) is shown.

Figure 3.8b shows a magnetic field scan performed with a MAT 253™ (Thermo Fisher Scientific, Bremen, Germany). Such scans are performed to evaluate the performance of the mass spectrometer by measuring single mass-to-charge ratios, for example, water (m/z 18), nitrogen (m/z 28) and oxygen (m/z 32) background values (background scan) and to check the gas cleanliness as well.[18] For other mass-to-charge ratios that can appear during a background scan see Table in the Appendix. As will be discussed later, the performance of the connected interface and the chromatographic system can be checked by such a scan, thus it should be carried out regularly. The height of background values depends on the application.

3.1.4 Ion Collection and Signal Pathway

It was mentioned in Chapter 1 that the first magnetic sector field mass spectrometers were equipped with a single Faraday cup detector. Nier later used a double collector Faraday cup, with which he was able to measure two ion beams simultaneously, for example, for CO_2 m/z 44

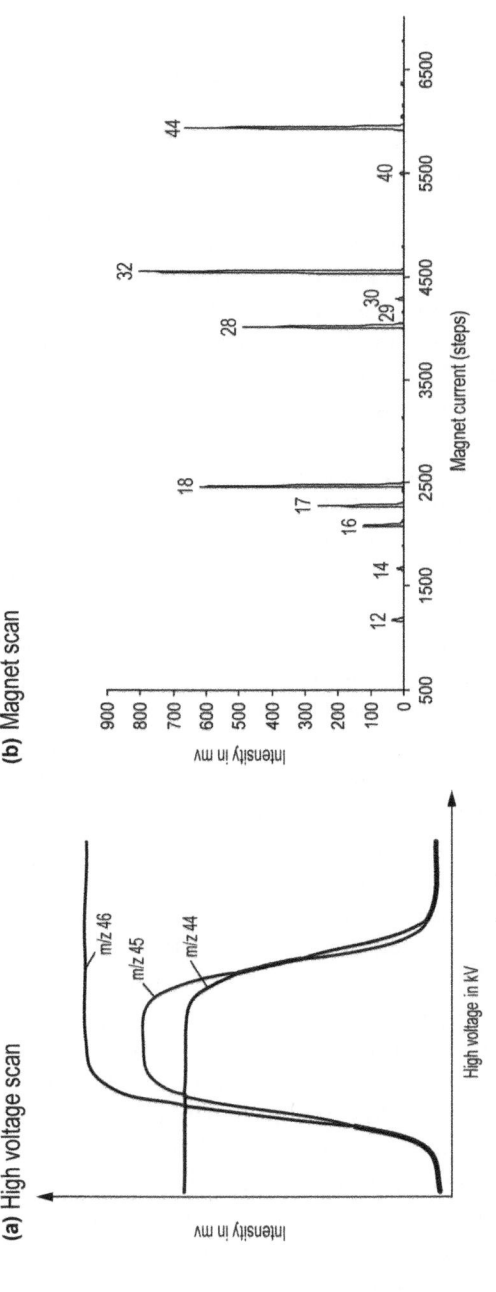

Figure 3.8 High voltage and magnet scan. a) High voltage or 'peak shape' scan of CO_2 carried out with a universal triple collector such as is used in a Delta V™ (Thermo, Bremen, Germany). The magnetic field is held constant during the scan. b) Magnetic field scan performed with a MAT 253™ (Thermo Fisher Scientific, Bremen, Germany). The measurement was carried out on the middle cup of a universal triple collector. (Peak designations represent m/z. For the individual m/z see Table in the Appendix.)

and 45, thus obtaining higher precision.[7,28] Nowadays, the majority of mass spectrometers are equipped with electron multipliers (EM) for ion detection, and IRMS instruments are probably the only analytical gas source mass spectrometers in commercial production that still employ Faraday cups (FC) for this purpose.[2,29] The reasons for keeping FC are explained in more detail below.

The number of cups has been successively increased over the decades and today between three and ten of them are positioned along the focal plane to detect multiple ion currents simultaneously. The major reason, of course, is that the ion currents are measured for *exactly* the same time intervals, so that ion source fluctuations cancel completely.[12] Furthermore, peak jumping and associated settling times are avoided. The collector system is aligned so that the ions exiting the magnetic sector field analyser strike the FC collector electrodes that are positioned along the focal plane of the mass spectrometer.[12]

A universal triple collector system is shown in Figure 3.9b. With such a collector arrangement it is possible to measure the mass-to-charge ratios of the isotopologues of CO_2 (44, 45, 46), N_2 (28, 29, 30), CO for oxygen determination (28, 29, 30), SO_2 (64, 65, 66), O_2 (32, 33, 34), SF_6 (127, 128, 129), N_2O (44, 45, 46), NO (30, 31).[12] The universal triple collector consists of a small middle cup and two outer wide cups. This is necessary because of the different dispersion ranges between nitrogen and the carbon dioxide isotopologues.[12] Nowadays this type of collector cup arrangement is the most widely used.

For the detection of hydrogen (H_2, m/z 2) and deuterium (HD, m/z 3) two collector cups are used. The arrangement of these cups depends on the focusing properties of the mass spectrometer. For a more detailed discussion of hydrogen isotope ratio measurements see Section 3.4.2.

Generally, FC are less sensitive compared to electron multipliers but are ideal for ion detection in IRMS instruments because: (i) Isotope ratio mass spectrometers require high signal stability to achieve high absolute precisions. The high counting rates required for high precision place IRMS signals in the normal detection range for FC, but would rapidly damage electron multipliers (SEM).[2] (ii) FC do not need gain adjustment for linear detection of very large ion currents.[2] In contrast, above 10^6 counts sec^{-1}, SEMs are increasingly non-linear. Up to 10^8 counts sec^{-1} one can apply a dead time correction, above this counting by SEMs is not possible any more. Hence, given the high counting rates required to achieve high precision SEMs are not an option.[18] (iii) FC are highly stable and rugged,[2] and (iv) rarely need replacement, compared to electron multipliers.[2] Additionally, (v) the response of this transducer is independent of the energy, the mass and the chemical nature of the ion.[2]

Instrumentation for Compound-specific Stable Isotope Analysis

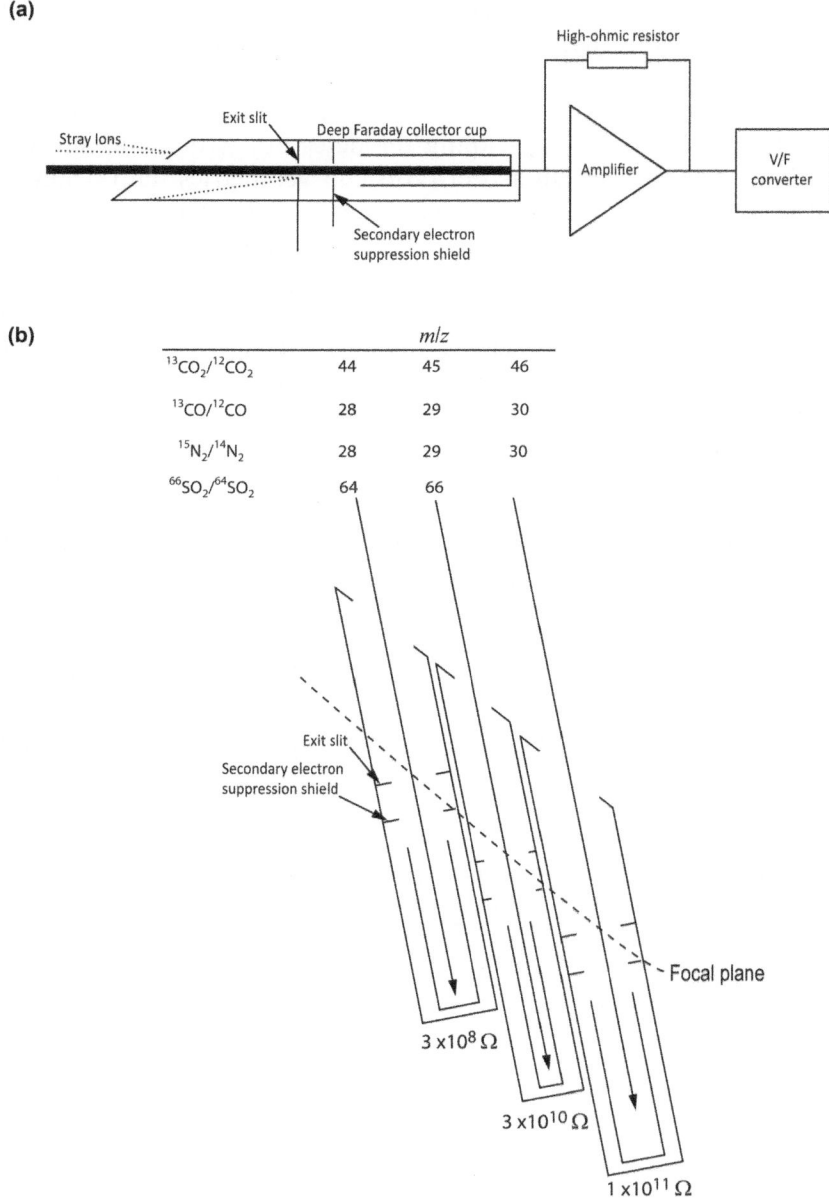

Figure 3.9 Collector system of an IRMS. a) Measuring channel consisting of an FC, an amplifier with a high-ohmic feedback resistor and a voltage-to-frequency converter. b) Diagram of a universal triple collector. Such a detector consists of a small middle cup and two outer wide cups. The feedback resistor values for the isotopologue mass-to-charge ratios listed in the upper part are shown.

Due to these characteristics, modern IRMSs still use FC for the detection of positively charged ions. The FC shown in Figure 3.9a is aligned so that ions exiting the analyser by the exit slit enter the cup and are discharged here. The design of the cup prevents reflected ions and secondary electrons from escaping the detector. In order to shield the cup from stray ions, high-aspect-ratio cup geometries (cup height/cup width) are used.[30] A preferred cup lining material is carbon. It produces very few secondary ions and allows for an efficient capture of ions, while minimizing losses due to scattering events.[12] A secondary electron suppressor plate, to which a small negative potential is applied, prevents sputtered electrons from escaping the cup. The collector electrode and the cup are connected to the ground via a high-ohmic resistor in the feedback loop (see Figure 3.9). Positively charged ions that strike the electrode are neutralized by a flow of electrons producing an electric current from ground through the high-ohmic resistor. The result is a voltage drop across the high-ohmic resistor according to Ohm's law ($U = RI$) proportional to the ion current entering the collector cup.[31] Each single collector cup possesses its own amplifier and the high-ohmic feedback resistors are matched to the individual abundances of the isotopes collected in the specific cup. Typical feedback resistor values matching the natural abundance of the light isotopes are shown in Table 3.1.

As an example, with a feedback resistor of $3 \times 10^8\ \Omega$ as it is applied for m/z 44 a 1 nA current produces a signal of 0.3 V.[32]

Table 3.1 Feedback resistor values matched to the natural abundance of the isotopologues measured with IRMS.

Measured gas	Mass-to-charge ratio (m/z)	Feedback resistor in Ohm
H_2	2	1×10^9
	3	1×10^{12}
CO_2	44	3×10^8
	45	3×10^{10}
	46	1×10^{11}
N_2	28	3×10^8
	29	3×10^{10}
	30	1×10^{11}
CO	28	3×10^8
	29	3×10^{10}
	30	1×10^{11}
SO_2	64	3×10^8
	66	1×10^{10}
O_2	32	3×10^8
	33	1×10^{12}
	34	1×10^{11}

Every single collector cup with its amplifier is connected to a voltage-to-frequency (VFC) or analogue-to-digital converter (ADC) that converts analogue currents into digital pulses, whereby signals are converted to frequencies at rates of up to 20 000 Hz V^{-1}.[18]

Each measuring channel (see Figure 3.9a) feeds an individual counter with the pulses for an operator-selected integration time (e.g., 250 ms). Each integration interval is followed by reading out the number of counts, and the ion current ratios are calculated.

Generally, small constant currents are added to the collectors, in order to generate electronic offsets.[33] In the case of carbon dioxide Merritt et al. reported, in absence of any ion current, offsets of 90–110 mV, 110–130 mV and 140–160 mV for m/z 44, 45, 46, respectively.[33]

Peterson and Hayes investigated the signal-to-noise ratios in mass spectrometric ion current ratios.[34] In the following, the contributions of different noise sources relevant for IRMS instruments are discussed on the basis of their publication. The overall noise of an ion current measurement σ_s is the sum of the ion beam noise or 'shot noise' denoted by the standard deviation σ_i, the noise of the transducer σ_s, noise introduced by signal conditioning such as amplification σ_c and noise introduced through the digitization process σ_d.[34]

An ion beam consists of a discrete number of ions and the ion current is given by $i = Nq_{ion}/t$, where N is the number of ions collected in a time interval t in seconds and q_{ion} is the charge in coulombs (C) carried by an ion.

The observation time t for the current measurement is, in the case of current follower amplifiers, given by 2τ, where the time constant of the amplifier is $\tau = RC$.[34] R is the value of the feedback resistor. For carbon measurements the capacitance values C of the electrometers are, for example, 470, 5, and 2 pF for m/z 44, 45, 46, respectively, so that the time constants are 0.14–0.2 s.[32] The capacitance values can be taken from the respective manufacturer manuals.

The number of ions hitting the collector per unit of time is a random process and time intervals between ions follow a Poisson distribution. When N ions are counted at the collector, the variance of the counting is $\sigma_N^2 = N$, so that the standard deviation of this process is $\sigma_N = \sqrt{N}$. Giving an example: when an average of 10^6 ions is measured per second the standard deviation for a 95% confidence interval would be $\sigma_N = 1000$ ions so that the number of ions hitting the collector within one second is in the range between 999 000 and 1 001 000. The variance in ion current i, the so-called shot noise, is then given by:

$$\sigma_i^2 = \left(\frac{\partial i}{\partial N}\right)\sigma_N^2 = \left(\frac{q_{ion}}{t}\right)^2 N \qquad (3.20)$$

Shot noise typically dominates overall noise of ion current measurements. In Chapter 5 this shot noise limit is used to calculate the amount of an element necessary to obtain a certain precision. Another source of noise is introduced by signal conditioning such as amplification σ_c. Here, the main source of noise caused by the high-ohmic resistor is Johnson noise:[30]

$$\sigma_{c(J)} = 2k_B RT/t \qquad (3.21)$$

where k_B is the Boltzmann constant; R the resistor value in Ω; T the temperature in K and t the integration time. Other sources of noise that contribute to σ_c are thermally induced variations. Typical temperature coefficients of the high-ohmic resistors are in the range of 200 ppm $°C^{-1}$. Therefore, the temperature has to be held constant at $\pm 0.01\ °C$, if a gain stability of 2 ppm should be reached.[30] This is achieved by putting the amplifiers in evacuated temperature-stabilized housings.[30] Johnson noise is particularly important for small ion currents, such as in the case of hydrogen isotope ratio determination. Modern microprocessor based data evaluation uses digital signals for processing and storage. Analogue detector signals, therefore, have to be converted into digital signals by an ADC. The transformation procedure of a continuous analogue signal to discrete steps in an ADC is also a source of noise (σ_d) that shall be explained briefly: an analogue signal is sampled by measuring the amplitude of the signal at fixed time intervals. This procedure is also called 'sampling'. The number of samples per second is the so-called 'sampling frequency' or 'sampling rate'. This means that the analogue–digital conversion is more precise, if the sampling frequency is high. The result of the sampling process is a pulse-amplitude modulation. The certain pulses are sampled and rounded to a discrete step. The rounding of a continuous signal to these discrete steps introduces noise, which is referred to as quantization error or "bit noise".[35]

The precision of a digitizer is expressed in terms of bits, where an N-bit board has $2^N - 1$ steps over a given range.[35] Sacks *et al.* gave the example that a 16-bit board has around 65 000 steps. If the board has a range of 0–10 V, then the step size Δ of the board is 10/$\sim 65\,000 = 0.15$ mV. IRMS systems are equipped with digitizers that deliver effective maximum digitization depths of 16 to 24 bits.[35] Sacks *et al.* investigated the effect of quantization noise on curve peak detection.[35] In Figure 3.10 the effect of digitalization is shown exemplarily for a Gaussian peak of 24, 16, 14, 12-bit resolution.[35] However, in modern instruments with 16- to 24-bit ADC boards this source of noise is negligible.

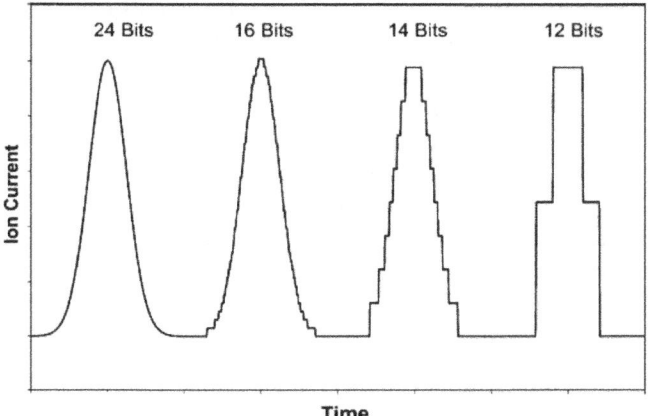

Figure 3.10 Gaussian signal collected by ADCs of various resolutions (24, 16, 14, and 12 bits) and quantization errors. Sacks et al. simulated a Gaussian signal collected by ADCs of various resolutions (24, 16, 14, and 12 bits) and resulting quantization errors. At 24-bit resolution, quantization error is not visible, and the peak appears as a smooth trace. At 16 bits, bit noise is evident primarily at the base of the peak. At 12-bit resolution, the signal is barely recognizable as a Gaussian shape.[35]
(Graphic reprinted with permission of Elsevier.)

3.1.5 Ion Corrections

As mentioned before, isotope ratios of light elements such as $^{13}C/^{12}C$, $^{15}N/^{14}N$, $^{18}O/^{16}O$ are not measured directly in the target compounds. Rather, the isotope ratio of isotopologues of simple gas species such as N_2, H_2, CO_2 for carbon and CO for oxygen after quantitative transformation are measured. The ion currents of these light gas isotopologues are detected on individual FC and the related isotope ratios or δ-values can be obtained by ion correction calculations. Using an IRMS instrument the isotope ratios $R(^hE/^lE)_c$ relative to an international standard $R(^hE/^lE)_{s,i-ref}$ are determined so that one can rewrite eqn (3.7) as

$$R(^hE/^lE)_c = \left(\left(\frac{^hi_c}{^li_c} \right) \Big/ \left(\frac{^hi_{s,i-ref}}{^li_{s,i-ref}} \right) \right) R(^hE/^lE)_{s,i-ref} \qquad (3.22)$$

where ^hi_c and ^li_c represent the ion currents of the heavy and the light isotopologue of the corresponding conversion gas, and $^hi_{s,i-ref}$ and $^li_{s,i-ref}$ are the ion currents of the heavy and the light isotopologue obtained by an international standard. Normally, these and the following correction calculations are carried out automatically by the

Table 3.2 Mass-to-charge ratios with corresponding isotopologues measured for $^{13}C/^{12}C$ isotope ratios and $\delta^{13}C$ determination.

Relevant m/z	Detected ionic species
m/z 44	$^{12}C^{16}O^{16}O^{+\bullet}$
m/z 45	$^{13}C^{16}O^{16}O^{+\bullet}$, ($^{12}C^{16}O^{17}O^+$, $^{12}C^{17}O^{16}O^{+\bullet})^a$
m/z 46	($^{12}C^{16}O^{18}O^{+\bullet}$, $^{12}C^{18}O^{16}O^{+\bullet}$), ($^{13}C^{17}O^{16}O^{+\bullet}$, $^{13}C^{16}O^{17}O^{+\bullet}$), $^{12}C^{17}O^{17}O^{+\bullet}$

^{a} $^{12}C^{16}O^{17}O^{+\bullet}$ and $^{12}C^{17}O^{16}O^{+\bullet}$ are listed as separate species due to the two times higher probability of the presence of ^{17}O. The same applies to m/z 46 species ($^{12}C^{16}O^{18}O^{+\bullet}$, $^{12}C^{18}O^{16}O^{+\bullet}$), ($^{13}C^{17}O^{16}O^{+\bullet}$, $^{13}C^{16}O^{17}O^{+\bullet}$).

instrument software and the user will not be aware of these black box procedures in the background. However, one should be aware of how these corrections are carried out. Furthermore, in terms of data consistency, identical correction procedures have to be used by different laboratories.[21] In the following we will focus on the ion corrections for carbon, nitrogen and hydrogen, the determination of corrections for oxygen measurements and sulfur are comparable with the ^{17}O correction explained below and will not be discussed in detail here.

Carbon. Measurements of carbon isotope abundances are based on the mass spectrum of CO_2. The isotopologues of CO_2 possess m/z ratios between 44 and 49 and can be relevant in the case of clumped isotope measurements. However, here we will only discuss the situation when the m/z ratios 44, 45 and 46 are measured. In Table 3.2, mass-to-charge ratios and the corresponding isotopologues of carbon dioxide are tabulated.

As can be seen, in Table 3.2, the m/z 44 only contains CO_2 molecules with the most abundant carbon and oxygen isotopes ^{12}C and ^{16}O, while m/z 45 consists of two isotopologues, the $^{13}C^{16}O^{16}O^{+\bullet}$ (93.5%) and smaller amounts of $^{12}C^{16}O^{17}O^{+\bullet}$ (6.5%).[21] Three isotopologues contribute to m/z 46.[36] $^{12}C^{18}O^{16}O^{+\bullet}$ is the most abundant and $^{13}C^{17}O^{16}O^{+\bullet}$ as well as $^{12}C^{17}O^{17}O^{+\bullet}$ have minor contributions of ($\sim 0.21\%$) and ($\sim 0.0036\%$), respectively.[21] Because of the 6.5% contribution of the $^{12}C^{16}O^{17}O^{+\bullet}$ isotopologue to the m/z 45 ion current, a correction is necessary.[21,36,37] In the literature this is often mentioned as the ^{17}O correction for $\delta^{13}C$ determination.[36]

In Chapter 2, the term fractional abundance ^iF was introduced. This term can be extended to molecules and isotopologues and the fractional abundance contribution of the involved isotopes to an isotopologue can be calculated by the following probability functions. For $^{44}CO_2$, $^{45}CO_2$ and $^{46}CO_2$ the contribution can be expressed by eqns (3.23), (3.24) and (3.25), respectively.

$$^{44}F = {}^{12}F^{16}F^{16}F \tag{3.23}$$

$$^{45}F = {}^{13}F^{16}F^{16}F + 2\,{}^{12}F^{16}F^{17}F \tag{3.24}$$

$$^{46}F = 2\,{}^{12}F^{16}F^{18}F + 2\,{}^{13}F^{16}F^{17}F + {}^{12}F^{17}F^{17}F \tag{3.25}$$

Dividing eqns (3.24) and (3.25) by (3.23) leads to the following isotopologue ratios (please note that we use the shorter expression ^hR_c and not the IUPAC recommended notation $R(^hE/^lE)_c$):

$$^{45}R_c = \frac{^{45}F}{^{44}F} = \frac{^{13}F^{16}F^{16}F}{^{12}F^{16}F^{16}F} + \frac{^{12}F^{17}F^{16}F}{^{12}F^{16}F^{16}F} + \frac{^{12}F^{16}F^{17}F}{^{12}F^{16}F^{16}F} = {}^{13}R_c + 2\,{}^{17}R_c \tag{3.26}$$

$$\begin{aligned}^{46}R_c &= \frac{^{46}F}{^{44}F} = 2\frac{^{12}F^{16}F^{18}F}{^{12}F^{16}F^{16}F} + 2\frac{^{13}F^{17}F^{16}F}{^{12}F^{16}F^{16}F} + \frac{^{12}F^{17}F^{17}F}{^{12}F^{16}F^{16}F} \\ &= 2\,{}^{18}R_c + 2\,{}^{13}R_c\,{}^{17}R_c + ({}^{17}R_c)^2\end{aligned} \tag{3.27}$$

The problem that occurs is that there is not enough information to calculate the three isotope ratios $^{13}C/^{12}C$, $^{18}O/^{16}O$ and $^{17}O/^{16}O$ by only two ion current ratios $^{45}R_c$ and $^{46}R_c$. In other words we have a system of two equations and three unknowns. Although in principle one could solve this problem by measuring the m/z 47 ion beam, this is not done routinely due to a lack of precision and accuracy caused by the very low abundance ratio of m/z 47 to m/z 44 ion currents ($\sim 0.0046\%$).[21] Instead, a further equation is introduced. The correction for the ^{17}O abundance of the m/z 45 ion beam is carried out by calculating the ^{17}O abundance from measurement of the $\delta^{18}O$ of the sample, which is obtained by determination of the ratio of the m/z 46 and m/z 44 ion currents.[21] The assumption was made and verified that in all terrestrial oxygen pools a proportional mass-dependent isotope fractionation between ^{17}O and ^{18}O exists, thus it is possible to correlate the ^{17}O and ^{18}O abundance. Due to the differential isotope ratio measurement, the correction has to be applied to the compound and the reference gas with the same procedure.

A possible relationship is the equation proposed by Craig in 1957,[38] which is known as the Craig correction:[36]

$$\left(\frac{^{17}R_c}{^{17}R_{i-ref}}\right) = \left(\frac{^{18}R_c}{^{18}R_{i-ref}}\right)^\lambda \tag{3.28}$$

where $\lambda = 0.5$ and ^{17}Rc and $^{17}R_{i-ref}$ are the $^{17}O/^{16}O$ isotope ratios of the component and the standard, respectively, and $^{18}R_c$ and $^{18}R_{i-ref}$ are the

$^{18}O/^{16}O$ ratios of the sample and the standard, respectively. It should be mentioned though that Craig did not use the definition given in eqn (3.27) himself because he applied an instrument that was equipped with a double instead of a triple FC arrangement.[21,38]

By using carbon dioxide, which was generated from Peedee belemnite (PDB) reference material (see Chapter 5) by using 100% H_3PO_4 at 25.3 °C, he determined the relation:

$$\left(\frac{^{45}R_{PDB}}{^{13}R_{PDB}}\right) = 1.0676 \tag{3.29}$$

In a similar manner, Craig gave an expression for ^{17}R and ^{13}R:

$$\left(\frac{^{17}R_{PDB}}{2^{13}R_{PDB}}\right) = 0.0338 \tag{3.30}$$

By introduction of the numerical values from eqns (3.29) and (3.30) in (3.31):

$$\delta^{13}C_c \approx \left(\frac{^{45}R_{PDB}}{^{13}R_{PDB}}\right)\left(\frac{^{45}R_c}{^{45}R_{PDB}} - 1\right) - \left(\frac{^{17}R_{PDB}}{2^{13}R_{PDB}}\right)\delta^{18}O \tag{3.31}$$

Craig obtained the expression (3.32) for correction of the $\delta^{13}C$-value by measuring CO_2 isotopologues.

$$\delta^{13}C_c \approx 1.0676^{45}\delta - 0.0338\delta^{18}O \tag{3.32}$$

It has already been mentioned that the value is related to PDB from 1957 and the numerical values given in this equation can strictly be used only for this standard. However, PDB was replaced by VPDB and the absolute value has been refined.[36]

Additionally, it turned out that the terrestrial relation λ between ^{17}O and ^{18}O is not exactly 0.5 but, rather, between 0.50 and 0.53.[37] Therefore, Santrock, Studley and Hayes proposed an iterative procedure. From eqn (3.28) the expression

$$^{17}R_c = K(^{18}R_c)^\lambda \tag{3.33}$$

can be generated, where K is a constant characteristic of the relationship between ^{17}R and ^{18}R that is given by

$$K = {}^{17}R_{s,i-ref}/({}^{18}R_{s,i-ref})^\lambda \tag{3.34}$$

Combining eqn (3.26) with (3.27) and substitution of eqn (3.33) for $^{17}R_c$ gives an expression for $^{18}R_c$ in terms of ion current ratios, K and λ:

$$-3K^2(^{18}R_c)^{2\lambda} + 2K^{45}R_c(^{18}R_c)^{\lambda} + 2^{18}R_c - {}^{46}R_c = 0 \qquad (3.35)$$

This equation has to be resolved iteratively for $^{18}R_c$ if λ differs from Craig's value of exactly 0.5. $^{17}R_c$ is then obtained from eqn (3.33) and finally $^{13}R_c$ can be calculated by using eqn (3.26).

Assonov and Brenninkmeijer compared different ^{17}O correction algorithms for CO_2 and found that the same data applied to different existing correction algorithms resulted in a bias of $\delta^{13}C$-values that exceeds the precision of modern IRMS instruments. The reasons for this bias have their origin in the various reported values for CO_2 evolved from NBS 19 by H_3PO_4 at 25 °C ($^{17}R_{VPDB-CO_2}$) (see Chapter 5) and λ, respectively.

Brand, Assonov and Coplen recently compared values for λ and found a value of 0.528 to be the best estimate for general use in the ^{17}O correction of CO_2 originating from the relevant most abundant oxygen pools.[21,39] Deviations from this value occur for non-terrestrial samples and in the case of mass-independent isotope fractionation such as the ^{17}O anomaly in ozone in the stratosphere that also affects CO_2 sampled from the atmosphere.[21,36]

Because of the potential inconsistencies in applied ion corrections, Verkouteren and Lee introduced a web-based tool to unify data treatment.[40] Brand et al. proposed a simplified (linearized) ^{17}O correction:

$$\delta^{13}C_c \approx {}^{45}\delta_{VPDB-CO_2} + \frac{^{17}R(^{45}\delta_{VPDB-CO_2} - \lambda^{45}\delta_{VPDB-CO_2})}{^{13}R} \qquad (3.36)$$

As mentioned, they suggested a λ-value of 0.528. The currently accepted ^{13}R ratio of 0.011180(28) for VPDB, gave a re-evaluated ^{17}R value of the evolved CO_2 of 0.000393(1), so that they employed a $^{17}R/^{13}R$ ratio of 0.03516(8). With eqn (3.36), $\delta^{13}C$-values with less than 0.010‰ deviation for normal oxygen bearing materials are obtained.[21] However, up to now, this correction is not implemented in any commercial software and thus not yet in widespread use.

Nitrogen. The situation in the case of nitrogen isotope ratio measurements by the isotopologues $^{14}N^{14}N$ with m/z 28 and $^{15}N^{14}N$, m/z 29 is straightforward because the measured ion beam ratios can be simply translated into isotope ratios (see Table 3.3).

Table 3.3 Mass-to-charge ratios with corresponding isotopologues measured for $^{15}N/^{14}N$ isotope ratio and $\delta^{15}N$ determination.

Mass-to-charge ratio of detected nitrogen isotopologue	Isotopologues
m/z 28	$^{14}N^{14}N^{+\bullet}$
m/z 29	($^{14}N^{15}N^{+\bullet}$, $^{15}N^{14}N^{+\bullet}$)
m/z 30	$^{15}N^{15}N^{+\bullet}$

Table 3.4 Mass-to-charge ratios with corresponding isotopologues measured for $^2H/^1H$ isotope ratio and δ^2H determination.

Mass-to-charge ratio of detected hydrogen isotopologue	Isotopologues
m/z 2	$^1H^1H^{+\bullet}$
m/z 3	($^2H^1H^{+\bullet}$, $^1H^2H^{+\bullet}$)

Here, even the isotopologue $^{15}N^{15}N^{+\bullet}$ can be neglected because of its low abundance and with the isotopologues from Table 3.3 the isotope ratio ^{29}R can be obtained easily by the following equation:

$$^{29}R_c = \frac{^{29}N_2}{^{28}N_2} = \frac{^{15}N^{14}N + {^{14}N^{15}N}}{^{14}N^{14}N} = 2 \times {}^{15}R_c \quad (3.37)$$

The factor 2 cancels out during calculation of the δ-value according to:

$$\delta^{29}N_c = \left(\frac{^{29}R_c}{^{29}R_{i-ref}} - 1\right) = \left(\frac{2 \times {}^{15}R_c}{2 \times {}^{15}R_{i-ref}} - 1\right) = \left(\frac{^{15}R_c}{^{15}R_{i-ref}} - 1\right) = \delta^{15}N_c \quad (3.38)$$

Hydrogen. The isotopologues of hydrogen with the corresponding mass-to-charge ratios are shown in Table 3.4. For the determination of hydrogen isotope ratios or δ^2H the mass-to-charge ratios 2 and 3 are measured, whereas the isotopologue $^2H^2H^{+\bullet}$ can be neglected.

As for nitrogen, in this case also no ion correction is necessary as the following equation shows:

$$^3R_c = \frac{2 \times {}^2H^1H}{^1H^1H} = 2 \times \frac{^2H}{^1H} \quad (3.39)$$

Again, the factor 2 vanishes in calculation of the δ-value. However, in case of hydrogen the formation of H_3^+ in the ion source has to be corrected for. The H_3^+ factor correction is described in detail in Section 3.4.2.

3.2 INLET SYSTEMS

One of the prerequisites of gas source IRMS is that the samples have to be converted into pure gases such as H_2, CO_2, N_2 and CO for hydrogen, carbon, nitrogen and oxygen isotope ratio analysis. In Section 3.1.2 the requirements on gas pressure to obtain linearity and sensitivity with a tight EI ion source have been described. The purpose of gas inlet systems is to limit the sample gas flow into the ion source and the pressure of a gas sample so that the system vacuum can be maintained between 5×10^{-6} to 10^{-8} mbar.[14] Another purpose of the inlet system is to introduce a sample and a calibration material such as a working or reference gas into the ion source under as nearly identical conditions as possible. This means that when an isotope ratio distortion during the introduction into the ion source cannot be prevented, the distortion should be the same for the sample as for the working gas.[12,14] Today, there are two common types of gas inlet systems for IRMS analysis. The first one is the so-called viscous flow or 'dual-inlet' system and the second, the 'continuous-flow system' (CF-IRMS).

3.2.1 Dual-inlet Systems

As we will see in Chapter 5, the measurement of pure substances that can be used as internal laboratory standards by dual-inlet is of enormous importance for CSIA. It is therefore necessary to be familiar with the fundamental background of this technique.

The dual-inlet principle was originally introduced by Murphey and co-workers[8] for the investigation of thermal gas diffusion.[2,12,41] Its incorporation in sector field mass spectrometry by McKinney et al. can be considered as the birth of high precision IRMS.[2,36] Sample introduction by dual-inlet requires an off-line sample preparation by closed-tube combustion with catalysts, subsequent gas purification (for example, removing water by cryogenic trapping) and further concentration in vacuum lines before the sample gas is introduced for measurement.[42,43] It is important to avoid interference by other gases during measurement.

The fundamental instrumentation has not changed since its introduction by Urey and McKinney in the late 1940s.[36] The general scheme of a dual-inlet system is shown in Figure 3.11. In principle, a dual-inlet system can be divided in two subsystems arranged in parallel that contain stainless steel tubes and all-metal design valves for the introduction of sample gas (left subsystem) and reference gas (right subsystem) under maximum inertness.[12] The gases are introduced into two separate welded-metal bellows (2–100 cm^3)[14] in which the gas pressure (10–1000 mbar)[14] can be adjusted automatically by electrically driven motors. The main part of the dual-inlet system is the 'changeover valve' shown in the middle of Figure 3.11. It consists of four valves that allow the inlet system to alternately switch within a couple of seconds between sample gas and reference gas, so that one gas enters the mass spectrometer's ion source while the other is directed to the waste line so that the flow is never interrupted. The employed reference gas is normalized to an internationally accepted standard (see Chapter 5). To avoid isotope composition changes, the sample gas and the reference gas bellows are connected to the changeover valve via capillaries (0.1–0.2 mm i.d.; 60–120 cm length),[36] which have a restriction (capillary crimp) near their ends.[12,44] At sufficiently high gas pressures (>50 mbar for hydrogen and >15 mbar for the other gases)[36] a viscous flow regime prevails in the capillary because the mean free path length of the gas molecules is short compared to the diameter of the capillary.[14] The restriction limits the gas flow and has a lateral dimension so that a molecular flow regime prevails behind the restriction. However, the restriction has the consequence that the gas in front of the crimp is isotopically heavier because the crimp has a higher conductance for the lighter isotopologue. The viscous flow results in a mass-independent flow of the isotopologues in the gas phase and prevents back-diffusion of the isotopically enriched gas at the crimp into the bellow reservoir.[28,36] Due to the fact that molecular flow prevails from the crimp via the source to the pump, no further isotope fractionation occurs.[28] The pressure that establishes the viscous flow conditions determines the minimum volume of gas that can be determined with high precision.[36]

With modern dual-inlet systems, in which computer-controlled adjustable welded-metal bellows are used for pressure matching and alternate introduction of sample gas and reference gas into the ion source via the changeover valve, highest precisions (about 0.01‰ for carbon isotope ratios and the other light elements) can be achieved.[36] This high precision is possible because the temperature dependence of electronic components, non-linearity or variations of the ion source or the magnetic field tend to cancel out by maintaining the same conditions for the sample as well as for the reference gas for a number of replicate measurements.[12] It is possible to measure even smaller gas volumes by

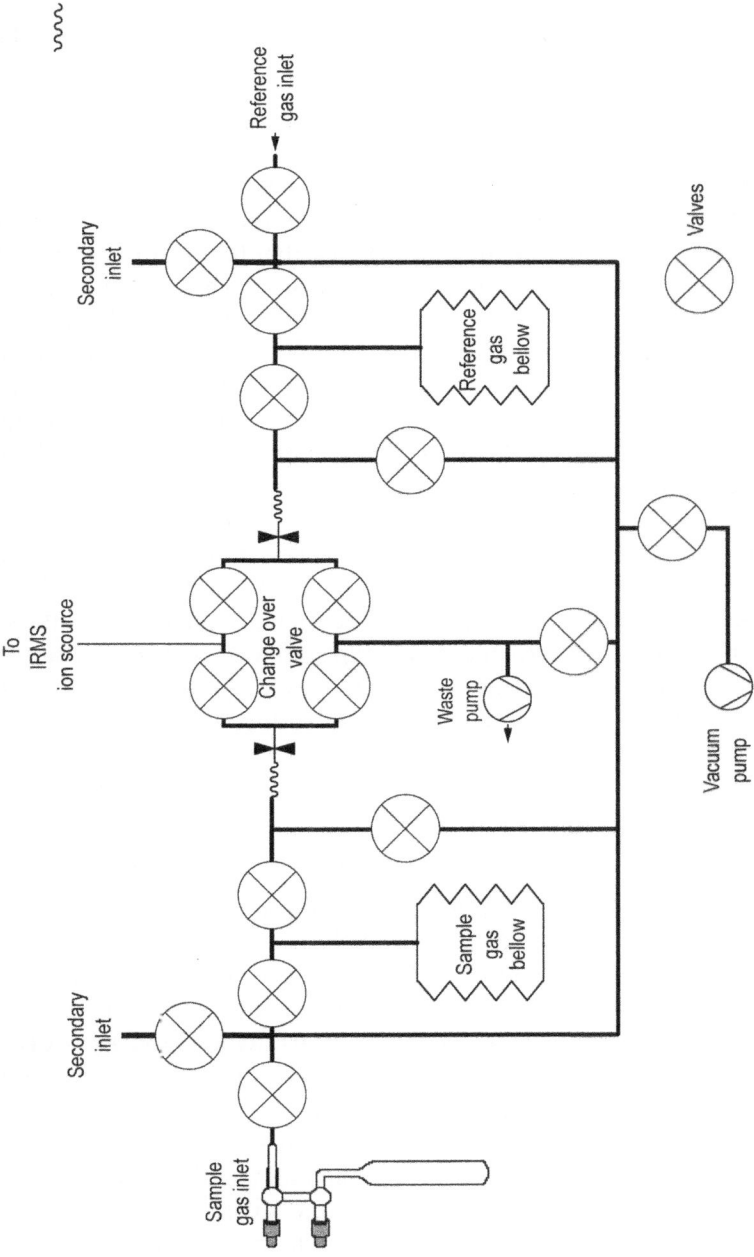

Figure 3.11 Simplified scheme of a dual-inlet system. For a detailed description, see text.

freezing a small volume in front of the capillary ('microvolume' technique). The physical limit of this volume or reservoir is ~250 μL. At operating pressures around 15 mbar a minimal sample size of 4 bμL gas is possible (1 bμL 'bar microlitre' is an amount of 1 μL gas at STP) compared to ~400 bμL without freezing.[36]

3.2.2 Continuous-flow Inlet Systems

It is obvious that the dual-inlet technique can never be used for hyphenation with separation techniques such as GC or LC which, as on-line techniques, introduce a high helium load into the ion source. In continuous-flow inlet systems the sample is transported within a helium carrier gas stream into the ion source.[14] Due to the use of helium as carrier gas, this principle is ideal for the coupling of IRMS with GC or elemental analysis (EA). As described in Chapter 1, there is a distinction between CSIA and BSIA. In the former, the effluent from an analytical column flows continuously through a conversion reactor, in which the separated compounds are transformed into a measurable conversion gas. This gas is introduced within the helium carrier into the ion source of the mass spectrometer. For BSIA, the whole sample is transformed first within, for example, an elemental analyser. Then the developed gases are separated on a GC column before they are introduced into the IRMS. For an illustration of these two types of sample preparation see Figure 1.3. In both cases, the flow of the carrier gas containing the conversion gas has to be limited when it is introduced into the IRMS ion source in order to provide a linear operation of the instrument (see also above).[14]

3.2.3 Open-split Systems

For the limitation of the gas flow into the ion source, 'open-split' systems have been developed that reproducibly transfer a constant stream of carrier gas into the ion source.[14] The open-split is also the transition point between atmospheric pressure and the vacuum of the mass spectrometer and it buffers gas pressure fluctuations, which have their origin in the conversion step (combustion or pyrolysis) of organic compounds into low molecular weight gases.[45] These fluctuations would lead to dynamic changes in the isotope fractionation during ionization and render accurate comparison between sample and standard impossible.[45]

Although the technical design between the different manufacturers and between CSIA and BSIA varies,[14] the principle of the open-split is realized in all of these. In the following we will discuss an open-split as it is typically used in CSIA. The necessary split ratios that have to be achieved vary for CSIA and BSIA. Typical carrier gas flow in capillary GC is between 1 and 2 mL min^{-1}. Compared to this relatively low carrier gas flow, the flow used in elemental analysers is much higher, i.e., between 80 and 120 mL min^{-1}.[14] Because of this, in CSIA split ratios between 2:1 and 5:1 are used, whereas the split ratios for BSIA via EA are almost two orders of magnitude higher.[14]

The restriction of the carrier gas flow into the ion source is achieved by a deactivated fused silica capillary of ~0.1 mm i.d. and ~1 m length in which a viscous flow regime prevails.[14] Under such conditions, the viscous flow follows the law of Hagen-Poiseuille:

$$\dot{V} = \frac{\pi r^4 \Delta p}{8 \eta l} \quad (3.40)$$

where $\dot{V}$ is the flow through the capillary, r the radius (here, e.g., ~0.1 mm), Δp the pressure drop across the capillary (here from ~1 bar to 10^{-6} mbar), η the dynamic viscosity of the gas and l the length of the capillary (0.7–2 m).

According to eqn (3.40), the flow is proportional to the 4th power of the capillary radius, and inversely proportional to the length of the capillary at a given pressure difference. The length of the transfer capillary was determined experimentally so that the source pressure is between 10^{-6} to 10^{-7} mbar and the carrier gas flow entering the ion source is 0.4 mL min^{-1}.[45] The transfer capillary can be introduced by a pneumatic actuator around 5 mm deep into the exhaust of a 0.32 mm i.d. capillary that delivers the converted effluent from a GC, EA or other preparation devices.[45,46]

A schematic open-split as it is used in continuous-flow instruments for CSIA is shown in the left panel of Figure 3.12. It consists of a glass tube (4 mm i.d.; 6 mm o.d.) with glued-in press fit connector, in which the carrier gas effluent capillary (0.32 mm i.d.) is inserted from the bottom. The restriction capillary to the IRMS can be moved up and down, which is referred to as open-split *out* and open-split *in*, respectively. In other words, if the open-split is *in*, the capillary 'sniffs' with this end at the carrier gas stream.[14] Any excess flow is flushed away so that the fixed split ratio prevails.[12] In the case of open-split *out*, the connection to the IRMS ion source is interrupted and the effluent is carried away by a permanent helium stream.[14] The permanent stream of helium flushes the open-split tube continuously so that other gases from the atmosphere are prevented from entering.[14]

In dual-inlet systems the main fraction of the sample gas is used to obtain non-fractional flow conditions and only around 10% of the

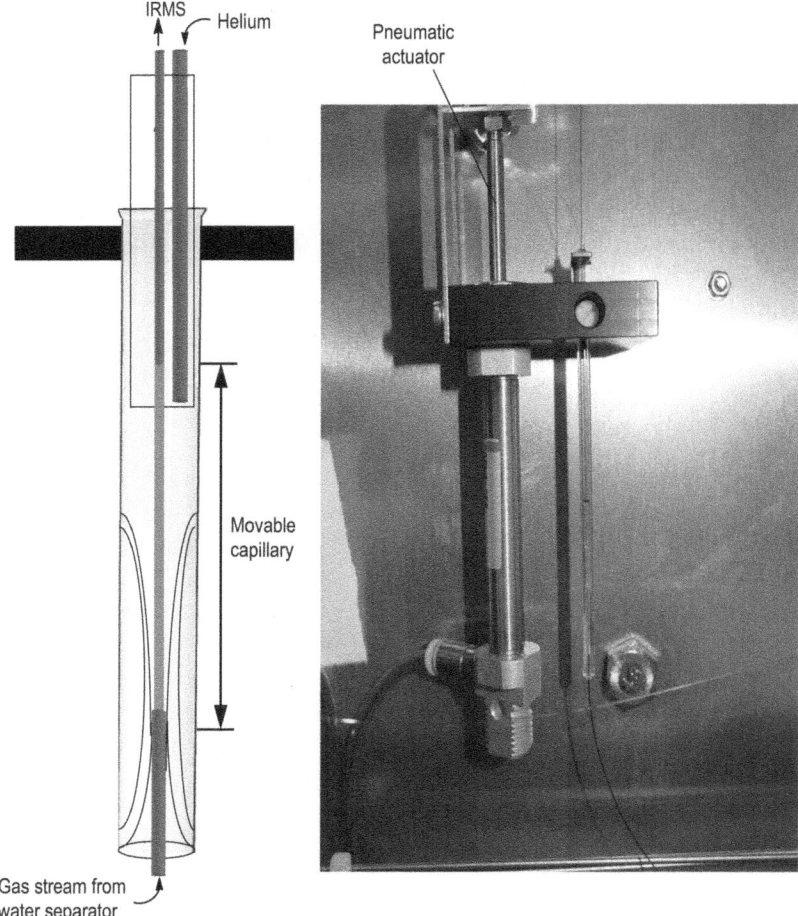

Figure 3.12 Open-split in an IRMS instrument. Left: schematic of an open-split inlet based on a sniffing capillary that can be moved by a pneumatic actuator into the incoming gas stream from the conversion oven via the water separator unit. A second capillary with a pure helium effluent prevents sample gas entering the mass spectrometer when the open-split capillary is in the upper position (open-split out). Furthermore, atmospheric gases are prevented from entering the mass spectrometer. Right: the open-split system of a GC interface.
(Thermo Fisher Scientific, Bremen, Germany; photo by Jens Laaks).

sample gas is actually measured.[14] In the case of continuous-flow the whole amount of a (converted) compound embedded in the carrier gas stream is available. Due to this the sample size required for an isotope ratio measurement is strongly (factor of 1 to 1000)[14] reduced by continuous-flow systems, down to the picomolar range.[12] Ratio linearity is important because in the case of GC or LC effluents the signal intensities of target compound peaks can vary considerably within or between chromatographic runs.[12] Whereas in dual-inlet IRMS the minimum sample size is determined by the pressure that determines the viscous flow conditions, the minimum sample size in continuous-flow is given by the overall signal to noise ratio.[14] This is discussed further in Chapter 5.

3.2.4 Reference Gas Inlet

It is obvious that the introduction of reference gas into a continuous-flow has to be made in a similar manner as the introduction of the sample. Reference gas pulses can be admitted to the ion source in various ways: (i) by a 6-port valve equipped with an injection loop,[2] (ii) by admittance from a changeover valve,[33] (iii) by a valve in front of the conversion oven,[47] or (iv) by a moving capillary switching device.[46]

A drawback of a 6-port valve is the short interruption of the carrier gas stream, resulting in background definition problems of the reference gas peak and reduced precision.[10] This effect can be reduced by moving capillary switching devices. Here we will discuss the reference gas inlet by such a device. Other methods for referencing are discussed in Chapter 5. Moving capillary switching devices will be found several times within this book. They are also used, for example, for heart-cutting in multi-dimensional GC (see Chapter 7).

Figure 3.13 shows a reference gas inlet on the basis of a capillary switching device. It consists of a small glass dome (1 mm o.d.),[10] in which a continuous stream of helium gas (5 mL min^{-1}) is introduced by a 0.1 mm i.d. fused silica capillary of 0.5 m length that reaches into the deepest part of the dome. A second sniffing capillary positioned further up in the glass dome with an internal diameter of 0.05 mm and a length of ~1.5 m is connected to the mass spectrometer (see Figure 3.13). A third fused silica capillary of 0.05 mm i.d. and 0.6 m length provides a flow of 1 mL min^{-1} CO_2, N_2, H_2 or CO reference gas to the glass dome. This capillary can be moved by a linear motion actuator to a position between the two other capillaries (sniffing point). This way, the reference gas pulses, such as are shown in the right panel of Figure 3.13, can be introduced into the mixing area of the dome, from where it can enter the sniffing capillary.[10,36]

Generally, modern interfaces for the coupling of continuous-flow techniques with IRMS include an open-split as well as a reference gas inlet. Often interfaces combine the open-split with the reference gas inlet, process an automatic dilutor to adjust sample peaks to that of the reference gas and can perform an automatic determination of the ratio

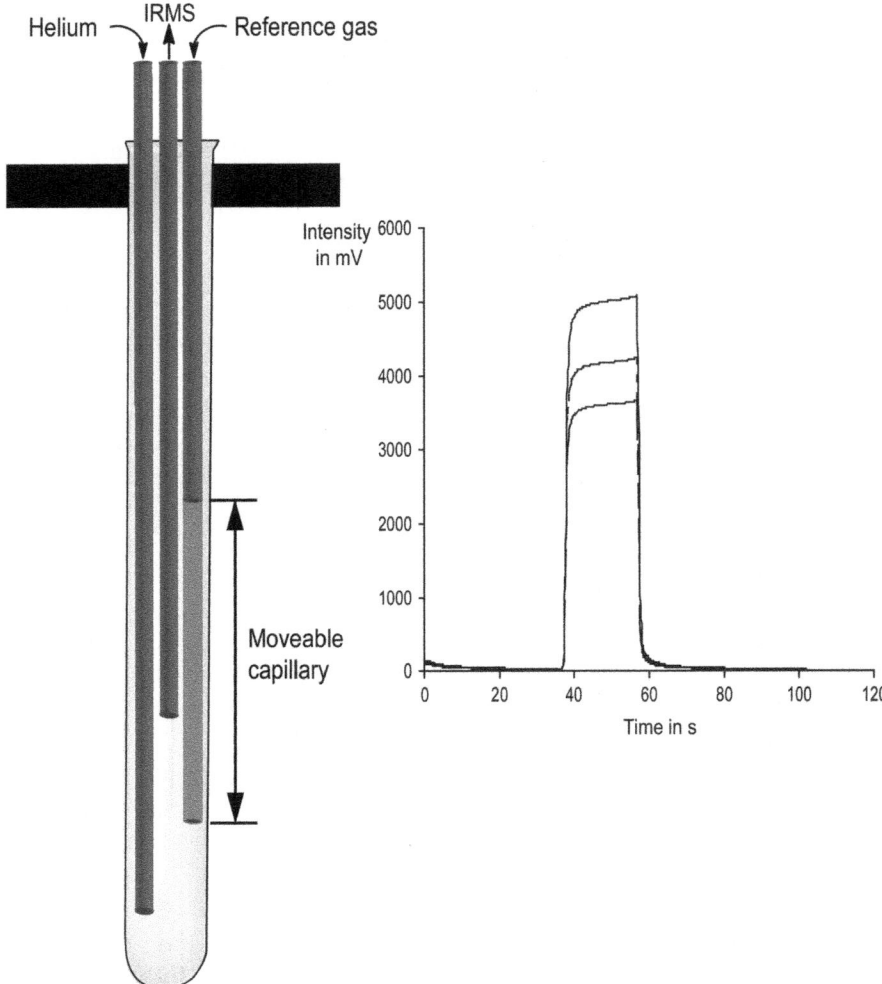

Figure 3.13 Reference gas inlet by a moving capillary switching device. Left: A schematic drawing of a moving capillary switching device. Right: reference gas pulses can be admitted by moving the capillary that delivers the reference gas to the glass dome. Note: The peaks are far from ideal. The ion source suffers from limited gas exchange performance. It can be improved by cleaning the ceramics close to the exit slit.

linearity L_R (see also discussion of L_R in Chapter 5). A manual change of the reference gas at the interface is no longer necessary because the last generation of IRMS interfaces contains several inlet ports for a variety of reference gases, which enables a faster changeover between different target elements. Other important features of such interfaces, such as units for conversion gas reduction, gas separation and drying, depend on the individual elements measured and will be discussed in the specific sections below.

3.3 PERIPHERAL DEVICES FOR CONTINUOUS-FLOW IRMS

Several peripheral devices that fulfil different purposes have been coupled to continuous-flow IRMS. The most prominent, the workhorse in SIA,[36] is EA-IRMS. Other devices have been developed, for example, for preconcentration of atmospheric gases,[48] or laser ablation for tree ring analysis,[49] to mention only two. The following section will provide a short glimpse of EA-IRMS. Similar to dual-inlet IRMS, EA-IRMS is important for CSIA because with this method internal laboratory standards can be measured and normalized to an internationally accepted reference material. Furthermore, CSIA is also still performed for specific target compounds by off-line separation and purification followed by EA-IRMS.

In 1983, Preston and Owens reported the coupling of an automatic nitrogen analyser (ANA) with an isotope ratio mass spectrometer.[50] Since this time, BSIA with elemental analysers such as the CNS analyser or ANSCA (automatic nitrogen, sulfur and carbon analyser)[14] has found a huge number of applications that still far exceeds the number of CSIA applications. Depending on the sample, several preparative steps, including dry freezing, grinding, aliquotization, etc., can precede the EA-IRMS measurement. Specific sample preparation protocols have been developed for a huge variety of samples and are available in literature. The prepared samples are weighed into tin or silver capsules[51] and introduced into an autosampler carousel, from which they fall into the conversion reactor.

Figure 3.14a–c shows the typical setups of a Flash EA-IRMS with different types of conversion reactors for the isotope measurement of different light elements. Figure 3.14a sketches the set-up for the determination of nitrogen and carbon. After introduction, the sample is immediately oxidized by an oxidation catalyst and the introduction of an oxygen gas pulse into the furnace. The filled reactor consists of quartz glass and is held at temperatures around 1000 °C at which the sample is completely converted to N_2, CO_2, O_2 and water.[53] The products of the oxidation process are transported in a helium gas stream through a reduction furnace that is filled with copper and is heated to a temperature of around 600 °C. In this reactor excess oxygen is scavenged and nitrogen oxide traces are reduced to elemental nitrogen. Subsequent to reduction, the gases are dried by passing a drying trap (e.g., filled with

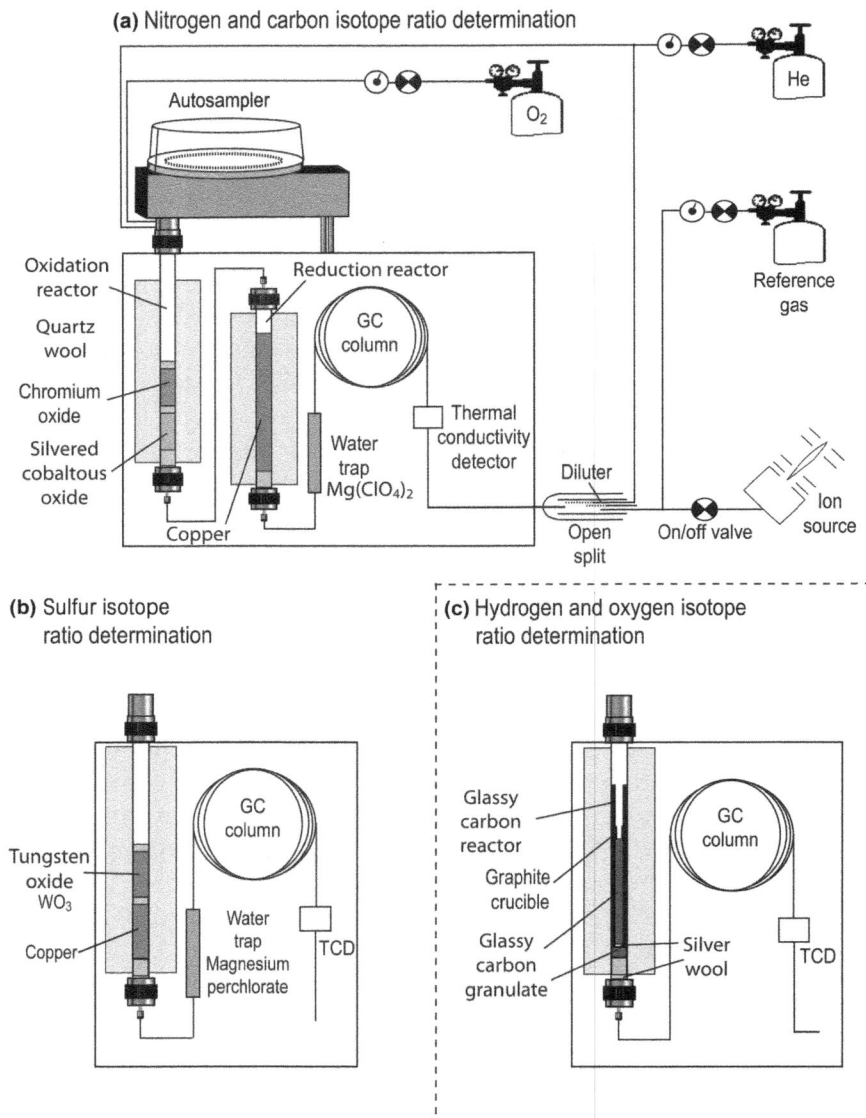

Figure 3.14 Elemental analyser setups for various target elements. The panels show the set-up of an elemental analyser in series with an interface and IRMS for the analysis of bulk samples of: (a) nitrogen and carbon; (b) sulfur. (c) a TC/EA-IRMS for bulk analysis of hydrogen and oxygen. The diagrams are based on a Thermo Fisher Scientific, Bremen, Flash Elemental Analyser.[52] It has to be mentioned that EA systems of other manufacturers possess similar but not identical components. In particular, the kind of combustion or high temperature conversion can be different; therefore we also refer to the specific manufacturer manuals for information. Generally a thermal conductivity detector (TCD) is used to measure the elemental composition of the sample.

$Mg(ClO_4)_2$) and are separated on a GC column incorporated in the elemental analyser. For the separation of the effluent gases N_2 and CO_2 the GC is equipped with a packed GC column (0.3–5 m length). Typically, a 3.5 m long Porapak PQ column (80/100 mesh) maintained isothermally at 40 °C is used. The separated gases are introduced in succession via an interface that contains an open-split and a reference gas inlet into the IRMS. Because of the misbalance of high C/N ratios of most organic materials, very small N_2 signals have to be measured along with large CO_2 signals. Therefore, some manufacturers have developed a dilutor that adds helium to the GC effluent for adjustment of the larger CO_2 peak. This allows the analysis of samples with very different C and N amounts with similar precision.[14] The analysis of sulfur, as shown in Figure 3.14b, is carried out separately since the oxidation reactors are different[53] and often Teflon™ instead of stainless steel packed columns are employed.

In addition to carbon, nitrogen and sulfur, high temperature conversion elemental analysers coupled to IRMS (TC/EA-IRMS) have been developed for the analysis of hydrogen and oxygen bulk sample isotope ratios (see Figure 3.14c). In high temperature or pyrolytic conversion the introduced samples are converted to H_2 and CO at temperatures between ~1100 °C and ~1450 °C.[53] At these high temperatures ceramic tubes lined with glassy carbon are employed. It is not possible to use quartz glass tubes, because CO can react with the quartz so that ^{18}O exchange occurs, thereby altering the isotope ratio.[14] After high temperature conversion, hydrogen and carbon monoxide are separated by a packed GC column and enter the IRMS via the open-split.

Well-defined protocols for the measurement of samples with EA have been developed and for quality assurance issues required for an appropriate determination of isotope ratios by EA-IRMS we refer to publications of Werner and Brand[36,54] and Qi and Coplen.[55,56]

3.4 CSIA BY GC-IRMS

From the previous discussion of IRMS instrumentation it can be concluded that a direct coupling with GC is not as straightforward as for organic mass spectrometry. As mentioned above conversion of organic compounds in low molecular gases such as CO_2, N_2, CO or H_2 for carbon, nitrogen, oxygen and hydrogen measurements, respectively, is necessary. For on-line conversion and proper gas introduction into the IRMS, special interfaces have been developed that convert separated compounds, eluting from the GC column, provide gas drying and control the gas flow into the ion source. Here, the continuous-flow inlet systems with open-splits discussed above offer ideal conditions for a coupling of GC with IRMS for CSIA. Due to the processes required to form the conversion gas it is not possible to apply a single conversion process for all elements. In general, two types of conversions are applied. The first type is a combustion (GC-C-IRMS) or oxidation interface for the conversion of carbon and nitrogen in organic compounds to carbon dioxide and nitrogen. Although the largest fraction of nitrogen from organic molecules is readily converted to N_2 in the interface, a few percent are oxidized further to nitrogen oxides. These have to be reduced to N_2 subsequent to the combustion process in an additional on-line reduction oven. The second type is referred to as high temperature conversion (GC-TC-IRMS) interface for hydrogen and oxygen isotope ratio determination. In this case an on-line reduction of organic compounds to hydrogen gas or carbon monoxide is used. Because of their fundamental differences the two types of conversion and the employed instrumental components are discussed separately in more detail in the following sections.

3.4.1 Carbon and Nitrogen Isotope Ratio Analysis

Compared with the other elements, CSIA of carbon is by far the most frequently used method.[51] The determination of nitrogen isotope ratios of individual compounds at natural abundance has a high potential and the applications are manifold. They reach from the investigation of nitrogen metabolism in plants[57] and humans to measuring gene expression, degradation of pesticides, to authenticity control of food products.[58] However, despite its high potential it is used much less frequently than carbon isotope ratio analysis because the measurement of $^{15}N/^{14}N$ isotope ratios of individual compounds is more challenging. From an instrumental viewpoint two more steps are required that need to be controlled properly, i.e., the additional reduction step and the subsequent removal of carbon dioxide by cryogenic trapping. Furthermore, the sensitivity of nitrogen isotope ratio analysis is inherently lower than that of carbon due to (i) the low amount of nitrogen present in organic molecules, (ii) the lower natural abundance of ^{15}N compared with ^{13}C (see Table 2.1), (iii) the fact that two atoms are necessary to form one N_2 molecule[59] (as an example, in the case of underivatized amino acids, the formation of one mole of nitrogen gas produces 4 to 22 moles of CO_2),[58] (iv) the ionization probability of N_2 is lower than that of CO_2 (see Figure 3.4), and (v) the high fraction of N_2 in air can result in high background values, even for very small gas leaks.[59]

Individual applications for CSIA of carbon and nitrogen in different fields are discussed in Chapter 6.

Figure 3.15 shows the scheme of a typical GC-IRMS system for the determination of carbon and nitrogen isotope ratios. In the first step, chromatographic separation of the analytes in a gas chromatograph takes place. Following separation, the analytes pass a T-piece or, if a second detector is involved, an X-piece prior to introduction into the combustion oven. In the oven, the compounds are completely oxidized by a catalyst and an oxygen-providing material (metal oxide) in a heated reactor (ceramic or quartz tube) to CO_2, N_2 and H_2O. Small amounts of nitrogen oxides (NO_x) that are formed during the combustion of nitrogen-containing compounds are reduced to N_2 in the subsequent reduction oven.[60-63] Water that would interfere in the ion source is removed by a Nafion™ membrane. In the case of nitrogen isotope analysis, additionally, the formed CO_2 is removed by a liquid nitrogen trap. Finally, the gas is introduced into the ion source via the open-split introduced above.

The Backflush System. The main task of the backflush system, shown in Figure 3.16, is to prevent the solvent from liquid injections reaching the oxidation oven since this could quickly exhaust or even exceed the

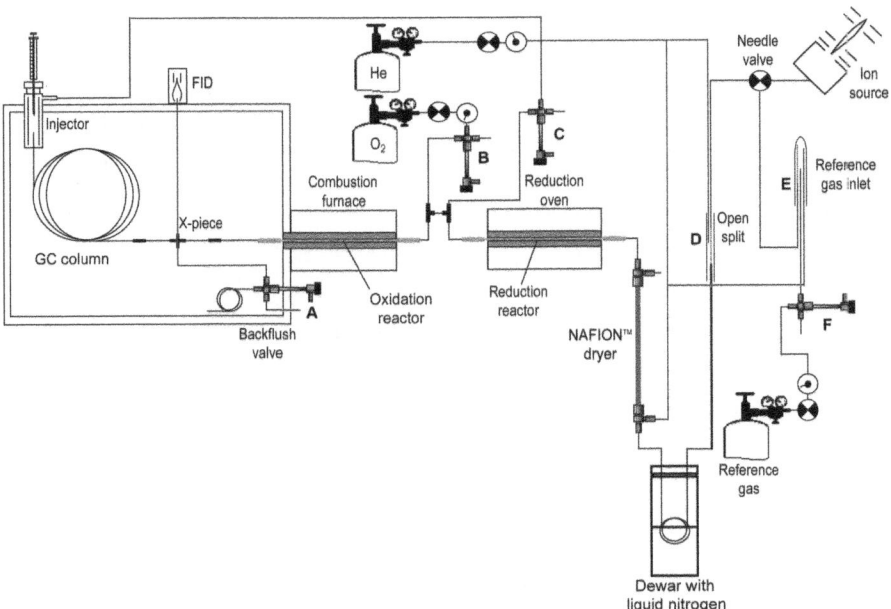

Figure 3.15 Set-up of an IRMS coupled to a gas chromatograph via a combustion interface for determination of carbon and nitrogen isotope ratios. For a detailed description, see text.

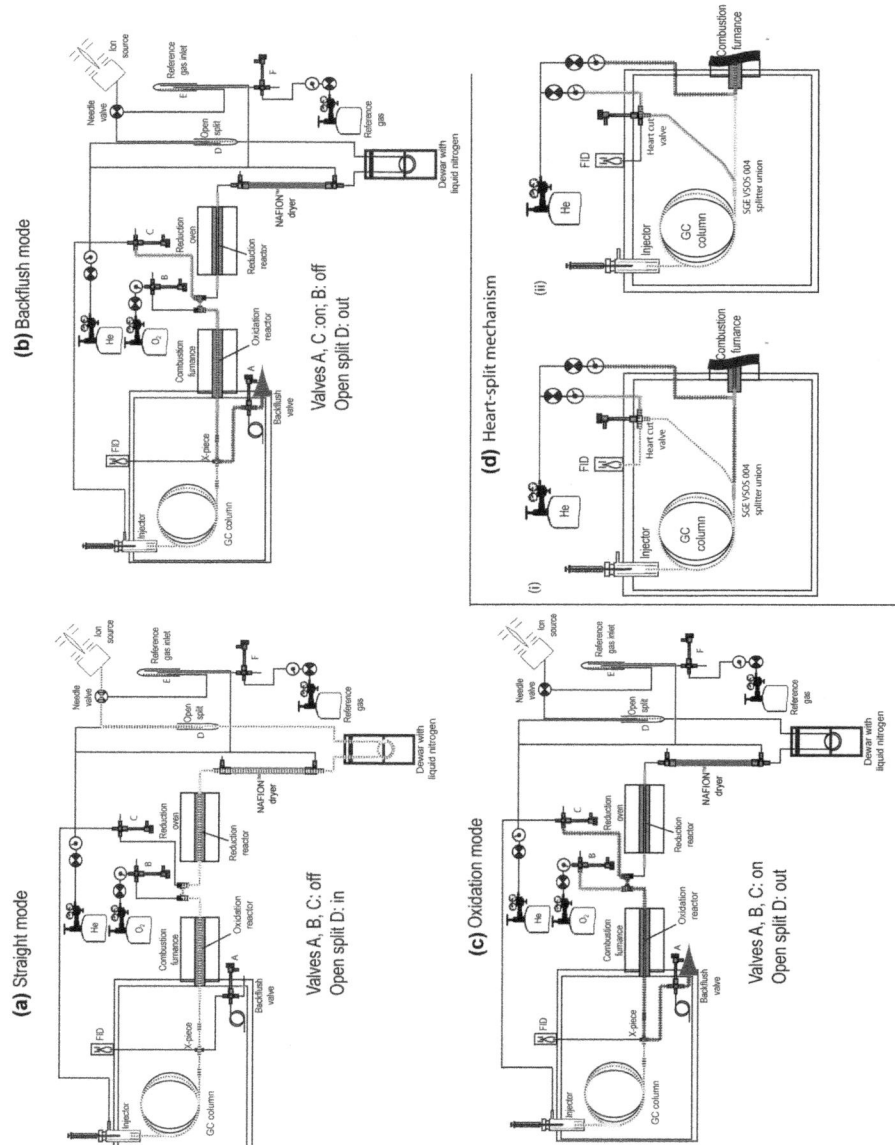

oxidation capacity.[11,61,64] Furthermore, the excess carbon can reduce the filament lifetime and incomplete combustion could result in contamination of the following interface parts and the IRMS. In addition to the excess carbon, higher amounts of water are developed which can saturate the Nafion™ membrane, preventing effective water removal.[33] Often chlorinated solvents such as dichloromethane or chloroform are used that yield high amounts of hydrochloric acid (HCl) within the combustion reactor, which could lead to corrosion of interface components. Due to all these reasons, it is mandatory to remove solvent peaks as far as possible via the backflush valve. With the backflush system it is also possible at any time during a run, to divert the effluent stream from the GC column to prevent nonessential sample components entering the reactor. Thus, isolated peaks from a complex chromatogram can be selectively introduced to the mass spectrometer.

The (GCC III) interface (Thermo, Bremen, Germany), for instance, possesses a backflush system as shown in Figure 3.16a–c. It consists of a pressure-actuated open T-valve. For the 'straight mode' shown in Figure 3.16a, the backflush valve (A) and the helium valve (C) are closed so that the GC column effluent can enter the combustion furnace and the interface. If the open-split is set to *in* the gas from the furnace can enter the ion source of the mass spectrometer. To avoid a dead volume in straight mode, the permanent exit side of the backflush valve is closed with a ∼1 m long, 0.1 mm i.d. deactivated capillary to obtain a small permanent bleeding vent. For the 'backflush mode' shown in Figure 3.16b the backflush valve (A) and the helium valve (C) are open so that a helium stream flows through the combustion reactor and further through the backflush valve vent. The column effluent is prevented from entering the combustion reactor and is removed through the backflush valve vent.

Figure 3.16 Backflush systems in GC-IRMS. Panels (a), (b) and (c) show the backflush system as it is used in Thermo Fisher Scientific instruments. (a) The valve positions and flow paths of column effluent (broken light gray line) are shown. Valves A, B, C are closed (off) so that GC, oven and interface are connected, which is referred to as 'straight mode'. (b) The so-called 'backflush mode' is shown, in which valves A and C are on (open) so that a helium stream (broken dark grey line) flows in the direction of the combustion reactor and through the backflush valve vent. The effluent from the column (broken light grey line) is prevented from entering the combustion reactor and is removed through the backflush valve vent. (c) The set-up for the oxidation mode is shown. In addition to the set-up in (b) valve B is open so that oxygen (broken regular gray line) can enter the oxidation reactor for regeneration but not the GC column and the mass spectrometer. (d) The 'heart-split mechanism' as it is used in IsoPrime-GC-V interface systems (for description see text).

An oxidation or reoxidation of the oxidation reactor can be carried out in the 'oxidation mode'. To that end, the backflush valve (A) inside the GC is opened so that the GC effluent is diverted to the vent. By opening the helium valve (C) and the oxygen valve (B), helium mixed with oxygen backflushes the oxidation furnace for reoxidation while pure helium passes over the reduction reactor.

In the case of the IsoPrime-GC-V interface (Elementar, Hanau) a so-called heart-cut mechanism has the tasks (i) to prevent solvent entering the combustion oven, (ii) to selectively divert interfering components or matrix, and (iii) to allow for reference gas injection without matrix co-elution from the column in the heart-split open mode. In Figure 3.16d the set-up of such a heart-split mechanism is shown in (i) heart-split open and (ii) heart-split closed mode. In this arrangement, helium is applied to both the heart-split valve and to the connector just in front of the oxidation reactor and at the heart-cut valve. The pressure of helium applied to the connector is higher than that applied to the heart-split valve. If the heart-split valve is open (i), this higher gas pressure leads to a backflush of helium towards the splitter union and further via the heart-split valve into the FID (flame ionization detector). The effluent from the column is thus prevented from entering the capillary towards the oxidation oven and is vented via the FID. If the heart-split valve is closed the gas pressure in the splitter rises until the column effluent pressure exceeds the helium pressure applied to the connector. Then, flow in the capillary between splitter and connector will reverse and the GC column effluent reaches the reactor.[65]

Combustion Oven Design for Carbon and Nitrogen. It has been mentioned earlier that a combustion step of the eluting analytes is necessary before compound-specific isotope ratio data of carbon and nitrogen can be obtained. Prerequisites for an on-line reactor are rapid and quantitative combustion of the molecules entering the reactor and concurrent minimum peak broadening,[45,66] long lifetimes[45] and minimal oxygen bleeding into the mass spectrometer. An incomplete combustion can lead to isotope fractionation resulting in low precision and accuracy.[33]

Merritt *et al.* thoroughly investigated materials and conditions in the oxidation reactor. As reactor material, quartz tubing can only be used up to 800 °C. At temperatures higher than 850 °C, quartz and cupric oxide (CuO) form an eutectic phase and the reactors failed at higher temperatures.[33] Therefore, they used non-porous alumina tubes (Al_2O_3) (0.5 mm i.d.; 1.5 or 3.0 mm o.d.; 30 cm length) and placed a copper or nickel and a platinum wire as catalyst with equal length inside (20 cm length; 0.1 mm diameter). Prior to the first use the Cu and Ni wires have to be oxidized in situ. The aluminium oxide tubes provide additional advantages with regard to durability and lower variability in outer and inner diameters even at high temperatures ($\leq 1300\,^\circ$C).[33] In commercially available oxidation reactors the wires are generally twisted. More than four metal wires cause flow restrictions that have their origin in the fact that the oxides occupy a larger volume within the reactor than the non-oxidized metal wires and the wires tend to break after several oxidation steps.[33,45]

The oxygen content of copper and nickel oxides is diminished by consumption during the oxidation of eluted compounds, column bleed and self-dissociation at elevated temperatures.[33] It was observed for CuO that at temperatures higher than 825 $^\circ$C the oxygen release by thermal decomposition is more important than the oxygen consumption by combustion. At temperatures >850 $^\circ$C the oxygen release results in short reactor life times. The thermal oxygen release from CuO is shown in Figure 3.17.

Because the metal oxides contain only a limited amount of oxygen, frequent regeneration is necessary either in situ by addition of small oxygen amounts into the carrier gas stream in front of the reactor or, after a certain time, via the backflush mechanism discussed above.[33,45,66] When oxygen is introduced permanently to the reactor, the O_2 addition has to be adjusted with care so that the filament of the IRMS will not be damaged or diminished in its lifetime.[33,45]

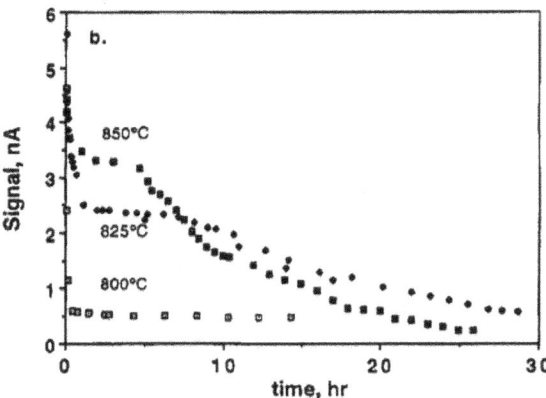

Figure 3.17 The dissociation of CuO at three different temperatures (800 $^\circ$C, 825 $^\circ$C, 850 $^\circ$C) by measuring the ion current at m/z 32 (oxygen) against time. The high oxygen release in the first 20–30 min starting with signals >30 nA are not included in this graph.[33]
(Graphic reprinted with permission of American Chemical Society.)

If CuO reactors are not reoxidized regularly, or if too much organic material passes through the reactor, spots of bare copper occur and sample components are not oxidized completely. Pyrolysis at the reactor inlet leads to visible residues of black soot. Reactors containing CuO that are operated at temperatures of around 800 °C need to be reoxidized every three days even when the reactors are not exposed to organic material.[33]

NiO behaves differently from CuO. It does not degrade thermally, even up to temperatures of >1300 °C and thus no increase of the m/z 32 ion current is observed. The lifetime of the NiO reactor is rather limited by the amount of organic material in the reactor and not by the self-dissociation.[33] For combustion at temperatures >1050 °C, oxygen has to be introduced regularly into the reactor, either continuously by the carrier gas stream or after each run to obtain accurate and precise isotope ratios. Very stable organic molecules such as methane need temperatures up to 1150 °C, achieved by using only NiO as oxidant.[33]

In summary, NiO based reactors provide several advantages compared with CuO: (i) better preservation of the oxidizing power, (ii) operability at temperatures above 1100 °C during continuous operation without loss of oxidation capacity, (iii) no requirement for an additional reduction reactor.[60]

In Thermo Fisher Scientific instruments for carbon and nitrogen the standard format has been a combustion reactor consisting of a nonporous alumina tube (320 mm length; 0.5 mm i.d.; 1.55 mm o.d.) filled with Pt, Cu and Ni wires of ~ 25 cm length.[67] Before first use, the reactor is oxidized with an oxygen flow of 0.5 mL min^{-1} at 940 °C for ~ 12 h.[68] The CuO and NiO represent the oxygen sources for a quantitative combustion of organic compounds to CO_2 and H_2O.

The oxygen in the reactor is generated by the reactions:

$$4CuO \rightleftharpoons 2Cu_2O + O_2 \qquad (3.41)$$

and

$$2NiO \rightleftharpoons 2Ni + O_2 \qquad (3.42)$$

The operating temperature is kept between 940–950 °C by a resistively heated Al_2O_3 furnace.[68] This type of reactor is, in principle, a combination of the CuO and NiO reactors investigated by Merritt *et al.* Here the NiO will release oxygen for the combustion only in the presence of organic compounds that immediately consume the formed oxygen, thus keeping the oxygen partial pressure extremely low. The developed Ni is

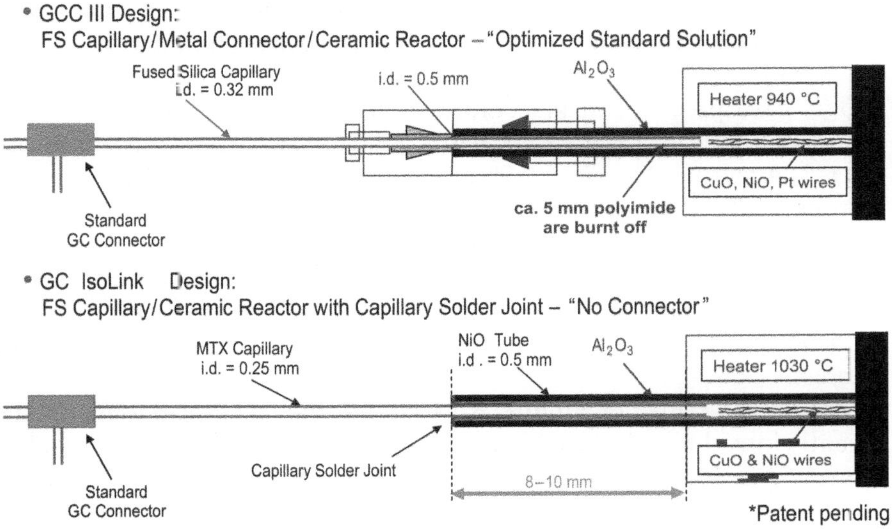

Figure 3.18 Comparison of the combustion oven design as it is used in GCC III™ interface (upper diagram) and the design as it is applied in the recently introduced GCIsoLink™ interface (lower diagram).
(Graphic with kind permission of Dieter Juchelka, Thermo Fisher Scientific, Bremen.)

reloaded with oxygen by the permanent oxygen loss of the CuO at the working temperature of 940 °C. Without organic compounds the reactor has to be reoxidized every 3–5 days. The reoxidation frequency depends on the compounds and samples measured. A connection between the GC column and the combustion oven is made by a non-coated fused silica column. This column is generally connected via a Valco™ connector with the alumina tube of the reactor. Use of connectors with a 0.5 mm i.d. bore hole to put the fused silica capillary into the hot zone of the oven is recommended, as illustrated in the upper panel of Figure 3.18. To that end, the polymer coating of the end of the capillary that is introduced into the reactor has to be burned off by using a lighter. The remaining soot of the coating has to be wiped away using an acetone or methanol soaked lint-free tissue. This and the introduction into the oven have to be carried out with great care because the remaining fused silica is very fragile.

The reactor performance strongly depends on the compounds to be oxidized as well as the measured element. Meyer *et al.* investigated the above-mentioned combustion reactor fillings for carbon and nitrogen isotope analysis of atrazine. They obtained accurate and precise carbon isotope measurements only with the Ni/NiO reactor operated at 1150 °C and with reoxidation after each chromatographic run for 2 min, providing a mean deviation $\Delta\delta^{13}C$ from dual-inlet measurements of -0.1–0.2‰ and a standard deviation (SD) of ± 0.4‰.[69] In contrast, CuO at 800 °C gave precise, but inaccurate values ($\Delta\delta^{13}C$ -1.3‰, SD 0.4‰), and CuO/NiO/Pt reactors at 940 °C gave inaccurate and imprecise data.[69] Accurate ($\Delta\delta^{15}N = 0.2$‰) and precise (SD ± 0.3‰) nitrogen isotope analysis was also possible with the Ni/NiO reactor at the same temperature but without the frequent reoxidation. This example shows that even for carbon isotope analysis, one cannot always use standard settings recommended by the manufacturers to generate reliable isotope data but must validate the performance for the specific compounds of interest.

The lower panel of Figure 3.18 shows the reactor design of the recently introduced GC IsoLink™ interface. The reactor consists of a Ni/NiO tube inside the Al_2O_3 tube, which is maintained at 1030 °C and is filled with Ni and Cu wires. In this design, the capillary from the GC connector is already fixed inside the alumina tubes.

In IsoPrime GC interfaces a quartz glass tube with a filling of copper oxide granules in the centre is used for carbon isotope analysis. The combustion process takes place at 850 °C. According to the manual the regeneration with oxygen overnight should be carried out only once because the capacity to hold oxygen diminishes. The quartz glass tubes can be filled with copper oxide by the user.[70]

A serious problem remaining in all reactor designs is the formation of salts with copper and nickel in the presence of chlorinated and fluorinated compounds. These poison the reactor resulting in a shortened lifetime and efficiency.[71,72]

Extra-column peak broadening by the combustion oven is of major concern because the high chromatographic resolution obtained can be altered or even destroyed by large tubing diameters, changes in tubing diameter or the connections and thus dead volumes.[73]

To this end, attempts have been made to develop zero-dead-volume combustion reactors on the basis of capillary columns connected via zero-dead-volume connectors but these have never been commercialized.[66,73] Goodman *et al.* developed a single-capillary interface design (SCID) consisting of a 2 m long deactivated fused silica capillary of (0.32 mm i.d.) containing two or three pieces of 30 cm long copper wires (0.1 mm o.d.). The furnace was maintained at 500 °C. The reactor capillary was embedded into a conventional quartz tube to hold it in place during operation.[73] With this interface Goodman *et al.* reached higher separation numbers and improved peak shapes compared with a packed conventional combustion interface.[73]

Ellis et al. developed a polyamide coated glass capillary as the furnace reactor. A disadvantage of applying polyamide coated fused silica capillaries is the low temperature tolerance of the polyamide. High temperatures as they are employed with ceramic or quartz tubes cannot be used. The coating of the capillary that traverses the hot zone of the furnace has to be removed.[66] Ellis et al. placed two copper and one nickel wire into the fused silica capillary combustion tube. The very fragile uncoated area of the capillary was placed in a ceramic tube for support.[66]

Reduction Oven. For the measurement of nitrogen isotope ratios a reduction oven is required subsequent to the oxidation oven to provide a quantitative conversion of nitrogen oxides N_xO_y formed in the combustion step to elemental nitrogen gas N_2. The reduction is achieved by passing the oxidation reactor effluent through a reactor filled with elemental copper wires or strands at 600–630 °C so that according to:

$$y\text{Cu} + N_xO_y \rightarrow y\text{CuO} + \frac{x}{2}N_2 \quad (3.43)$$

copper oxide is formed from the Cu wires or granula.

IsoPrime nitrogen interfaces possess a regeneration membrane allowing hydrogen gas to penetrate the membrane and infuse into the carrier gas stream upstream of the reduction furnace for regeneration by copper oxide reduction.

Additionally, the reduction reactor acts as a scavenger for surplus oxygen bleeding from the oxidation reactor, which prevents oxygen from entering the ion source. It is also advantageous to employ the oven for carbon isotope analysis, when nitrogen- or sulfur-containing compounds are present. The minor by-products of the combustion, NO_x and SO_x are corrosive to the IRMS and $^{14}N^{16}O_2^+$ is an isobaric interference for the isotopologue $^{12}C^{18}O^{16}O^+$ with m/z 46 measured in the minor ion beam for ion correction during carbon isotope analysis.

Isobaric Interferences for Carbon Measurements. Proton donating compounds present within the IRMS ion source can lead to proton-transfer reactions that can cause isobaric interferences. For example, this is the case with methane or water, which react with CO_2 and produce a bias of the minor ion beam (m/z 45) and thus with $^{13}C^{16}O_2^+$ due to the following reactions:

$$CO_2 + CH_4^{+\bullet} \rightarrow CO_2H^+ + CH_3^\bullet \quad (3.44)$$

$$^{12}C^{16}O_2^{+\bullet} + H_2O \rightarrow {}^{12}C^{16}O_2H^+ + HO^\bullet \quad (3.45)$$

These isobaric interferences will lead to an overestimation of the abundance of the heavy isotope ^{13}C. Therefore, in the case of methane measurements it has to be ensured that a quantitative conversion of methane to CO_2 in the combustion oven is provided. Water is the inevitable by-product of the combustion of organic compounds and has to be eliminated as far as possible in the interface after the combustion oven, as discussed in the following.

Different systems for on-line water removal can be applied. Trapping can be carried out by directing the effluent through $Mg(ClO_4)_2$, by placing a capillary behind the combustion oven through a dry ice acetone slush or liquid nitrogen[74] or, most commonly,[58] by using a countercurrent Nafion™ membrane dryer. The membrane-based dryer has the advantage that it avoids the common problems of cryogenic drying, including clogging by ice formation or cold spots. Additionally, it possesses a low dead volume, a low resistance to flow and is continuously self-regenerative and resistant to chemical degradation.[74] The construction and operation of a Nafion™ membrane countercurrent dryer was investigated in detail by Leckrone and Hayes.[74,75]

Figure 3.19 shows a typical Nafion™ dryer unit. It consists of a glass or stainless steel tube (4 mm i.d.; 5 mm o.d.), in which a Nafion™ membrane tube (0.3 mm i.d.; 0.5 mm o.d.; 20 cm length) is connected tightly to the fused silica capillary coming from the reduction oven and the fused silica capillary that leads to the open-split. A countercurrent flow of dry helium with a flow of 3–4 mL min^{-1} is maintained between the Nafion™ membrane and the glass or stainless steel tube.

Nafion™ is a copolymer consisting of tetrafluoroethylene (Teflon®) and perfluoro-3,6-dioxa-4-methyl-7-octane-sulfonic acid (see Figure 3.19), which has the characteristic property of being permeable to water. This is caused by the sulfonic acid groups that can take up 13 water molecules, which makes the membrane hygroscopic and water permeable.[74] Low molecular weight polar compounds such as ketones, alcohols, aldehydes, and water-soluble ethers can also penetrate the membrane but gases such as carbon dioxide, nitrogen and carbon monoxide cannot and will remain within the Nafion™ tube, from which they are transported to the open-split. The driving force for the transport through the membrane is the humidity gradient between the tube interior and the countercurrent helium stream. The dry helium stream removes the water so that no equilibrium is reached and continuous water removal is obtained. Leckrone and Hayes found that a Nafion™ dryer operated at ambient temperature can remove water from a saturated stream with up to 99.96% efficiency, produces a dew point as low as −62 °C, and is about as effective a drying agent as $CaSO_4$. The cooling

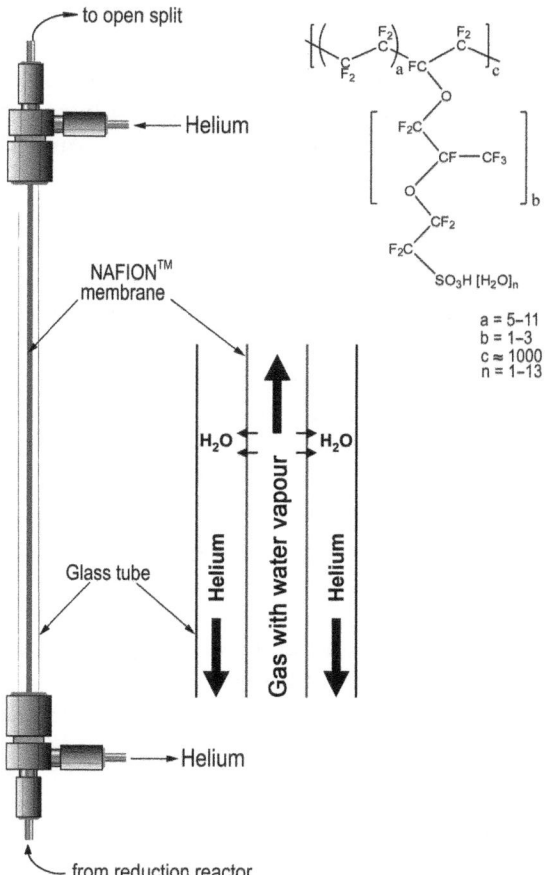

Figure 3.19 Scheme of a typical Nafion™ dryer unit. Gas containing water vapour flows through a Nafion™ tube, which is purged by a countercurrent flow of dry helium on the outside. The difference in chemical potential of water across the membrane drives the water from carrier to purge.

of the Nafion™ dryer would enhance the water removal so that at an operation temperature of 0 °C water is removed with an efficiency of 99.997%, yielding a dew point of −80 °C, comparable with a $Mg(ClO_4)_2$ or a dry ice/acetone cryogenic trap efficiency. However, the drying efficiency at room temperature is already sufficient to obtain reliable isotope ratios.[74] Nafion™ tube material should be stored in the dark because it will be destroyed by UV light. Within the glass tube this is generally not a problem because UV light cannot penetrate the glass. A colour change of the tubing from light brown to a darker brown colour can be an indicator of its deterioration.[33]

Isobaric Interferences for Nitrogen Measurements. During the oxidation process of organic compounds carbon dioxide is formed, which produces CO^+ fragments by unimolecular decomposition of $CO_2^{+\cdot}$ in the ion source. $^{12}C^{16}O^+$ and $^{13}C^{16}O^+$ are isobaric interferences to $^{14}N_2^{+\cdot}$ and $^{14}N^{15}N^{+\cdot}$ that are also measured on m/z 28 and 29, respectively. Therefore, in nitrogen isotope analysis, carbon dioxide has to be removed from the carrier gas effluent before it enters the ion source. This can be achieved by chemical or cryogenic trapping. The CO_2 is removed by immersing the fused silica capillary tubing (0.32 mm i.d.; 1–1.5 m length) carrying the effluent stream in a dewar filled with liquid nitrogen (see Figure 3.15). The cryogenic trap is placed between the water separator and the open-split. The cryogenic trap cannot be used permanently because CO_2 can clog the capillary depending on the carbon content of the determined compounds and matrix. A rule of thumb is that around 20 measurements are possible. To remove the CO_2, the open-split is set to *out*, so that the CO_2 cannot enter the ion source when the capillary is removed from the dewar.

As in carbon isotope analysis, protonation reactions with hydrogen-bearing materials such as water can lead to the formation of N_2H^+ and thus bias the low abundance minor ion beam measured at m/z 29.[36] Thus, water removal as described above is also essential in nitrogen isotope analysis.

3.4.2 Hydrogen Isotope Ratio Analysis

Hydrogen shows the largest variations ($-700‰$ (electrochemically produced hydrogen) to $+100‰$)[18] in the ratio of its stable isotopes because of strong isotope fractionation caused mainly by the huge relative mass difference of 2. Due to these large variations it is very attractive to measure isotope ratios of hydrogen.[36,76] However, this is complicated by the low natural abundance of ^2H of only 0.0156% (ocean water)[18] compared with 1.1% for ^{13}C.

As we will see in Chapter 6, the knowledge of hydrogen isotope ratios is important to elucidate the origin of compounds or clarify possible transformation reactions in combination with other element isotope ratios (dual isotope ratio plots).

For hydrogen compound-specific isotope analysis by GC-IRMS several attempts have been made to overcome the associated technical problems. The preferred gas species for isotope ratio analysis of hydrogen is molecular hydrogen H_2. For the on-line analysis of organic compounds it is necessary to convert hydrogen-bearing organic molecules quantitatively into hydrogen gas within the carrier gas stream. Two approaches can be followed to achieve this.[77] The first is a combustion, which is followed by a subsequent reduction of the generated water in order to produce molecular hydrogen.[78,79] The second approach applies pyrolysis or catalysis in order to produce hydrogen.[77]

Tobias and Brenna demonstrated CSIA for hydrogen by using an on-line combustion micro-reactor, which was filled with CuO and held at 850 °C. The oxidation reactor was followed by a reduction reactor filled with nickel metal held at 950 °C.[78,79] An inherent problem in combustion/reduction systems is oxygen bleeding from the combustion reactor, which diminishes the capacity of the reduction reactor.[45,77] Hence, the second approach has become more popular, where reductive metal catalysts or pyrolysis in carbon conditioned alumina tubes is applied.[77] This approach is nowadays exclusively used in commercial instruments.

A quartz tube filled with chromium granules as catalyst held at 1000 °C is used in the IsoPrime™. A conditioning of the furnace filling is not necessary once the system has reached its operational temperature.

Burgoyne and Hayes, however, showed that a quantitative conversion by pyrolysis can be achieved without metal reductants by using a carbon-lined non-porous alumina tube reactor heated to temperatures $>1440\,°C$.[77] At such high temperatures, high yields of H_2 can be obtained according to eqn (3.46) by pyrolysis of hydrocarbons.[80]

$$C_nH_xO_y \xrightleftharpoons{\text{HT or Py}} \frac{x}{2}H_2 + yCO + (n-y)C \qquad (3.46)$$

Additionally, solid carbon black or, in presence of oxygen, CO will be formed.[77] In Figure 3.20 it is shown that the pyrolysis of propane results mainly in the formation of extremely stable methane at temperatures below 1400 °C. This methane is in equilibrium with elemental carbon and hydrogen gas according to:

$$CH_4 \rightleftharpoons C + 2H_2 \qquad (3.47)$$

At temperatures above 1430 °C the equilibrium constant of reaction 3.47 exceeds 10^3 and conversion of CH_4 to H_2 exceeds 99.9% for $p_{CH_4} \leq 10^{-3}$ atm.

Resistance heaters based on Nichrome as they are applied for carbon cannot be used because they cannot reach the necessary temperatures of $>1400\,°C$.[45] Heating elements in pyrolysis ovens are thus based on silicon carbide and can reach operation temperatures up to 1500 °C.[45] The high temperatures employed are above the transformation temperature of quartz so that alumina has to be used.[45] However, it was

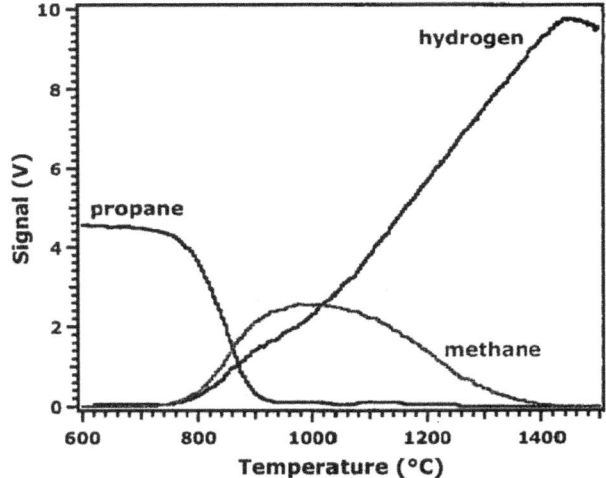

Figure 3.20 Generation of hydrogen from propane. Ion currents representative of propane (m/z 44), methane (m/z 16), and hydrogen (m/z 2) as a function of furnace temperature.[77]
(Graphic reprinted with permission from the American Chemical Society.)

observed by Burgoyne *et al.* that at temperatures above 1470 °C a loss in signal yield was observed, which can be explained by a higher porosity of the alumina tubes at these high temperatures.[77] Thus, an optimal temperature of 1450 °C is recommended.[45,77] At this temperature, prolonged residence times for quantitative pyrolysis of organic compounds to hydrogen, achieved by lowering the carrier gas flow, is required.[45,82] Since the tube diameter is fixed in a given interface, it is recommended to limit the carrier gas flow to 1.5 mL min^{-1} for analyte separation. Otherwise incomplete conversion can cause incorrect isotope ratio data results.[76,83] Under these conditions the alumina tubes have to be replaced every two to four weeks. Improved peak shapes and reproducibilities are obtained after a layer of graphite has been deposited inside the tube, covering the aluminium oxide surface. For conditioning the reactor tubes by graphitization a flow of pure methane (5–10 min) can be applied.[77] Instead of methane gas also two or three injections of 1 μL of an organic solvent,[45] for example, iso-octane, via the GC injector can be carried out at a split ratio of 10 : 1.[81]

Figure 3.21 shows a set-up of a typical high temperature conversion or pyrolysis interface. In contrast to carbon and nitrogen measurements a reduction oven as well as a Nafion™ dryer is not necessary because minor traces of water are converted to CO and H_2 gas.[45] The measurement of halogenated compounds containing chlorine or fluorine is

Instrumentation for Compound-specific Stable Isotope Analysis

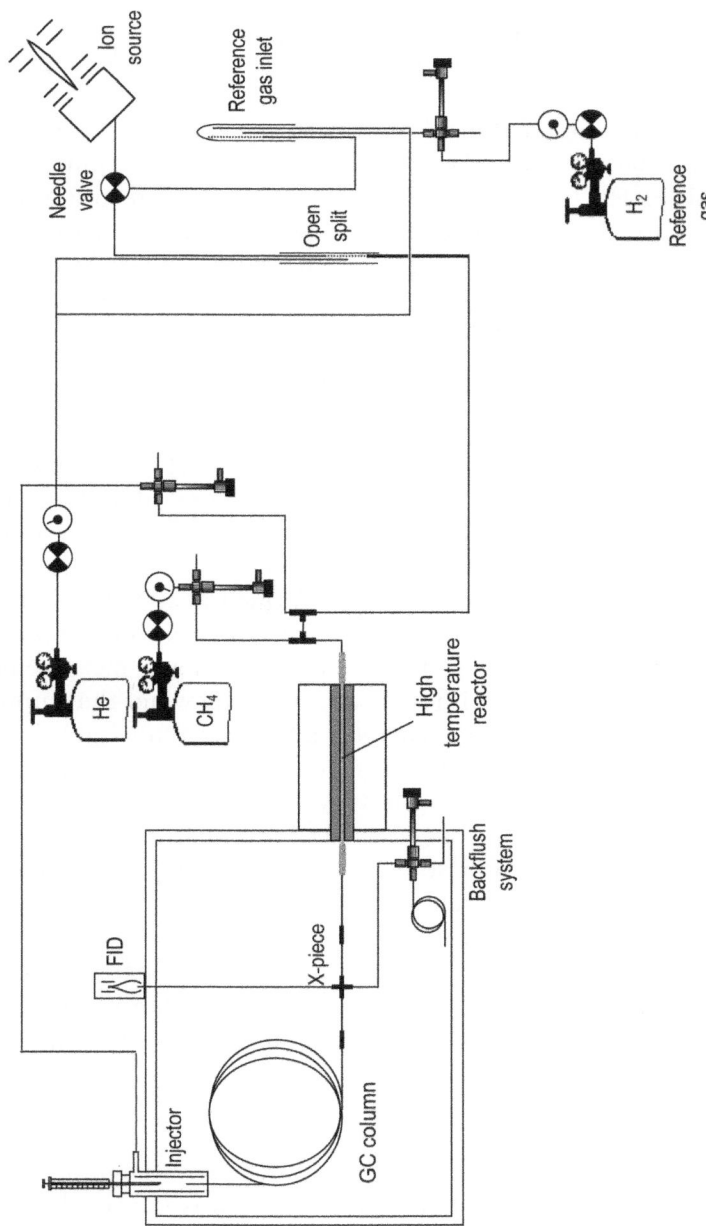

Figure 3.21 Set-up of a GC high temperature conversion system for the analysis of hydrogen isotope ratios. Note that a water removal device is not necessary. For the regeneration of the high temperature reactor see text.

problematic because the pyrolysis of such compounds forms acids such as HCl and HF inside the oven.[45] Besides the destruction of subsequent fused silica columns following the oven such species lead to corrosion and damage to the IRMS. Additionally, isotope fractionation between H_2 and HX can result in inaccurate hydrogen isotope ratio measurements.

As has already been mentioned, molecular hydrogen gas is measured for the determination of the $^2H/^1H$ ratio. To that end, the ionic species $^1H^1H^{+\bullet}$ and $^1H^2H^{+\bullet}$ are measured at m/z 2 and m/z 3, respectively. As we saw in Section 3.1.5, an ion correction is not necessary. However, in the case of hydrogen another problem caused by an isobaric interference occurs. Inevitably, the triatomic ion $^1H_3^+$ is formed in the ion source according to the following ion molecule gas phase reaction:[12,36]

$$^1H_2^{+\bullet} + {}^1H_2 \rightarrow {}^1H_3^+ + {}^1H^\bullet \tag{3.48}$$

The formed $^1H_3$ ion interferes with the m/z 3, $^1H^2H^{+\bullet}$, leading to an overestimation of the isotope ratio $^2H/^1H$. In natural abundance measurements of $^1H^2H^{+\bullet}$ from organic compounds the $^1H_3^+$ can account for as much as 5–30% of the m/z 3 ion current.[83] Another source for $^1H_3^+$ is tungsten oxide, which can be formed at the surface of the filament by O_2 or H_2O.[12] This interference is important temporarily when an IRMS is switched from carbon to hydrogen measurements because the hydrogen reacts with the oxide, forming water that supports the formation of $^1H_3^+$ ions. Therefore, ion currents and isotope ratios need some time (one hour or more) to stabilize and reach a state without further interfering ions with m/z 3 formed due to this process.[12]

In continuous-flow, the ion source is tuned to maximum linearity, whereby a large extraction potential is used (see Section 3.1.2). Because of this, ion–molecule reactions are minimized by shorter residence times in the ion source and thus the amount of $^1H_3^+$ formed is reduced.[36] Nevertheless, a correction for the formed $^1H_3^+$ is mandatory. According to eqn (3.48) the reaction constant is proportional to the amount of $^1H_2^{+\bullet}$ and $^1H_2$:

$$[^1H_3^+] \propto [^1H_2^{+\bullet}][^1H_2] = K[^1H_2]^2 \tag{3.49}$$

The proportionality constant K in this equation is the so-called 'H_3-factor', which is expressed in ppm mV^{-1}. In laboratory practice, the value of K is usually determined by measuring the $(m/z\ 3)/(m/z\ 2)$ ion current ratio in a standard on/off procedure, in which the hydrogen gas pressure is successively increased. Least-squares regression is then used to obtain K as its slope by using the following relation:[83]

$$^3R = \frac{^3i}{^2i} = \frac{(^3i_{1\text{H}^2\text{H}^{+\cdot}} + {^3i_{1\text{H}_3^+}})}{^3i_{1\text{H}_2^{+\cdot}}} = \frac{^3i_{1\text{H}^2\text{H}^{+\cdot}}}{^2i_{1\text{H}_2^{+\cdot}}} + K \times {^2i_{1\text{H}_2}} = {^{*3}R} + K \times {^2i_{1\text{H}_2}} \quad (3.50)$$

Here, ^3R is the measured ion current ratio, $^{*3}R$ is the corrected $^1\text{H}^2\text{H}^{+\cdot}/^1\text{H}_2^{+\cdot}$ ratio, ^2i and ^3i are the ion currents measured at m/z 2 and m/z 3, and $^2i_{\text{H}_2^{+\cdot}}$, $^3i_{1\text{H}^2\text{H}^{+\cdot}}$, $^3i_{1\text{H}_3^+}$ are the ion currents caused by the involved ionic species. For the regression between five and ten hydrogen pressure steps should be used in the standard on/off procedure. With K the corrected isotope ratio $^{*3}R$ can be calculated. The typical determination of the H_3-factor by a standard on/off procedure is visualized in Figure 3.22.[83]

In modern instruments, K values are typically between 5–30 ppm mV^{-1}, which is equivalent to 5–30‰ nA^{-1}.[83] This value should be stable to ± 0.1 ppm mV^{-1} within 24 h and should be measured daily before a measurement series.[81] Modern IRMS software automatically corrects all data acquired for $^1H_3^+$ contribution after 'H_3-factor' determination. Generally it can be said that the H_3-factor should be low for precise hydrogen isotope ratio measurements. In case of continuous-flow analysis factors such as (i) the dynamic conversion process of eluting compounds to hydrogen, (ii) large variations in the partial pressure of H_2 caused by different amounts of the eluting compounds, and (iii) the helium carrier gas itself can change the mechanism of $^1H_3^+$ development as well as its magnitude.[84] For a more comprehensive and detailed discussion of different corrections for $^1H_3^+$ we refer to the publications by Sessions et al.[83,85]

The presence of helium affects the measurement of hydrogen isotope ratios in several ways.[2] First, ^{4}He might form $^4\text{He}^{2+}$, which adds to m/z 2.[18] Second, helium gas used in the lab can contain around 1 ppm ^{3}He, which forms $^3\text{He}^{+\cdot}$ in the source and is measured as an isobaric interference in addition to $^1\text{H}^2\text{H}^{+\cdot}$ at m/z 3. However, the constant continuous intake of ^{3}He and formation of $^3\text{He}^{+\cdot}$ can be corrected for by background subtraction.[2,86]

Third, a more severe problem that affects the ^{2}H/^{1}H ratio measurement is connected to the abundance sensitivity. The abundance sensitivity is defined as the contribution of an ion current at mass m to the neighbour mass $m+1$.[12] In particular, the high helium load in continuous-flow applications leads to the formation of $^4\text{He}^{+\cdot}$ by collisions in the ion source and along the flight path.[76] Collisions of the $^4\text{He}^{+\cdot}$ ions during acceleration result in an energy loss of the ion beam.

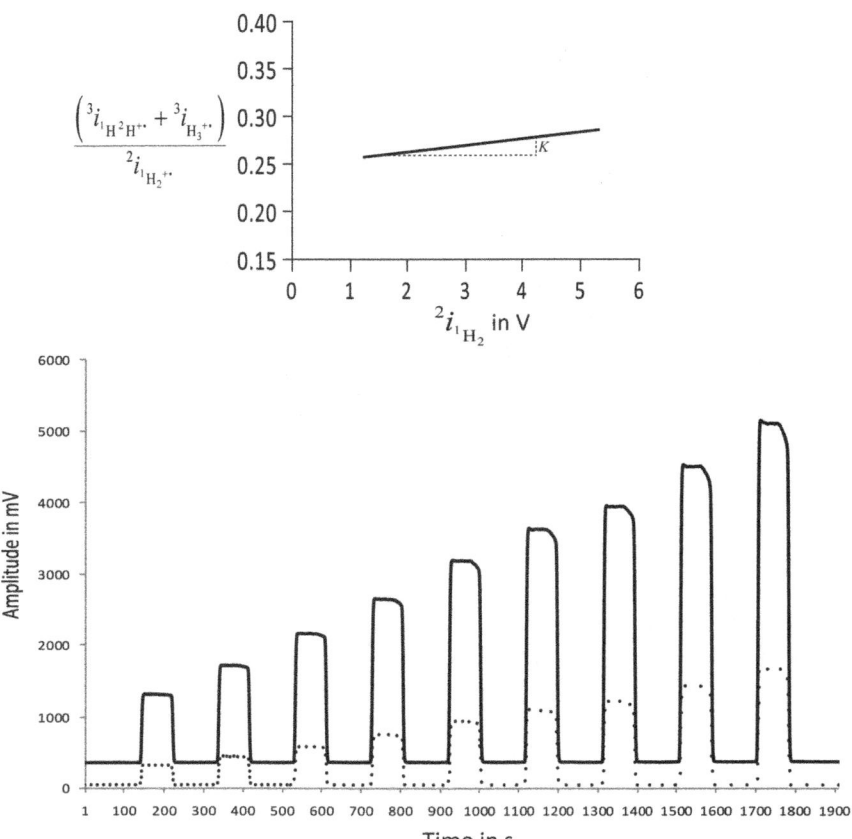

Figure 3.22 Schematic illustration of H_3-factor determination used to calculate the value of K. Bottom: the standard on/off procedure (9 steps) for hydrogen, in which the black line represents the m/z 2 and the dashed line the m/z 3 trace. Top: a typical linear relationship 'H_3-plot' used to calculate the value of K.

Additionally, the resulting ion beam is around six orders of magnitude larger than the $^1H^2H^{+\bullet}$ ion beam.[45] The huge m/z 4 ion beam of the $^4He^{+\bullet}$ tails into the m/z 3 channel for $^1H^2H^{+\bullet}$ determination and even saturates this channel.[12] However, an undisturbed m/z 3 is mandatory in order to achieve the sensitivities required for $^1H^2H^{+\bullet}$ determination. One way to reduce the formation of $^4He^{+\bullet}$ is the application of differential pumping (see Section 3.1.1). Further technical solutions to reduce the $^4He^{+\bullet}$ ion current on the m/z 3 collector have been investigated. These include:

(i) The use of inert carrier gases other than helium, such as neon, argon and xenon.[2] However, these gases are not suitable for the task.

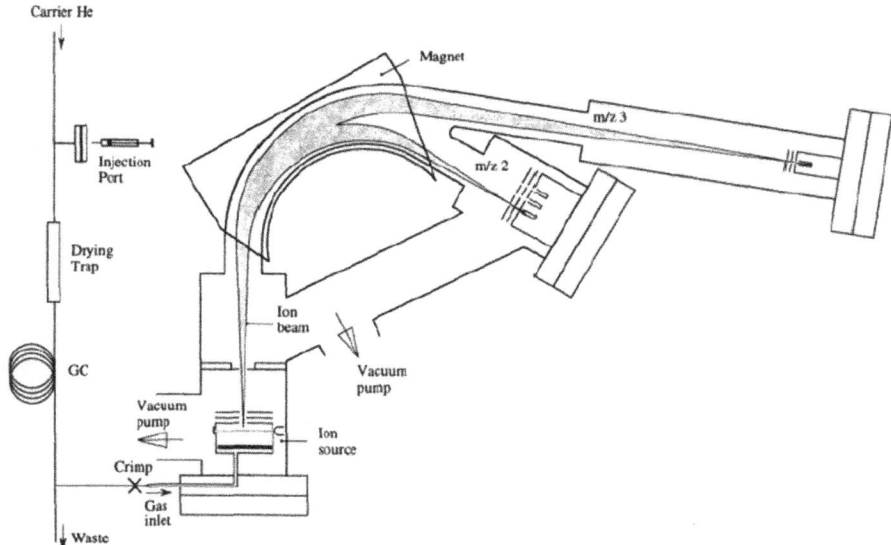

Figure 3.23 Diagram of the mass spectrometer with gas inlet system used by Prosser and Scrimgeour. The mass spectrometer principally matches a commercially available Sercon 20-20™. The mass spectrometer shows large dispersion for m/z 2 and m/z 3. The m/z 4 beam is not deflected sufficiently, so that $^4\text{He}^{+\bullet}$ hits the flight tube and not the collector cup.
(Graphic reprinted with permission from the American Chemical Society.)

Due to their high mass, optimal chromatographic conditions cannot be obtained.[2] In comparison to helium, they have very high densities and low ionization potentials, so that high-intensity ion beams are formed, which rapidly damage ion optical components and insulators.[2,78] Furthermore, Ne and Xe are expensive.[2]

(ii) Prosser and Scrimgeour used a mass spectrometer, shown in Figure 3.23, providing a high dispersion of m/z 2 from m/z 3, and causing the m/z 4 to hit the flight tube walls. Thus, an interference on the m/z 3 channel by $^4\text{He}^{+\bullet}$ is reduced due to the larger dispersion. This principle is still used in Sercon 20-20 IRMS instruments.[87]

(iii) Tobias and Brenna used a palladium filter system (PFS), in which a Pd membrane is employed prior to the ion source. This selectively restrains helium while allowing hydrogen gas to penetrate it.[78,79] As membrane material, an alloy of Pd with 25 wt-% Ag at 555 °C was used, which was permeable to $^1\text{H}_2$, $^1\text{H}^2\text{H}$, and $^2\text{H}_2$. Thus, only hydrogen without helium carrier gas can enter the ion source and no $^4\text{He}^+$ is formed.[79] However, the applicability of this system has been questioned and it has never been commercialized.

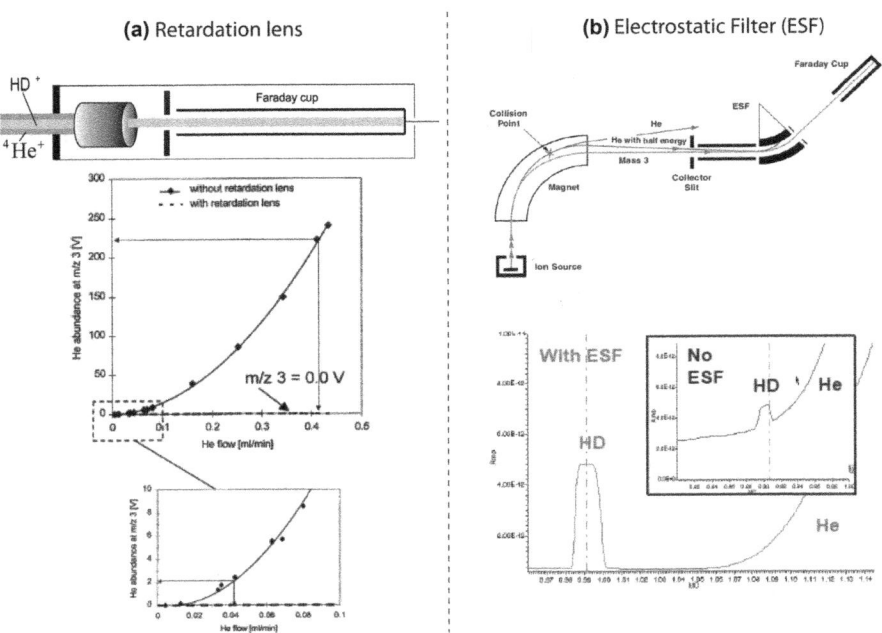

Figure 3.24 Ion optical devices used to prevent m/z 3 channel interference by $^4\text{He}^+$ ions. a) Retardation lens in front of the m/z 3 collector cup. It can be taken from the graphs below the diagram that without using the lens (solid line), the low energy tail of $^4\text{He}^{+\bullet}$ superimposes the m/z 3 ($^1\text{H}^2\text{H}^+$). With retardation lens (dashed lines along the abscissa) a complete suppression of the interfering signal is possible. (Lower diagram reprinted with permission of Wiley InterScience.[76]) b) Use of an electrostatic sector filter (ESF) to separate the $^4\text{He}^{+\bullet}$ ion beam from the m/z 3 collector cup. In the lower panel, the reduction of the $^4\text{He}^{+\bullet}$ tail with the ESF is shown.[88] (Graphic with kind permission of Elementar (Hanau, Germany).)

(iv) Another way to exclude $^4\text{He}^{+\bullet}$, is the use of ion optical devices such as a retardation lens in front of the m/z 3 collector cup, or an electrostatic filter (ESF) as shown in Figure 3.24a and b, respectively. Hilkert *et al.* applied a collector design with a retardation lens directly in front of the Faraday cup of the m/z 3 channel.[76] The retarding voltage for the lens is taken from the source potential by a voltage divider chain. The energy of ions allowed to pass the energy filter is 600 eV, corresponding to a retarding voltage of 2400 V (3 kV nominal energy).[76] With this arrangement used in ThermoFisher instruments an almost complete suppression of the $^4\text{He}^{+\bullet}$ tail is possible (see Figure 3.24a).[76]

The ESF shown in Figure 3.24b is used in IsoPrime™ mass spectrometers. The ESF provides similarly efficient suppression of the $^4\text{He}^{+\bullet}$

interference as a retardation lens. The ESF is set up to reject any ions that have lost a part of their original source energy because the flight path of such ions is altered and they hit the filter.[88]

3.4.3 Oxygen Isotope Ratio Measurement

The main application for oxygen isotope measurements is the determination of $^{18}O/^{16}O$ isotope ratios of water, carbonates, silicates, O_2 and CO_2 in air.[61] For compound-specific applications in combination with GC, the method principle was introduced in 1994.[61] However, in the last 15 years only a few applications have been reported, which are discussed in Chapter 6.

As for hydrogen, the isotope ratio analysis of oxygen also relies on pyrolysis or thermal conversion of organic compounds. As shown in eqn (3.46) a quantitative conversion to carbon monoxide CO is used. At the pyrolysis temperature (1280 °C), oxygen exchange between CO and aluminium oxide of the reactor can occur. This exchange is prevented by using alumina pyrolysis tubes coated with platinum and filled with nickel wires.[76,89–91] For proper reduction conditions 0.6–0.8 mL min^{-1} of an auxiliary gas (1% H_2 in He) are introduced continuously into the reactor. The reactor is conditioned by injection of 1 μL *iso*-octane prior to the measurements.[92] The $^{18}O/^{16}O$ isotope ratios are determined by measuring *m/z* 28 and *m/z* 30, and the $\delta^{18}O$-values are calculated against CO reference gas. For the necessary ^{17}O-correction see Section 3.1.5.[61] Carbon monoxide handling is not without danger: apart from its toxicity it is flammable and the coated reactor has to be grounded because the ceramic parts of the reactor will be conductive at temperatures >1000 °C so that the inner Pt coating can be in electrical contact with the heater.[92] Thus, if oxygen isotope analysis is aimed for in a stable isotope laboratory, further safety precautions need to be considered.

3.5 CHROMATOGRAPHIC ASPECTS OF CSIA

3.5.1 Separation Basics

In the following, some basic principles of chromatography will be introduced, which are necessary for the discussion of GC and LC in combination with IRMS. Furthermore, this section should help newcomers in the field who are not familiar with separation science to understand the basic principles and operations.

According to IUPAC, chromatography is defined as a physical and chemical method of separation in which the components to be separated are distributed between two phases, of which one is a stationary phase, while the other, mobile phase is a fluid that moves in a definite direction. In chromatography, the mobile phase can be a liquid, a gas or a supercritical fluid, while the stationary phase can be a liquid, a gel or a solid.

Chromatographic methods can be classified according to the nature of the stationary and the mobile phases. In first instance, one can distinguish between adsorption chromatography and partitioning chromatography. In adsorption chromatography, the separation process is based on formation of equilibria between the mobile phase and the surface of a solid adsorbent material. If the mobile phase is a gas, the method is termed *gas-solid chromatography* (GSC). Here, either packed capillary columns made of glass or steel or so-called *porous layer open tubular columns* (PLOT) are used. PLOT and packed capillary columns, filled with linked porous polymers, inorganic or organic molecular sieves, silica gels or aluminium oxide are mainly applied for the separation of gases or volatile low molecular weight organic compounds such as alkanes. A prominent example in the context of IRMS is the separation of gases within an elemental analyser (see Section 3.3). These columns have typical internal diameters between 2–6 mm and lengths between 0.2–5 m. The particle size of the stationary phase material is generally between 60/80 mesh (250–177 μm) and 100/120 mesh (149–125 μm). PLOT columns are capillary columns with a thin coating of solid phase adsorbent particles immobilized on the inner wall.

If a liquid mobile phase is used in combination with an adsorptive solid stationary phase, the term *normal phase chromatography* (NPC) is used. In this type of chromatography, silica gel or aluminium oxides are often used as stationary phases, so that the stationary phase is more polar than the mobile phase (organic solvent).

Partitioning chromatography is based on the successive equilibrium formation between a mobile phase and a chemically immobilized liquid. The GC equivalent of this separation technique is called *gas-liquid chromatography* (GLC). In this process a polymeric liquid is bonded covalently to the surface of a fused silica capillary (FS) or stainless steel capillary (SS), which is called a wall coated open tubular column (WCOT).

Liquid-liquid chromatography (LLC) is in most cases *reversed phase chromatography* (RPC). Here, the mobile phase – typically a mixture of water and an organic solvent is more polar than the stationary phase, which is often a C_2- to C_{18}-phase or a polymer immobilized on a silica gel core particle. In particular, for *high temperature, high performance liquid chromatography* (HT-HPLC), more temperature-stable core materials such as titanium oxide or zirconium oxide are frequently used (see Section 3.6).

If equilibrium chromatography is assumed, the basic principle of chromatographic separation of individual components in a mixture can be described by the successive formation of multiple equilibria of individual compounds between the mobile phase and the stationary phase. Each equilibrium can be expressed by the distribution constant $K_d = c_S/c_M$, where c_S is the compound concentration in the stationary phase and c_M, its concentration in the mobile phase.

For partitioning chromatography, the ratio of mobile phase volume to that of the stationary phase is the phase ratio β. For open tubular columns it is given by $\beta = d_c/4d_f$, where d_c and d_f are the internal diameter of the column and the stationary film thickness in mm, respectively. If the distribution constants of the compounds present in a sample differ sufficiently, a separation of the single compounds is possible. The higher the distribution constant the greater is the affinity of the compound to the stationary phase.

Figure 3.25 shows an example of the generic separation of two compounds. Under ideal chromatographic conditions the recorded peaks of eluted compounds follow a Gaussian peak shape (see Figure 3.25b). The retention time t_R of a compound is defined as the time between sample injection and the appearance of the compound's peak maximum. In GC, t_R is proportional to the length of the used column, the internal diameter (i.d.) of the column, the film thickness d_f as well as the flow F of the mobile phase. The flow (in chromatography often denoted as flow rate) F is generally given in mL min^{-1}. Under constant conditions (injection, oven temperature programme, flow conditions) t_R of a compound is constant. Therefore, the identification of peaks can be carried out by injection of the pure compound, which should result in the same retention time. If a detector such as the IRMS is used that gives no structural information on the compound, retention time matching is the only means to ensure correct peak designation. Therefore, laboratory standards have to be included between sample runs to ensure retention time stability. The same requirement is valid for liquid chromatography.

In the case of constant mobile phase flow F, the retention time is proportional to the retention volume V_R, which is given by:

$$V_R = t_R F \tag{3.51}$$

The retention volume gives information about how much mobile phase has passed the column until half of a compound has been eluted (strictly speaking only for symmetric peaks).

Compounds that are not retained by the stationary phase pass the column in the holdup time t_M, which is equal to the residence time in the

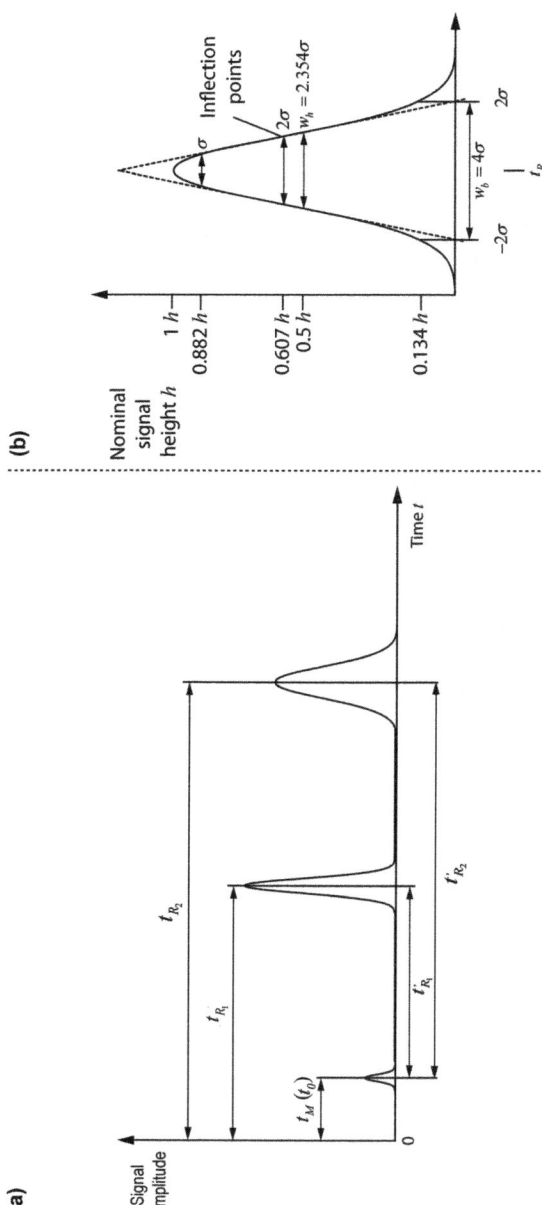

Figure 3.25 Basic chromatographic parameters relevant in CSIA. a) Exemplary separation of a two-component mixture showing the relevant retention parameters, such as the holdup time t_M, the total holdup time t_0, the retention time t_R and the net retention time t'_R. For details see text. b) Widths of a Gaussian peak at different heights as a function of the SD of the peak. Here, w_h is the full width at half height (fwhm) and w_b the baseline or base width of the peak.

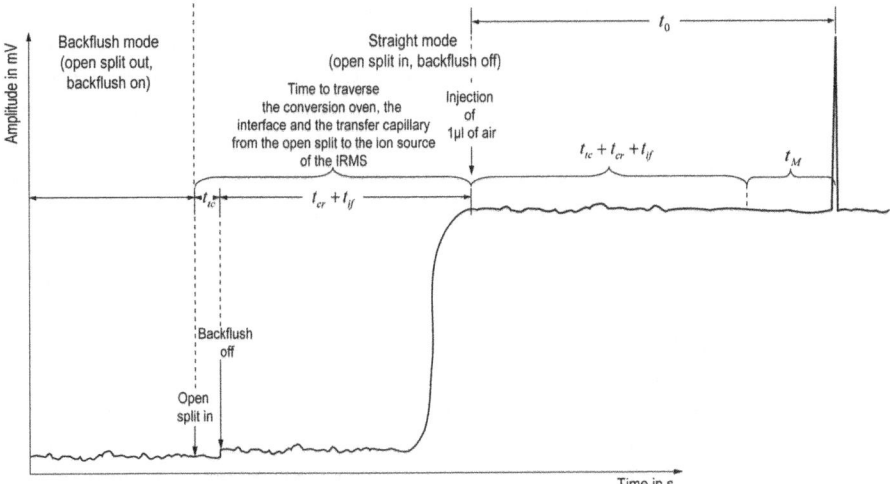

Figure 3.26 Total holdup time in GC-IRMS. The artificially drawn graph shows how the different parts of the conversion oven and the interface contribute to the total holdup time t_0.

mobile phase. By dividing the length of the separation phase (column length) L by the holdup time, the average linear velocity $\bar{u}$ in cm s^{-1} of the mobile phase can be determined. Generally, the carrier gas outlet flow in mL min^{-1} is $F = 60\bar{u}\pi r_c^2$. In the case of GC the carrier gas compressibility correction factor j has to be taken into account (see GC monographs).

The holdup time can be estimated by $t_M \approx 0.5 L d_c / F$ or measured by injection of an unretained compound. In LC, uracil or thiourea are often used for this purpose. In GC-IRMS the holdup time can be measured by injection of 1 μL air and monitoring the mass trace for argon (m/z 40). The holdup time is a feature of the analytical system. In GC- or LC-IRMS systems the total holdup time t_0 consists of the time of the analyte in the mobile phase t_M, the residence time in the conversion reactor t_{cr} as well as in the interface t_{if} and the transfer capillary to the IRMS ion source t_{tc}. Thus, it is given by: $t_0 = t_M + (t_{cr} + t_{if} + t_{tc})$. The knowledge of this time is very important for method development and also for troubleshooting purposes. We therefore recommend determining these times for a given set-up prior to measurements.

Figure 3.26 shows the components of the total holdup time. In addition, the amplitude level during the different modes of the interface is shown. For the explanation of the different modes, see Figure 3.16 in Section 3.4.1.

In order to evaluate the retention behaviour of the stationary phase, a correction of the retention time by the total holdup time is necessary. This corrected retention time is the adjusted or net retention time t'_R, which can be calculated by $t'_R = t_R - t_0$. Accordingly, a holdup volume V_0 can be defined, which can be used to calculate the adjusted retention volume $V'_R = V_R - V_0$. In GC, the adjusted retention volume has to be corrected by a compressibility correction factor j, which compensates for the carrier gas expansion along the column from the entrance, with inlet pressure p_i, to the column exit, with outlet pressure p_o (for more detailed information see text books on GC). The adjusted retention volume is directly proportional to the distribution constant K_d via $V'_R = K_d V_S$, where V_s is the volume of the stationary phase.

Since, in chromatography, retention times rather than retention volumes are typically measured and K_d might not be known a priori, a more practical expression for the quantification of a compound's retention and for the comparison between stationary phases is the retention factor k. The retention factor (formerly termed capacity factor k') is defined by:

$$k = \frac{t_R - t_0}{t_0} = \frac{t'_R}{t_0} \tag{3.52}$$

This parameter is a measure of how much longer a compound stays in the stationary phase than in the mobile phase, so that a compound elutes earlier when k becomes smaller. The distribution constant K_d is related to the retention factor k by multiplication with the phase ratio β so that

$$K_d = k\beta \tag{3.53}$$

A measure of the ability of a stationary phase to separate two compounds 1 and 2, of which compound 2 is the more strongly retained, is the separation factor α. The separation factor is a function of the retention factor of both compounds:

$$\alpha = k_2/k_1 \tag{3.54}$$

By convention, the larger retention factor is in the numerator. According to this expression, two compounds can only be separated if $\alpha > 1$ is obtained for the separation system. If $\alpha = 1$, a complete co-elution of the compounds occurs. The separation factor is directly related to the difference in free energy $\Delta(\Delta G^\circ)$ for the interactions of the two compounds in the system by:

$$\Delta(\Delta G^\circ) = -RT\, ln[(K_d)_2/(K_d)_1] = -Rt\, ln\, \alpha \tag{3.55}$$

where R is the universal gas constant and T the temperature in K.

In this section we have so far only used the so-called kinetic model that does not take into account the width of peaks. For further discussion, the theoretical plate model has to be used, which relates the chromatographic separation with the theory of distillation. According to this theory, a chromatographic column can be hypothetically divided into so-called theoretical plates. As already mentioned, the separation process is a result of the formation of multiple equilibration steps between the stationary phase and the mobile phase. Each equilibration step is represented by a theoretical plate and the overall number of theoretical plates is N. Adapted from the theory of distillation, the height equivalent of a theoretical plate (HETP) or H can be determined by dividing the number of theoretical plates by the length L of the separation phase. In chromatography, nowadays the terms plate number for N and plate height for H are more common and generally recommended. A chromatographic column is more efficient if the plate number N is high or H is low. The elute is also called 'band'. In an ideal case the eluted peaks possess a Gaussian shaped profile following:

$$h = h_{max}\, exp\left[-(t - t_R)^2 / 2\sigma^2\right] \quad (3.56)$$

where h is the peak height of the compound, h_{max} the peak height at the peak maximum, t the time and σ^2 the peak variance.

Figure 3.25b shows a Gaussian shaped peak. The peak variance is a fundamental parameter giving information about the compound dispersion (band broadening) in the chromatographic system including column and extra-column effects. The latter are more pronounced in IRMS systems due to the extra-column volume from conversion reactor and interface. The peak variance is inversely proportional to the plate number, so that $\sigma^2 = 1/N$. It is related to the plate height via the equation $H = \sigma^2/L$. The standard deviation σ describes the spread of the compound molecules in a band. The band broadening is also a function of the retention time so that

$$N = \left(\frac{t_R}{\sigma}\right)^2 \quad (3.57)$$

The plate numbers N in LC are typically between $0.5\text{--}5 \times 10^4$. In GC, typical values for N are between $10^4\text{--}10^6$ for an open tubular column and $10^3\text{--}10^4$ for a packed column.[93]

For a Gaussian peak profile, 95.5% of the compound molecules in an eluting band are within $\pm 2\sigma$ around the mean that is represented by the retention time t_R. As shown in Figure 3.25b, this corresponds to the baseline or base width of the peak. The baseline or base width w_b is measured at the base of the peak by drawing tangents to the deflection points of the Gaussian curve, so that $w_b = 4\sigma$. Therefore, eqn (3.57) can be written as:

$$N = \left(\frac{4t_R}{w_b}\right)^2 = 16\left(\frac{t_R}{w_b}\right)^2 \tag{3.58}$$

It is generally easier to measure the full width at half height (fwhm) w_h for a peak, which corresponds to 2.354σ. N is then given by:

$$N = \left(\frac{2.345 t_R}{w_h}\right)^2 = 5.54\left(\frac{t_R}{w_h}\right)^2 = 2\pi\left(\frac{t_R H}{A_{peak}}\right)^2 \tag{3.59}$$

where A_{peak} is the peak area. By combining eqns (3.57) and (3.59) the standard deviation of the peak and thus the peak broadening can be calculated according to:

$$\sigma = A_{peak}/H\sqrt{2\pi} \tag{3.60}$$

In CSIA it is useful to know the various contributions that lead to peak broadening. The ideal situation for a peak profile is that only the separation in the column has an influence on peak broadening and it is independent of other system components.[93] However, this is rarely achieved in real life, in particular not in CSIA, where conversion reactors and other parts of the interface between the chromatographic system and the IRMS are contributing to overall peak broadening. All extra-column band broadening contributions are additive in their variances. For a typical GC-IRMS system the total peak variance then is represented by:

$$\sigma_{peak}^2 = \sigma_{col}^2 + (\sigma_{inj}^2 + \sigma_{con}^2 + \sigma_{int}^2) \tag{3.61}$$

where σ_{peak}^2 is the total peak variance observed in the chromatogram, σ_{col}^2 the peak variance contributed by the column, and the extra-column band broadening contributions are given in brackets, such as the variance introduced by the injection σ_{inj}^2 and the variances that have their origin in the conversion reactor σ_{con}^2 and the interface σ_{int}^2 including all kinds of tubings used for connections up to the IRMS. For LC-IRMS the contributions to extra-column band broadening are comparable.

In this context, the difference between the flat-topped shaped reference gas peaks and the Gaussian shaped compound peaks needs to be discussed briefly, because this difference might not be directly obvious to

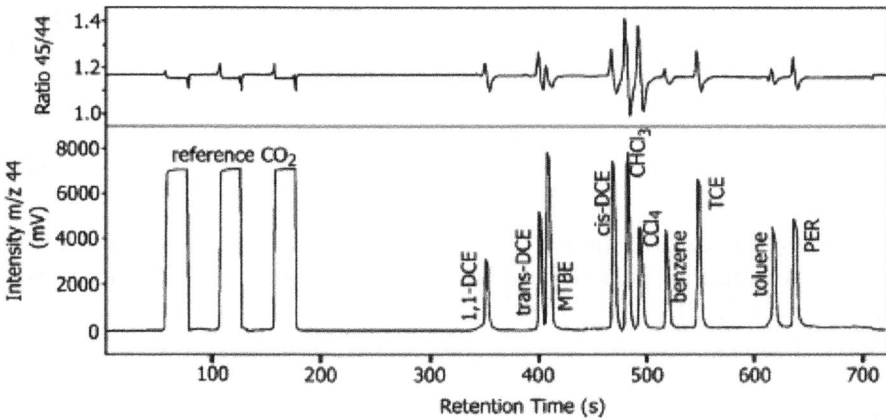

Figure 3.27 Exemplary GC-IRMS chromatogram for a mixture of volatile organic compounds (VOCs). At the beginning of the chromatogram three flat-topped CO_2 reference gas peaks can be seen. In contrast to this, the separated VOCs show Gaussian shaped peaks. In the upper part, the isotope ratio trace ^{45}R is shown. The S-shaped 45/44 ratio signal (isotope swing) is caused by the inverse isotope effect in GC (see discussion in text). The spikes at the peak flanks of the reference gas peaks have their origin in different amplifier time constants.[61]
(Chromatogram is taken from Zwank et al.[94] reprinted with permission of American Chemical Society.)

all readers and is a question frequently asked by students and on separation science conferences.

Figure 3.27 shows a typical GC-IRMS chromatogram, with three flat-topped reference gas peaks at the beginning of the chromatogram and approximately Gaussian shaped compound peaks. The flat-topped peak shape has its origin in the continuous direct introduction via the helium stream into the ion source for several seconds (typically 20–30 s) (see Section 3.2.4). In contrast to this, an injected sample with a certain introduction band width gives a transient signal, which is affected by diffusion processes (longitudinal diffusion, eddy diffusion) and mass transfer between the mobile and the stationary phases.[47,71] The same behaviour is valid for a compound separated by LC-IRMS.

Finally, it should be noted that in reality peaks are seldom perfectly Gaussian shaped; rather, they tail to some extent, i.e., they show an asymmetric peak profile. Models such as that of Foley and Dorsey[95] have been developed to describe such situations. However, a discussion of these models is beyond the scope of this book and we need to refer to more specialized texts.[93,96]

As we will see in the following, a sufficient resolution R_S for precise and accurate isotope measurements is of the utmost importance in CSIA. Chromatographic resolution R_S is defined as:

$$R_S = \frac{2(t_{R_2} - t_{R_1})}{w_{b_1} + w_{b_2}} \approx \frac{t_{R_2} - t_{R_1}}{w_{h_1} + w_{h_2}} \tag{3.62}$$

where t_{R_1} and t_{R_2} are the retention times of two adjacent peaks, w_{b_1}, w_{b_2} the corresponding peak widths, and w_{h_1}, w_{h_2} the peak half widths (see also Figure 3.25).

The peaks are resolved if the sum of the half widths of the peaks is smaller than the difference of the retention times ($R_S > 1$). For two equally sized Gaussian shaped peaks, a resolution R_S of 0.5 corresponds to a peak overlap of $\sim 16\%$, a value of 1.5 ($w_b = 4\sigma$) is said to be *baseline* resolution, in this case there is a peak overlap of $\sim 0.2\%$. $R_S = 2$ results in 0% overlapping so that the space between the two peaks is equal to the peak width at baseline wb and the peaks are completely separated. In order to relate resolution with chromatographic parameters that can be influenced by the user, the following equation has been derived:

$$R_S = \underbrace{\left(\frac{\sqrt{N}}{4}\right)}_{(i)} \underbrace{\left(\frac{\alpha - 1}{\alpha}\right)}_{(ii)} \underbrace{\left(\frac{k_2}{k_2 + 1}\right)}_{(iii)} \tag{3.63}$$

where k_2 is the retention factor of the second peak. R_S incorporates both efficiency and separation.

Term (i) in eqn (3.63) characterizes the dispersion in the chromatographic system and is mainly affected by the column conditions. An increase in N has the effect that the peak bands become narrower so that a better separation can be achieved by conservation of their relative position to each other. As has already been discussed, the number of theoretical plates N is directly proportional to the length of the column L and an increase in N can be achieved by using a longer analytical column. However, the resolution is only proportional to the square root of the number of theoretical plates. A doubling of the column length improves the resolution only by a factor of 1.4 with the disadvantage of a doubling of the retention time. It is therefore important to carefully weigh up the benefit of longer columns. In LC, an increase of N can be achieved by using smaller particle size packing materials but the related increase in pressure has to be taken into account.

Term (ii) describes the selectivity of the chromatographic system. If α increases, two peaks move apart from each other so that the resolution increases. α and K_d are affected by the parameters that change the compound's equilibrium between the sample in the mobile phase and the stationary phase: these are, for example, the kind (composition) of the stationary phase, the temperature and, more pronounced in LC, the composition of the mobile phase. Term (ii) has generally the biggest effect on the resolution and it shows the importance of choosing the best stationary phase material for an individual separation problem. The resolution can also be improved by term (iii), by using a thicker stationary phase (or more stationary phase) and lower column temperatures. However, this effect is only pronounced for early eluting peaks.

Isotope effects in chromatography are a common phenomenon in separation science. In GC, ^{13}C-enriched or deuterated standards tend to be eluted earlier than their non-labelled isotopologues. In LC, the effect is not as pronounced as in GC and the direction of the effect (i.e., stronger or weaker retardation of the heavy isotopologue) is less predictable. Many examples of such isotope effects have been reported in the literature.[98–102]

Such chromatographic isotope effects for liquid-vapour phase equilibria have their origin in small differences in vibrational energy shifts in association with the condensation of different isotopologues.[103] Depending on the predominating interactions in the condensed phase, the intramolecular vibrations can be shifted to lower or higher frequencies.[13,104] The associated energy shifts are greater for molecules containing the lighter isotope[103] so that in the case of attractive interactions, solutes containing the heavy isotope possess slightly higher energies than non-substituted species. Due to this, isotopically heavier species are slightly less retarded and, consequently, have shorter retention times.[13] The magnitude and direction of the time displacement depend on the compound properties and on chromatographic parameters such as flow, temperature, and the nature (polarity) of the stationary phase.

In the following, the importance of a total separation ($R_S=2$) in CSIA will be discussed in the context of chromatographic isotope effects. The so-called inverse isotope effect, as it occurs in GC, is shown in the lower left panel of Figure 3.28.[13,105] The heavier isotopologue of a target analyte is eluted earlier from the column than the lighter one and after on-line conversion in the oxidation reactor this separation is maintained for the separately detected m/z 45 ($^{13}CO_2$) and 44 ($^{12}CO_2$) isotopologues. In the case of carbon, the m/z 44 signal in general follows the m/z 45 signal by 60–150 ms.[13,97] In the upper left panel the corresponding isotope ratio of the m/z 45 to m/z 44 is shown, which indicates the passage of CO_2 through the ion source by the rise and fall of the m/z 45 to m/z 44 isotope ratio.[13] The S-shaped curve,[97,106] or bimodal

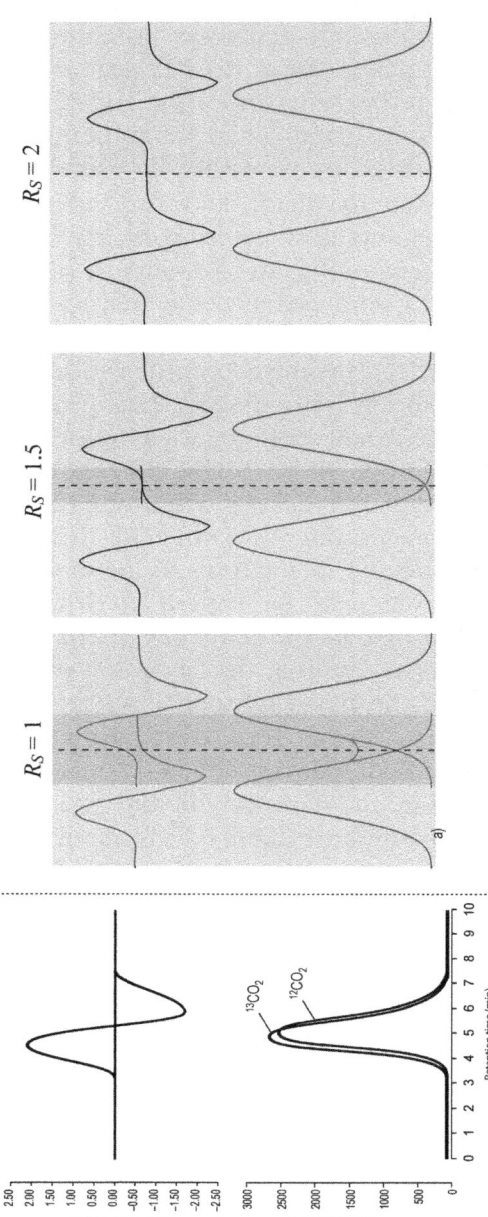

Figure 3.28 Isotope effect in chromatography. As an example, displacement between $^{13}CO_2$ and $^{12}CO_2$ that causes the S-shaped 45/44 ratio signal in the upper graph ('isotope swing') is shown. In the right panel, the effect of resolution on the separation of the isotope signals of two adjacent peaks is highlighted. Clearly $R_S = 1$, often acceptable in quantitative analysis, is insufficient to fully separate the isotopologues of both peaks. Hence, a biased isotope ratio may be obtained for both compounds. (Graphic was modified after Rautenschlein et al.[97])

oscillation[107] (often denoted as 'isotope swing') shows an increasing isotope ratio in the first part of the eluted peak, which indicates that the peak is strongly enriched in ^{13}C-containing molecules, whereas the second part of the peak is depleted in ^{13}C-containing molecules.[13] In contrast, the m/z 46 to m/z 44 ratio does not show this phenomenon, because the oxygen originates from the post-GC combustion.[11,13] Because peak areas of both ion traces 44 and 45 are needed for the isotope ratio determination, adjacent overlapping component peaks can alter the δ-values of both compounds.[108] Goodman et al. investigated the detrimental effects of peak overlapping on isotope data precision and accuracy.[109] They reported that even a small extent of peak overlap can have a dramatic effect on isotope ratios although precision measured with repeated injections was excellent. This effect was even more pronounced for the smaller peak of a mismatched pair.[109] To sum up, there is no substitute for a good separation in CSIA and total peak separation ($R_S \approx 2$) is an essential requirement for achieving accurate compound-specific isotope ratio measurements.[66] Especially, for complex matrices this can be challenging and careful sample preparation has to be applied (see Chapter 4).

3.5.2 Peak Detection in CSIA

For the determination of isotope ratios it is necessary to integrate the relevant m/z traces of the obtained peaks. The 'peak definition', which is the determination of the peak start and end time is generally performed on the basis of the most abundant m/z trace of an ion beam, which gives the highest signal-to-noise ratio. In case of Thermo software, the instantaneous rate of change of the signal is determined using a five-point linear regression for each data point (see Figure 3.29).[13] The start and end of a peak can be determined by a user-defined slope threshold value. The top of the peak is determined by the sign change of the slope. Irrelevant, small peaks can be excluded by choosing an amplitude threshold value.

Due to the isotope effects observed in chromatography, the mass traces for, e.g., m/z 44 and m/z 45 are time shifted. To avoid systematic errors, a correction of this time shift is necessary. This correction is carried out by the software, by separate examination of the precise peak maxima through fitting parabolas to nine points at the top of the individual peaks for the individual ion traces.[13]

The baseline level or 'background' can have a strong influence on the isotope ratio determination and it is therefore necessary to correct the integrated ion currents by subtraction of the background from the total signal at each point. In IRMS, the background has two components: an electronic offset (see Section 3.1.4), usually referred to as baseline, and a chemical part that represents the ion current on the m/z of interest when

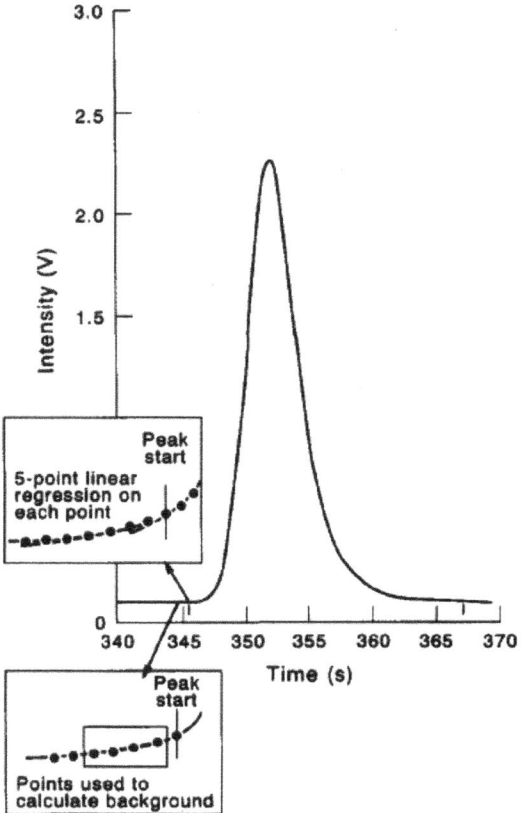

Figure 3.29 Procedures used to determine peak start times and background levels. The graphic shows the five-point linear regression on each point, which is used to determine the peak start.
(The graphic is adapted from Ricci et al.[13] and reproduced with permission of the American Chemical Society.)

there is no reference gas or sample peaks introduced to the mass spectrometer (the blank).[12] This chemical part includes residual gases in the ion source or combusted column bleed. Different background situations can be observed in CSIA:

(i) For isothermal separations a stable background is typically obtained.
(ii) An increasing background occurs during temperature rises leading to increased column bleeding, and on increasing flanks of unresolved complex matrix.
(iii) A decreasing background can be observed during column cooling or on decreasing flanks of unresolved matrix.

(iv) In case of compound co-elution, background increase and decrease can occur.

Most frequently, the 'individual' background algorithm is used. In this algorithm, the background for each peak is determined by averaging the five points immediately preceding the start of the peak (see Figure 3.30).[13] Although neglected sometimes in practice, the application of background algorithms has to be evaluated carefully by the user for individual problems. In our experiments, the 'Basefit' background algorithm showed best results for steep slopes (as observed, for instance, in programmed GC or HT-HPLC measurements).

3.5.3 Carrier Gas, Carrier Gas Flow and GC Columns for CSIA

In GC, generally, inert gases of low viscosity and favourable solute diffusivity, such as hydrogen and helium, are commonly used. However, the selection of the kind of carrier gas used in GC-IRMS is predetermined by the mass spectrometer and the conversion process. Hence, helium is used exclusively for GC separation. Although other carrier gases have been discussed, especially in combination with hydrogen isotope ratio analysis, all such approaches have failed (see Section 3.4.2). It is recommended to use ultrapure (He 5.0; 99.999%) or research grade (He 6.0, 99.9999%) helium as carrier gas. In Table 3.5 the optimal carrier gas flows for various internal column diameters from a chromatographic viewpoint are given. Under these conditions the linear gas velocity is in the optimum of the van Deemter curve.

Although GC-IRMS has revolutionized the measurement of atmospheric trace gases by using PLOT columns,[72] we will focus here on separation of organic compounds and not on atmospheric trace gases such as carbon dioxide or nitrogen oxides. For more information on chromatographic separation of these we refer to the literature.[48,72] For organic compounds, generally WCOT columns are employed and will be discussed here.

Table 3.5 Optimal carrier gas flows for various internal column diameters.

Internal column diameter in mm	Linear gas velocity in $cm\,s^{-1}$	Carrier gas flow in $mL\,min^{-1}$
0.18	30–45	0.7–0.9
0.25	30–45	1.3–1.8
0.32	30–45	2.2–2.8
0.53	30–45	6.0–7.9

Data adapted from Ref. 110.

Except for carbon isotope ratio measurements, the necessary analyte amounts are on the high side for capillary column chromatography. In particular, for the compound-specific isotope analysis of hydrogen, 10 to 100 times higher absolute amounts injected on the column are necessary than for carbon.[76] In the case of oxygen and nitrogen, the low elemental abundance in organic molecules requires higher amounts to be injected.[45] To give an example, for a precise isotope ratio measurement of trichloroethylene (TCE), around 1 nmol of carbon is required on the GC column, compared to 10 nmol of hydrogen. Translated into a TCE concentration for a 1 µL on-column injection, this means 66 mg L^{-1} for carbon and 1375 mg L^{-1} for hydrogen. Such high amounts easily exceed the capacity of conventionally used capillary columns, leading to overloading of the column, as indicated by deteriorating peak shapes (broadening, fronting, peak splitting) and reduced resolution. The analyte capacity of capillary columns mainly depends on the internal column diameter and on film thickness.

Most commonly used column lengths are 30 m and 60 m. The internal diameter of columns can vary between 0.1 mm (microbore capillary) and 0.53 mm (megabore). The internal diameter of capillary columns is proportional to the capacity. Mostly, narrow bore capillary columns with an internal diameter of 0.25 or 0.32 mm are used. As discussed above, the selectivity term in eqn (3.63) has the highest influence on the peak resolution. It is therefore important to choose the optimum stationary phase for the set of target compounds that have to be separated. The number of different separation phases is high and the abbreviations given by the various suppliers can confuse users. In Table 3.6, the most common types of stationary phase and corresponding columns offered by different suppliers as well as an overview of the typical compound classes separated by these columns are given.

New columns and columns that have been stored for longer periods of time should be conditioned (for example, overnight) before the first run, requiring that the column is connected to the injector so carrier gas can flow through the column. The adjusted carrier gas flow should be checked by a flow meter. During conditioning it is advisable to disconnect the column from the FID and the backflushing system to prevent possible contamination. A recommended value for the conditioning temperature is often provided by the manufacturer. This temperature value should be high enough to remove non-volatile residues but low enough not to reduce the lifetime of the column. In the case of new columns, solvent residuals from production, and in the case of stored columns, possible residues that have been accumulated through the storage or fragments from column degradation, are removed.

Table 3.6 Commercially available stationary phases and their application areas.

Stationary phase material as well as polarity of capillary columns for gas chromatography	Commercially available columns with the corresponding stationary phase	Compounds and compound classes typically separated with these columns	Minimum and maximum temperature limits in $°C^{a}$ [93]
100% dimethylpolysiloxane nonpolar phase	Equity-1, SPB-1, DB-1, HP-1, AT-1, Optima-1, ZB-1, Rtx-1, BP1, CP-Sil 5 CB, OV-1, SE-30	Aliphatic, olefinic and aromatic hydrocarbons; long chain fatty acids and alcohols; waxes	–60–325/420
5% phenyl, 95% dimethylpolysiloxane nonpolar phase	Equity-5, SPB-5, DB-5, HP-5, AT-5, Optima-5, ZB-5, Rtx-5, BP5, CP-Sil 8 CB, SE-52, OV-5	Aliphatic, olefinic and aromatic hydrocarbons; long chain fatty acids and alcohols; waxes	–60–325/420
6% cyanopropylphenyl, 94% dimethylpolysiloxane medium polar phase	SPB-624, DB-624, DB-VRX, AT-624, Optima-624, ZB-624, Rtx-624, BP624, CP-Select 624 CB, OV-624	Volatile priority pollutants (chlorinated hydrocarbons, BTEX)	20–280
35% phenyl, 65% dimethylpolysiloxane medium polarity	SPB-35, DB-35, HP-35, AT-35, ZB-35, Rtx-35, VF-35, OV-35, Optima-35	Pesticides, drugs, PCBs, phenols	40–300/340
50% phenyl, 50% dimethylpolysiloxane medium polarity	SPB-50, DB-17, HP-50, AT-50, Optima-17, ZB-50, CP-Sil 24 CB, OV-17, Rxi-17	Pesticides, PAHs, ethylchloroformates of organic acids	40–325/390
14% cyanopropylphenyl, 86% dimethylpolysiloxane medium polarity	Equity-1701, DB-1701, AT-1701, Optima-1701, ZB-1701, Rtx-1701, BP10, CP-Sil 19 CB, VF-1701, OV-1701	Alcohols, pesticides, fatty acid methyl esters	–20–280
Proprietary modified dimethylpolysiloxane phase intermediate polarity	VOCOL, DB-502.2, HP-VOC, AT-502.2, Rtx-502.2, Rtx-Volatiles	Trihalomethanes, light hydrocarbons, aromatics	
Modified polyethylene glycol polar phase	Carbowax Amine, CAM, AT-CAM, Stabilwax-DB, CP-Wax-51	Amines	40–250

Table 3.6 (*Continued*).

Stationary phase material as well as polarity of capillary columns for gas chromatography	Commercially available columns with the corresponding stationary phase	Compounds and compound classes typically separated with these columns	Minimum and maximum temperature limits in °C[a] [93]
Polyethylene glycol polar phase	SUPELCOWAX, DB-WAX, AT-WAX, Optima-WAX, ZB-WAX, Rtx-WAX, BP20, CP-Wax 52 CB, AT-AquaWax, Stabilwax, HP-Innowax, Carbowax®, Omegawax, AT-FAME, FAMEWAX	Fatty acid methyl esters, flavours, aldehydes, amines, ketones	20–250
50% cyanopropylmethyl, 50% phenylmethyl polysiloxane polar phase	SPB-225, DB-225, AT-225, Optima-225, Rtx-225, BP225, CP-Sil 43 CB, OV-225	Fatty acid methyl esters, flavour compounds	40–260/280
Polyethylene glycol nitroterephtalate polar phase	SPB-1000, DB-FFAP, Optima-FFAP, ZB-FFAP, BP21, CP-FF,AP CB, HP-FFAP, OV-351	Fatty acid methyl esters, carboxylic acids, flavours	40–250/260

[a]These values can vary slightly among manufacturers, thus one should always consult their specifications. For film thicknesses above one micrometre, the upper limits are around 20 °C lower. Values behind the slash are the maximum limits for the high temperature versions.

Column bleeding can be of major concern in GC-IRMS, because it can diminish the conversion oven capacity,[45] can cause ghost peaks as well as an increased baseline, and thus can result in reduced sensitivity. Also, peak detection can be altered by huge column bleed. Thick film columns can overcome peak capacity problems. However, thick film columns tend to bleed more than thin film ones. Thick film columns generally possess lower temperature limits than thin film columns. Column bleeding is more pronounced for polar stationary phases such as polyethylene glycol (PEG). More recently, polar columns have become available that show low bleed provided they are handled correctly. Although the amount of column bleed is low nowadays, it is generally recommended to employ low bleed columns (often indicated by the synonym 'MS' mass spectrometry). Due to the rigorous demands on low bleed in GC-IRMS, further developments of low bleeding columns such as stationary phases based on sol gel-processes or ionic liquids may become of particular interest in this area in the future.

3.5.4 Multidetection

Effluent splitting at the end of the analytical column prior to the conversion reactor enables simultaneous detection of the separated compounds via another type of detector such as a FID or an additional organic mass spectrometer (MS).[111–113]

The FID is a universal GC detector for combustible effluents. It is used for general purpose and routine analysis. It possesses a rather high sensitivity for almost all organic compounds but has negligible response to water and gases such as carbon dioxide, carbon monoxide, nitrogen, oxygen, argon and helium.

Flame ionization detectors can be used for method development, determination of backflush times to cut off solvent peaks, as well as for quantification. Because only non-structural information is available by the mass flow dependent FID it is comparable to an IRMS.[10] In Thermo instruments, a Valco™ x-splitter is used to connect the analytical column, the conversion reactor, the backflush valve and the FID. Generally, a 0.1 mm i.d. fused silica capillary is used to connect the splitter with the FID. In IsoPrime instruments, the GC column effluent is split by a splitter union (SGE™ VSOS 004) with two fused silica capillaries (250 µm i.d.). One capillary is connected to the end of the quartz furnace tube. The other is connected to a so-called 'heart-split' valve (SGE MOVPT 1/100 needle valve). A long, fused silica capillary (40 cm; 320 µm i.d.) connects the 'heart-split' valve to the FID. The heart-split valve can be used to regulate the transfer to the FID.

For system troubleshooting, the FID can be used to determine possible problem sources. This can be of particular importance in isotope analysis of nitrogen and oxygen because interference by a hydrocarbon that could affect, for example, the conversion process in the reactor might otherwise not be noticed. On the other hand, carbon dioxide is not detected by the FID, thus, for example, in the case of headspace injections from natural or mineral aqueous solutions, an early eluting CO_2 peak is observed by the IRMS but not by the FID.

Meier-Augenstein, Brand and co-workers coupled an ion trap (IT-MS) for simultaneous molecular mass detection with a GC-IRMS. The effluent from the GC is split by a Valco™ cross-piece and is then introduced via an open-split into the ion source of the IT-MS.[111] It is often found in literature that 90% of the effluent goes into the combustion furnace for isotope ratio measurement to avoid compromising the sensitivity in isotope analysis, while the remaining 10% effluent is transferred to the organic MS.

3.6 LC-IRMS

Before we start with the discussion of coupling LC with IRMS, we would like to note that early on H. Urey introduced an (off-line) ion chromatographic step for separating potassium and lithium isotopes using zeolithe materials.[114,115] During the Second World War, the scope of this technique was widened to separate transuranium elements.[116] This is another example of the wide range of Urey's interest in solving scientific problems by new technologies.

It was pointed out in the previous sections that not all analytes are amenable to GC because either they are not sufficiently volatile or they are thermally labile so that they tend to decompose during injection or separation. For some of these analytes, derivatization and thus a transformation into more volatile and temperature stable molecules are possible. However, derivatization can be laborious, and the problems involved in derivatization in GC-IRMS will be discussed in Chapter 4. Even if a derivatization with valid corrections and without any isotope fractionation is possible, it represents an additional expenditure, which can be avoided by application of LC in combination with an IRMS (LC-IRMS or HPLC-IRMS). Furthermore, low amounts of purified complex biochemical molecules such as proteins, nucleic acids, polar lipids and carbohydrates can be measured and, in such cases, the LC-IRMS coupling provides an exceptional advantage with regard to sample amount and sample preparation.[117,118]

An alternative approach to LC-IRMS is the off-line preparative separation of analytes by LC with subsequent bulk analysis of the collected fractions to produce isotope measurements of individual compounds.[118] This approach is mainly restricted by current sample-size requirements for EA-IRMS analysis,[118] is more tedious than an on-line hyphenation and errors can be introduced due to the required sample handling. To sum up, LC use can greatly enhance the areas of research in which CSIA may be routinely used.[119]

As required for gas source IRMS instruments, an LC interface has to provide a complete and reproducible conversion of the compound from the LC effluent into a measurable gas followed by a quantitative separation of this gas from the eluent.[2] Most importantly, organic modifiers must be removed quantitatively for carbon isotope analysis.

Instrumentation for Compound-specific Stable Isotope Analysis 135

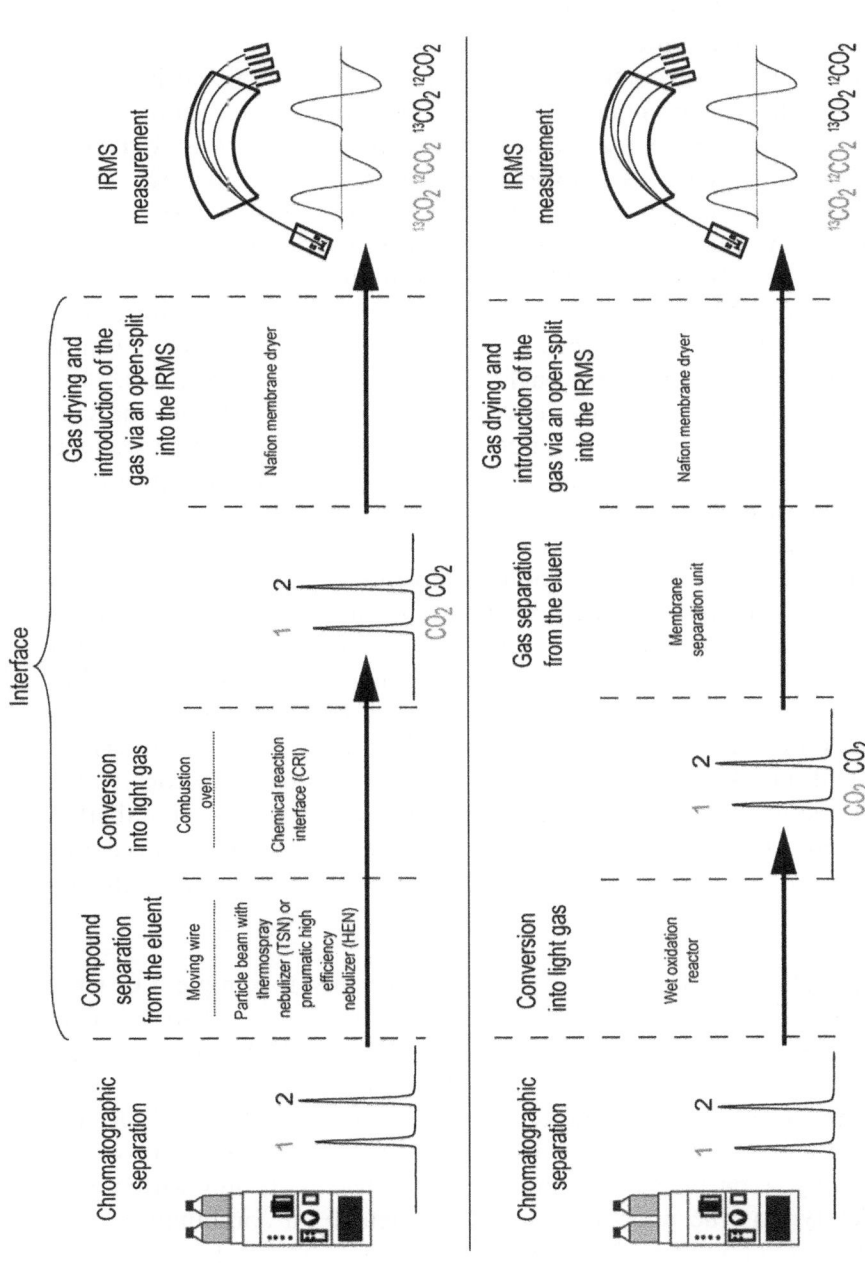

Figure 3.30 Schematic overview of the two approaches used for LC-IRMS interfaces. Top: an interface based on a moving wire (MW) or CRI. Bottom: the basic concept of a wet chemical combustion interface.

The solvent carbon would be combusted together with the target analyte and thus exhaust the oxidation capacity and overwhelm the measurand signal. Thus, a reliable and reproducible solvent removal prior to combustion is the most critical step in LC-IRMS hyphenation.[2]

Two principle approaches have been followed as depicted in Figure 3.30. In the first approach, represented in the upper diagram of Figure 3.30, the organic eluent is removed directly after chromatographic separation followed by a conversion of the remaining analytes into a measurable conversion gas. Subsequently, the gas is dried by a Nafion™ membrane dryer and introduced into an IRMS via an open-split. Two interface types follow this approach, the moving belt or moving wire (MW)[117,118,120] and the chemical reaction interface (CRI).[106,121] The second approach, represented in the lower diagram of Figure 3.30 contains a wet chemical combustion step to oxidize the whole effluent from the chromatographic system, including analytes and all other carbon-containing compounds, to water and gases such as nitrogen oxides and carbon dioxide. The carbon dioxide produced is then separated by a membrane separation unit into a dry helium gas stream, further dried and transferred into the IRMS. The commercially available wet chemical combustion interfaces are based on this principle.[122] In this approach, organic solvents, modifiers or buffers will also be converted and can lead to incorrect carbon isotope ratio values. Therefore, only fully aqueous eluents with inorganic buffers can be employed, whereas in the CRI and MW approach organic solvents and buffers can be used. In the following, the three interface types will be discussed in more detail. For carbon isotope ratio measurements at natural abundance, as well as for ^{13}C-labelled compounds using LC-IRMS with the wet chemical combustion interface we also refer to two detailed reviews by Godin and co-authors.[1,123]

Chemical Reaction Interface. The first attempts to couple LC with IRMS have been carried out by Abramson and co-workers.[117] This approach is based on a microwave discharge CRI, which was developed as a universal interface for GC and LC for different types of mass spectrometers in 1982 by Markey and Abramson.[106,119,124,125] Originally, the LC-CRI was developed for an MW interface,[126] which was later replaced by a particle beam interface.[1,127] The first step within this interface type is the removal of the LC eluent by a combination of a thermospray nebulizer (TSN) working with a flow of 1 mL min^{-1} connected to a longitudinal membrane gas diffusion cell (GDC) in which the analytes are separated from the eluents by a countercurrent helium stream and a subsequent momentum separator for final desolvation.[1,128] The combination of TSN and GDC has the commercial name universal interface (UI) and has been distributed by Vestec.[106,119,125] Jorabchi et al. used a new configuration with a pneumatic high efficiency nebulizer (HEN).[129] An advantage of the CRI compared with an MW interface discussed below is that it does not contain moving parts.[128]

After desolvation the target molecules are transformed into gaseous molecules such as CO_2, O_2, H_2O and NO, using a microwave-induced helium plasma, which is spiked with a reaction gas such as O_2 or SO_2. For carbon isotope ratio analysis, O_2 and SO_2 have been used as reaction gases to prevent a conversion of organic carbon to CO instead of CO_2 and for extending filament lifetimes.[130] SO_2 as reactant gas resulted in the formation of metastable ions which lead to interferences.[119]

The CRI has the advantage that both $\delta^{13}C$ and $\delta^{15}N$ can be measured.[119] When measuring $\delta^{15}N$ no additional CO_2 trap, as is required for combustion interfaces in GC, is necessary because with the CRI, $^{15}NO^+$ is generated and can be monitored without CO^+ and N_2^+ isobaric interferences.[119] However, Teffera and co-workers were not able to measure $^{15}N/^{14}N$ because a large signal at m/z 32 originating from the O_2 reactant gas spreads into the m/z 31 mass position that would be used to monitor the $^{15}NO^+$.[119]

Instabilities of the CO_2 production in the microwave-induced plasma cause the production of substantial amounts of CO, NO_2 (m/z 46), and $C_2H_5O^+$ so that with analyte amounts higher than 20 mg, these by-products cause interferences leading to incorrect carbon isotope ratios. Due to these difficulties, the CRI has not found widespread use apart from a few metabolic applications.[106,119,121]

Moving Belt or Moving Wire Interface. The first attempts to couple LC to mass spectrometry (MS) or FID date back to the late 1960s.[1,128] For a coupling with detection systems that are sensitive to organic solvents, for example, LC eluents, so-called moving transport (moving belt or wire) interfaces had been developed.

Generally, these interfaces employ a metal wire (stainless steel, nickel), on which the LC effluent containing the target analytes is continuously deposited. The wire is reeled off from a feed spool and subsequently passes the following stages: (i) cleaning oven, (ii) coating assembly, (iii) drying step, and (iv) combustion reactor, before it is wound up on a collect spool. Figure 3.31 illustrates a system that was developed at Finnigan in Bremen by Peter Dobberstein and Willi Brand[117] with later

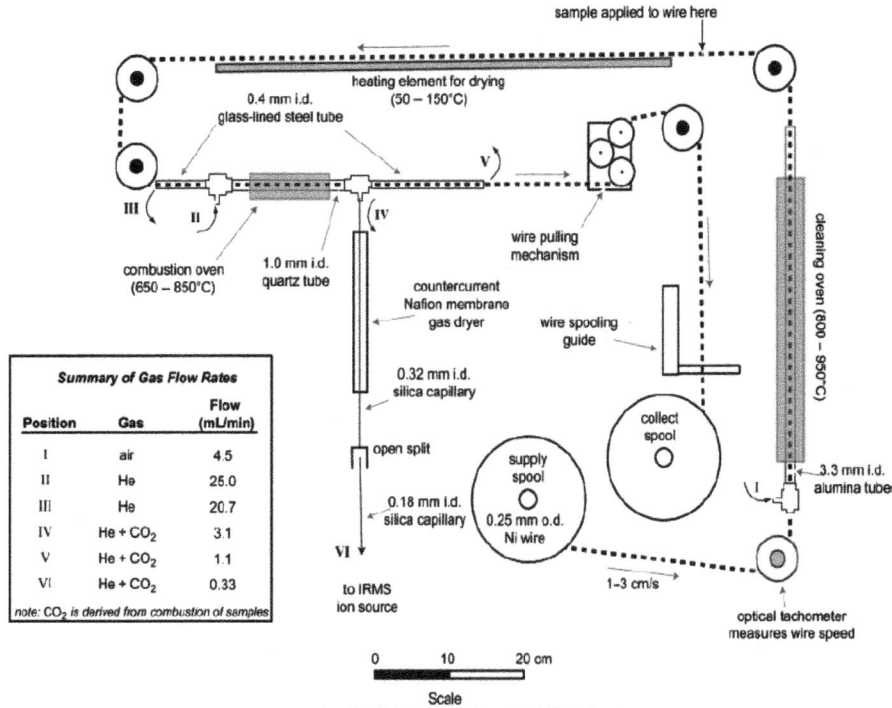

Figure 3.31 Schematic drawing of a moving wire system according to Sessions.[118] The scaled drawing shows the different segments such as cleaning oven, the coating assembly (sample applied to the wire), the drying oven as well as the combustion reactor. The generated CO_2 gas is transported via a helium stream through a Nafion™ membrane dryer and an open-split into the IRMS.
(Graphic reprinted with permission of the American Chemical Society.)

operating modifications by Alex Sessions.[118] The individual parts have the following purposes:

(i) The cleaning oven removes organic impurities and surface oxidation and is held at 800–1000 °C.
(ii) Within the coating assembly a portion of the eluent is deposited on the wire ($\sim 1/10$ of the eluent flow).
(iii) In the drying step the solvent is completely evaporated at around 150–200 °C, whereas the involatile target analytes remain on the wire.
(iv) The combustion oven consists of a tube containing CuO held between 650 and 850 °C. The sample compounds are quantitatively converted to CO_2, which is introduced via a helium gas

stream through a Nafion™ gas dryer and an open-split into the ion source of an IRMS.[117,131]

Caimi and Brenna employed a modified Pye Unicam MW interface that was originally intended for a hyphenation of LC and FID.[117,157] The sensitivity of this wire-based transport system is limited by the capacity of the wire. In a modified version with a belt, this was improved by up to 20%.[2] Their improved system deposits the eluent via an aerosol spraying unit. With the modified instrument, they obtained a limit for high precision isotope ratio measurements of about 3 pg of analyte on-column for flow injection and LC modes.[131] For carbon isotope ratio determination of fat-soluble analytes they demonstrated a precision and accuracy of <2.5‰.[131]

Brand and Dobberstein reported a linear range between $0.1-2\,\mu g\,\mu L^{-1}$ with $\pm 0.5‰$ at a concentration $>0.5\,\mu g\,\mu L^{-1}$. The detection limit was determined to be 167 pg ^{13}C (17 pg ^{13}C on the wire after eluent split).

The MW interface was never commercialized. More recently, Sessions and co-workers constructed a new MW interface[118] by modifying the prototype developed by Brand and Dobberstein.[117] The modified interface was used as a device for bulk sample analysis of carbon and showed accuracy better than 0.5‰ for nearly all dissolved analytes (lipids, proteins, nucleic acids, sugars, halocarbons) and a 1000-fold improved sensitivity compared with existing elemental analysers.[118]

Wet Chemical Combustion Interface. The commercially available LC-IRMS interfaces exclusively apply wet chemical combustion for the conversion of organic compounds into measurable gaseous compounds. The function of the interface is to convert separated components individually and quantitatively into CO_2 for IRMS without isotope fractionation. The principle is based on a chemical reaction as it is also used to convert organic compounds to CO_2 in total organic carbon (TOC) analysers.[22] A similar principle is employed in a TOC-IRMS coupling developed by Gilles St. Jean.[132]

Nowadays, there are two commercial interfaces for LC coupling on the market, the LC-IsoLink™ (Thermo Fisher Scientific, Bremen, Germany) and the LiquiFace™ (Elementar Analysensysteme, Hanau, Germany) introduced in 2004 and 2010, respectively.[122,133]

Figure 3.32 is a schematic diagram of a wet chemical combustion interface as realized in the LC-IsoLink (Thermo Fisher Scientific, Bremen, Germany). The HPLC effluent containing the separated compounds is mixed via a T-piece with an oxidation reagent of $M_2S_2O_8$

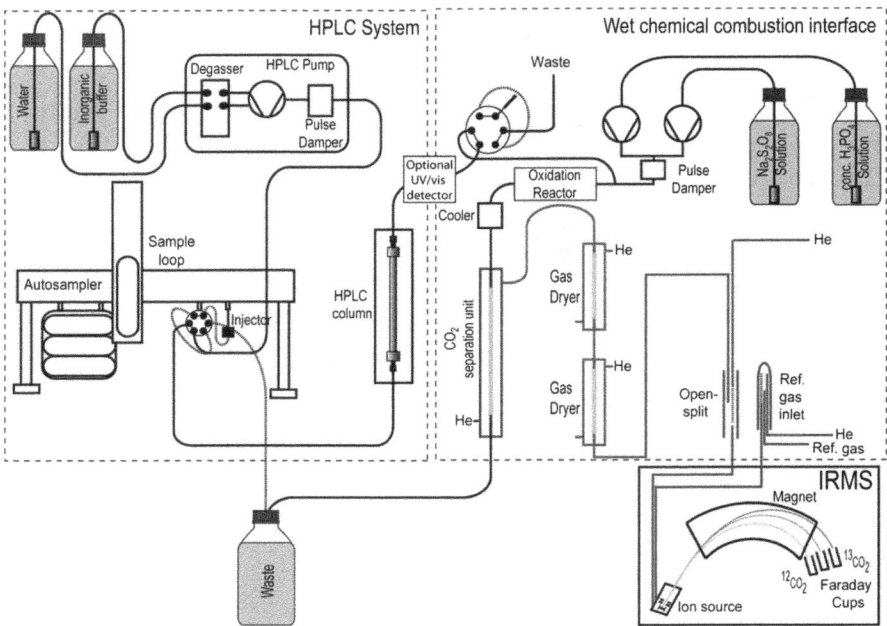

Figure 3.32 Principle of the LC-IRMS interface based on wet chemical combustion. Subsequent to chromatographic separation, the HPLC column effluent is mixed with oxidation reagent (sodium peroxodisulfate) and concentrated acid (phosphoric acid) via a T-piece before flowing into an oxidation reactor, which is held at $T = 99.9\,°C$. The developed CO_2 is separated from the aqueous phase by a membrane separation unit. The CO_2 permeates the membrane and is transported by a dry helium gas stream through two Nafion™ gas dryer units. The dried gas is introduced via an open-split into the IRMS ion source. The interface also possesses a reference gas inlet for the introduction of reference gas pulses into the ion source.

($M = Na^+$, K^+, NH_4^+) and a concentrated acid (conc. H_3PO_4). These two reagents are pumped separately by two head pumps. Following this, the mixture flows through a capillary oxidation reactor that is held at a temperature of around $100\,°C$. Within the reactor the components in the effluent stream are oxidized to CO_2, water and, if nitrogen is present, to N_2 and a variety of nitrous oxides (N_xO_y) including nitrate and nitrite. After oxidation, the eluent is cooled, and the formed CO_2 is separated from the liquid phase by passing through a gas separation membrane, which is surrounded by a countercurrent flow of dry helium. Note that the membrane is permeable for any gas, not just CO_2. Hence, O_2, H_2O, N_2O, N_2 and more can get through and reach the mass spectrometer.[18] Before the CO_2-containing helium stream enters the ion source of the

IRMS via an open-split, the gas is dried in two on-line Nafion™ gas dryer units.

Oxidation Reactor. For the combustion step mostly sodium peroxodisulfate ($Na_2S_2O_8$) at a concentration of $\sim 0.8\,mol\,L^{-1}$ is applied as oxidation reagent. Sodium peroxodisulfate dissociates in water according to:

$$Na_2S_2O_8 \rightleftharpoons 2Na^+ + S_2O_8^{2-} \qquad (3.64)$$

$$S_2O_8^{2-} + 2e^- \rightleftharpoons 2SO_4^{2-} \qquad (3.65)$$

Peroxodisulfate anions $S_2O_8^{2-}$ are strong but relatively stable oxidants with a standard reduction potential of $E° = 2.01$ V for the reaction in eqn (3.65).[134,135] Carbohydrates (here given by an average sum formula CH_2O) are oxidized according to the gross equation:[122]

$$4S_2O_8^{2-} + 2CH_2O + 2H_2O \rightarrow 8SO_4^{2-} + 2CO_2 + 8H^+ \qquad (3.66)$$

The main oxidative species in this reaction are sulfate radicals ($SO_4^{\bullet -}$). The sulfate radical is one of the strongest oxidants known, with a standard reduction potential of $E° = 2.47\,V$;[136] and reaction rates for the peroxodisulfate ion are 10^3 to 10^5 times slower than for the sulfate radical.[135] Sulfate radicals can be formed from peroxodisulfate anions through activation by heat, UV light, and transition metals such as silver according to:

$$S_2O_8^{2-} \xrightarrow{T,UV,metalions} 2SO_4^{\bullet -} \qquad (3.67)$$

At elevated temperatures such as those used in the wet combustion process, peroxodisulfate decomposition rates increase.[135] Sulfate radicals are highly reactive electrophilic reagents with a short half life. They preferably remove electrons from organic molecules producing organic radical cations.[135] Thus, reaction rates decrease for compounds containing nitro groups or an increasing number of ether bonds that destabilize the cationic intermediate. Moreover, the thermal activation of peroxodisulfate is strongly pH-dependent. Although high temperatures and low pH (<2) result in increased radical generation, recombination of generated radicals will dominate and radical-to-compound reactions are suppressed.[137,138] Therefore, an addition of acid subsequent to oxidation would be preferable, but is not realized in the commercial interface systems.

The acid catalysed reaction and the formation of oxygen can be explained by the following reactions:

$$S_2O_8^{2-} + H^+ \rightleftharpoons HS_2O_8^- \qquad (3.68)$$

$$HS_2O_8^- + H_2O \rightleftharpoons H_2SO_5 + HSO_4^- \qquad (3.69)$$

$$H_2SO_5 + H_2O \rightleftharpoons H_2O_2 + H_2SO_4 \qquad (3.70)$$

$$H_2O_2 \rightleftharpoons H_2O + 1/2 O_2 \qquad (3.71)$$

where, in the first rate-determining step (eqn (3.68)) peroxodisulfate is protonated to form peroxomonosulfuric acid which hydrolyzes further to hydrogen peroxide. Subsequently, hydrogen peroxide thermally decomposes under oxygen production.[137,139]

Various inorganic ions, such as chloride, carbonate, or bicarbonate ions can react with sulfate radicals.[135] In particular, high chloride concentrations can lead to reduced reaction rates for compound mineralization. However, in the coupling with LC, this could only be a potential problem when using ion chromatography for the separation of organic anions. Even then, separation of the interfering inorganic anions is achieved routinely. Hence, only when using the interface for direct injection of aqueous samples (i.e. without a separation column) for bulk analysis of small sample volumes, might the interfering anions be relevant.

The length of the oxidation reactor is a potential source for extra peak broadening. To minimize the extra peak broadening, the reactor consists of a low dispersion knitted tubing.

This has the advantage that the mobile phase flow direction changes by 180° as it passes from one serpentine to another which results in an extensive radial flow that greatly reduces the dispersion. Furthermore, the heat exchange between the heater and the capillary is more effective.[116,140]

Although the wet chemical oxidation interface has been shown to produce reliable carbon isotope data in many cases, it is particularly useful in the comparison of sample series that are all treated in the same way.

Membrane Separation Unit. As mentioned before, the developed CO_2 is separated from the liquid phase by a membrane through pervaporation.[65]

Instrumentation for Compound-specific Stable Isotope Analysis 143

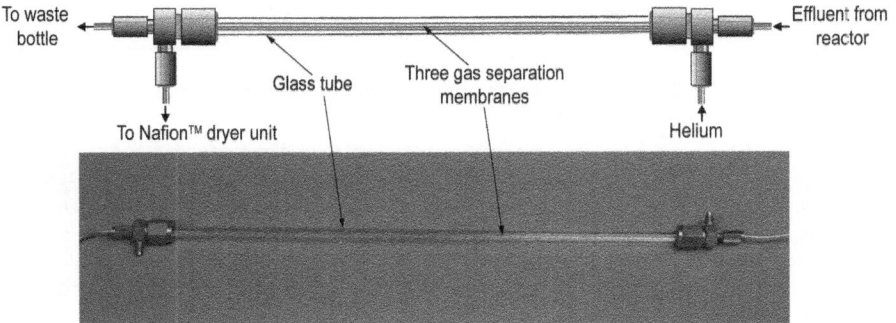

Figure 3.33 Diagram and photograph of a gas separation unit as used in an LC-IRMS. The cooled effluent from the reactor passes through three gas separation membranes. The membranes are stretched inside a glass tube through which a stream of dry helium gas flows in the opposite direction to the effluent. The CO_2 passes through the membrane from the effluent into the countercurrent He gas stream.
(Photo by Dorothea M. Kujawinski.)

The separation unit, shown in Figure 3.33, consists of three membrane tubings with excellent gas permeability for CO_2 (and other gases), at the ends of which three silica capillaries are inserted for connection. The liquid passes through the three membrane tubings and a helium stream flows along the exterior of the membrane. Due to the concentration and pressure gradient, CO_2 permeates the membrane and is transported away with the helium stream. The membrane consists of polyperfluoroethylenoxide with a thickness of approximately 40 µm and an inner diameter of the tubing of approximately 210 µm.

The acidification of the LC column effluent promotes the formation of CO_2 by shifting the equilibrium,

$$CO_2(aq) + H_2O \rightleftharpoons HCO_3^- + H^+ \quad (3.72)$$

to the left-hand side and the high ionic strength of the solution leads to an additional salting-out effect that enhances the CO_2 transfer into the gas phase.[122]

In addition to CO_2, O_2 generated via reaction (3.71) is able to diffuse through the transfer membrane into the helium carrier gas and is transferred into the mass spectrometer, where the high oxygen content (typical signal size 20–35 V for m/z 32) can lead to detrimental effects on measurement accuracy and precision as well as on filament lifetime.[141] The high oxygen content in the ion source can enhance space charge effects. In addition, high oxygen contents can lead to oxidative reactions with the hot filament and other surfaces in the source. The consequences

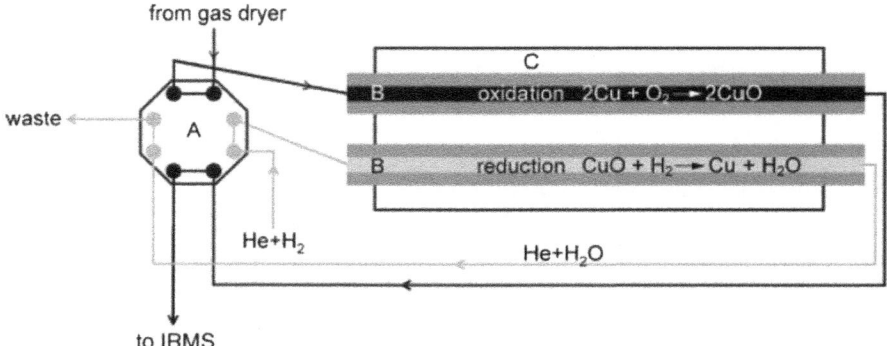

Figure 3.34 Schematic of a reduction reactor used as oxygen scavenger for LC-IRMS. The reduction reactor consists of an eight-port valve (A), two ceramic tubes filled with copper wire (B) and a heating reactor (C). The gas coming from the gas dryer passes through the 580 °C hot reduction reactor, in which the oxygen is fixed. It is possible to switch between the two reactors. While one reactor is in use the other is regenerated by a hydrogen gas stream.[141]
(Reprinted with permission of Wiley InterScience.)

of such uncontrolled reactions are ion source tuning problems and CO_2 background fluctuations.[141]

Hettmann et al. therefore developed a reduction reactor as oxygen scavenger, which they implemented in a commercial system (for a schematic of the set-up see Figure 3.34).

The unit consists of two reactors filled with copper wires that fix the oxygen. While one reactor is in use the other is reduced by a hydrogen gas stream.[141]

Separation Columns for LC-IRMS. Commercially available interfaces for LC-IRMS cannot be operated in combination with chromatographic separations driven or supported by organic modifiers. Thus, stationary phases need to be compatible with pure aqueous mobile phases to ensure reproducible separations. Also column bleed, i.e., loss of carbon from the stationary phase, should be as low as possible. Common columns for ion chromatography or RPC, which are the two main separation methods reported in literature, fulfil these requirements, as long as they are operated under the conditions recommended by the manufacturer.

Sulfonated polystyrene divinylbenzene materials have been used with pure water or acidic eluents for the analysis of amino acids, sugars, ethanol, glycerol and small organic acids by anion exclusion,[142,143] ligand exclusion[144–146] and ligand exchange.[147] Sulfonate anchor groups have also been used to separate amino acids and sugars via cation exchange.[122,148]

Pure anion exchange separations have been performed on a common polymer stationary phase with quaternary ammonium groups as exchanging functional group for aminosugars,[149] glycerol and carbohydrates.[133,150,151] Using $NaNO_3$ as a pusher in anion exchange chromatography, low concentrated NaOH eluents (1 mM) were used to achieve a separation of acidic carbohydrates similar to that in gradient mode with higher concentrated NaOH solutions (up to 100 mM) although column specifications were the same. Thus, problems associated with solvation of CO_2 and transformation to carbonate in highly concentrated basic solutions were avoided.[133,150]

Another application of anion exchange chromatography has been reported for the determination of $\delta^{13}C$-values of nucleotide triphosphates separated on columns packed with polyethylene imine bound to deactivated silica.[152]

Godin et al. used a two-dimensional chromatographic technique to separate amino acids.[153] In this work a C_{30} reversed phase column was installed behind a cation exchange column and baseline separation for 11 out of 15 amino acids was achieved. Based on this work, McCullagh et al. used a mixed mode separation column to analyse amino acids.[107] The stationary phase consists of C_{12} chains with embedded anionic groups allowing for ion exchange and hydrophobic interactions. Amongst others, Smith et al. used the same analytical column to achieve baseline separation of almost all amino acids from protein hydrolysates.[154] Unfortunately, analysis time for a single chromatographic run was more than three hours.

Contrary to ion chromatography, operating reversed phase columns under pure aqueous conditions can be difficult due to an impending 'phase collapse'. However, the use of organic modifiers is fully excluded. Godin et al. reported that even 0.01% (v/v) methanol or acetonitrile increases the CO_2 background to 40% of the scale of the IRMS, prohibiting a measurement within the dynamic range of the instrument.[153] Nowadays, polar embedded groups in between the hydrophobic chains keep the stationary phase wetted with eluent allowing the use of pure water as the mobile phase. Some of these stationary phases have been reported in the literature, such as a silica-based amide-embedded C_{16}-column for the analysis of bentazon[155] or other polar modified C_{18} columns for the analysis of small organic acids.[156]

Another way to use RP chromatography with pure aqueous eluents is to increase the temperature of the mobile and stationary phases. The static permittivity of water decreases with temperature, rendering its elution strength similar to that of a methanol–water mixture.[157] Consequently, in HT-HPLC temperature gradients can replace organic solvent gradients. As a first example of this promising option for LC-IRMS, Godin et al. applied temperatures up to 170 °C to column and eluent to separate hydrosoluble fatty and phenolic acids on a very hydrophobic porous graphitic carbon column.[143] Commercially available columns are often not intended to be used at these high temperatures. The silica support as well as the bonded phase might be rapidly hydrolyzed, increasing CO_2 background and deteriorating separation efficiency of the column. Hence, a careful investigation of the magnitude of column bleed by monitoring of the CO_2 background and its effect on $\delta^{13}C$-values is mandatory prior to HT-HPLC method development. Godin et al. used a procedure of injecting CO_2 reference gas during a blank HT-HPLC-IRMS run.[143] It was shown that precision was high ($\geq 0.3‰$) and accuracy was achieved as long as the CO_2 background was stable and not drifting.[143] Additionally, analyte degradation on the column at high temperatures should be carefully checked by comparison of peak areas as well as accuracy of $\delta^{13}C$-values.[143]

Zhang et al. investigated four different stationary phases for their applicability in HT-HPLC-IRMS. Under isothermal and temperature gradient conditions, the column bleed of XBridge C18 (up to 180 °C), Acquity C18 (up to 200 °C), Triart C18 (up to 150 °C), and ZirchromPBD (up to 150 °C) showed no influence on the precision and accuracy of $\delta^{13}C$-values.[158]

Due to the flow limitations of the LC-IRMS interfaces (total flow $\sim 700\,\mu L$) selection of column dimensions is an important factor in method development. As pointed out by Godin et al. and McCullagh et al. column diameter should be as low as possible to increase signal height to analyte mass ratio, thus increasing sensitivity.[143,159] With a lower internal diameter of the column, the optimum flow rate according to the van Deemter equation is within the restrictions of the interface. On the other hand, reducing the column i.d. from 2 mm to 1 mm results in about four times smaller sample loading capacity.[143] In order to improve resolution, McCullagh et al. suggested the use of longer columns with an optional flow gradient at the end of the chromatographic run to decrease peak width.[159] Although resolution can be further enhanced by using smaller particle sizes in the sub-2 μm range (Ultra-High Pressure Liquid Chromatography) this has not been reported yet for LC-IRMS but might be a promising tool in the future.

3.7 WEB FINDINGS

Mion Chrom project ('mionchrom')
An open-source project developed for interpretation of results from GC-IRMS. This project is currently specialized for stable carbon isotopes ($^{13}C/^{12}C$):

- Auto-integrate folders of ASCII files and calculate isotope ratios ($\delta^{13}C$ and $\delta^{18}O$ values).
- Explore the chromatograms with zoom in, centre view, integrate manually, in a clean layout free of buttons and minimal movement of cursor.
- Pre-process the data to remove noise, export the results with changes in *.png, *.xls and *.txt.
- Save and retrieve work like manual integration and peak identification.

http://sourceforge.net/projects/mionchrom/

REFERENCES

1. J.-P. Godin, L.-B. Fay and G. Hopfgartner, *Mass Spectrom. Rev.*, 2007, **26**, 751–774.
2. J. T. Brenna, T. N. Corso, H. J. Tobias and R. J. Caimi, *Mass Spectrom. Rev.*, 1997, **16**, 227–258.
3. K. Sakaguchi-Soder, J. Jager, H. Grund, F. Matthaus and C. Schüth, *Rapid Commun. Mass Spectrom.*, 2007, **21**, 3077–3084.
4. C. Aeppli, H. Holmstrand, P. Andersson and O. Gustafsson, *Anal. Chem.*, 2010, **82**, 420–426.
5. A. Amrani, A. Sessions and J. Adkins, *Geochim. Cosmochim. Acta*, 2008, **72**, A20–A20.
6. A. O. Nier, *Rev. Sci. Instrum.*, 1940, **11**, 212–216.
7. A. O. Nier, *Rev. Sci. Instrum.*, 1947, **18**, 398–411.
8. C. R. McKinney, J. M. McCrea, S. Epstein, H. A. Allen and H. C. Urey, *Rev. Sci. Instrum.*, 1950, **21**, 724–730.
9. A. O. Nier, W. R. Eckelmann and R. A. Lupton, *Anal. Chem.*, 1962, **34**, 1358–1360.
10. W. A. Brand, *Adv. Mass Spectrom.*, 1998, **14**, 661–686.
11. W. A. Brand, *J. Mass Spectrom.*, 1996, **31**, 225–235.
12. W. A. Brand, in *Handbook of Stable Isotope Analytical Techniques*, ed. P. A. De Groot, Elsevier, Amsterdam, 2004, vol. 1, pp. 874–906.

13. M. P. Ricci, D. A. Merritt, K. H. Freeman and J. M. Hayes, *Org. Geochem.*, 1994, **21**, 561–571.
14. I. T. Platzner, K. Habfast, A. J. Walder and A. Goetz, *Modern Isotope Ratio Mass Spectrometry*, J. Wiley, Chichester; New York, 1997.
15. W. Demtröder, *Experimentalphysik 1 Mechanik und Wärme*, Vierte, neu bearbeitete und aktualisierte Auflage edn., Springer-Verlag Berlin Heidelberg, Berlin, Heidelberg, 2006.
16. M. L. Vestal, *Chem. Rev.*, 2001, **101**, 361–375.
17. S. Asche, A. L. Michaud and J. T. Brenna, *Curr. Org. Chem.*, 2003, **7**, 1527–1543.
18. W. Brand, *Personal Communication*.
19. *National Institute of Standards and Technology Electron-Impact Cross Sections for Ionization and Excitation*, http://physics.nist.gov/PhysRefData/Ionization/molTable.html.
20. Y. K. Kim, W. Hwang, N. M. Weinberger, M. A. Ali and M. E. Rudd, *J. Chem. Phys.*, 1997, **106**, 1026–1033.
21. W. A. Brand, S. S. Assonov and T. B. Coplen, *Pure Appl. Chem.*, 2010, **82**, 1719–1733.
22. G. R. Peyton, *Mar. Chem.*, 1993, **41**, 91–103.
23. S. Taya, *Nuclear Instruments & Methods*, 1978, **152**, 399–405.
24. T. W. Burgoyne and G. M. Hieftje, *Mass Spectrom. Rev.*, 1996, **15**, 241–259.
25. H. Wollnik, *J. Mass Spectrom.*, 1999, **34**, 991–1006.
26. W. G. Cross, *Rev. Sci. Instrum.*, 1951, **22**, 717–722.
27. J. H. Gross, *Mass Spectrometry : a textbook*, Springer, Berlin; London, 2004.
28. K. Habfast, ed., *Advanced Isotope Ratio Mass Spectrometry I: Magnetic Isotope Ratio Mass Spectrometers*, J. Wiley, Chichester; New York, 1997.
29. D. W. Koppenaal, C. J. Baringa, M. B. Denton, R. P. Sperline, G. M. Hieftje, G. D. Schilling, F. J. Andrade and J. H. Barnes, *Anal. Chem.*, 2005, **77**, 418A–427A.
30. M. E. Wieser and J. B. Schwieters, *Int. J. Mass Spectrom.*, 2005, **242**, 97–115.
31. D. A. Skoog, F. J. Holler and S. R. Crouch, eds., *Principles of Instrumental Analysis*, 6th edn., Thomson Brooks/Cole, Belmont, CA; United Kingdom, 2007.
32. D. A. Merritt and J. M. Hayes, *Anal. Chem.*, 1994, **66**, 2336–2347.
33. D. A. Merritt, K. H. Freeman, M. P. Ricci, S. A. Studley and J. M. Hayes, *Anal. Chem.*, 1995, **67**, 2461–2473.

34. D. W. Peterson and J. M. Hayes, in *Contemporary Topics in Analytical and Clinical Chemistry*, ed. D. M. Hercules, G. M. Hieftje, L. R. Snyder and M. E. Evenson, Plenum, New York, 1978, pp. 217–252.
35. G. L. Sacks, C. J. Wolyniak and J. T. Brenna, *J. Chromatogr. A*, 2003, **1020**, 273–282.
36. R. A. Werner and W. A. Brand, *Rapid Commun. Mass Spectrom.*, 2001, **15**, 501–519.
37. J. Santrock, S. A. Studley and J. M. Hayes, *Anal. Chem.*, 1985, **57**, 1444–1448.
38. H. Craig, *Geochim. Cosmochim. Acta*, 1957, **12**, 133–149.
39. S. S. Assonov and C. A. M. Brenninkmeijer, *Rapid Commun. Mass Spectrom.*, 2003, **17**, 1007–1016.
40. R. M. Verkouteren and J. N. Lee, *Fresenius J. Anal. Chem.*, 2001, **370**, 803–810.
41. B. F. Murphey, *Physical Reviews*, 1947, **72**, 834–837.
42. W. W. Wong and P. D. Klein, *Mass Spectrom. Rev.*, 1986, **5**, 313–342.
43. D. L. Hachey, W. W. Wong, T. W. Boutton and P. D. Klein, *Mass Spectrom. Rev.*, 1987, **6**, 289–328.
44. R. E. Halsted and A. O. Nier, *Rev. Sci. Instrum.*, 1950, **21**, 1019–1021.
45. A. L. Sessions, *J. Sep. Sci.*, 2006, **29**, 1946–1961.
46. D. A. Merritt, W. A. Brand and J. M. Hayes, *Org. Geochem.*, 1994, **21**, 573–583.
47. W. Meier-Augenstein, *Rapid Commun. Mass Spectrom.*, 1997, **11**, 1775–1780.
48. W. A. Brand, *Isot. Environ. Health Stud.*, 1994, **31**, 277–284.
49. B. Schulze, C. Wirth, P. Linke, W. A. Brand, I. Kuhlmann, V. Horna and E. D. Schulze, *Tree Physiology*, 2004, **24**, 1193–1201.
50. T. Preston and N. J. P. Owens, *Analyst*, 1983, **108**, 971–977.
51. Z. Muccio and G. P. Jackson, *Analyst*, 2009, **134**, 213–222.
52. *ThermoFinnigan High Temperature Conversion Elemental Analyzer (TC/EA) Operating Manual. Ident. No. 112 76 01, Issue 11/2001.*
53. S. Benson, C. Lennard, P. Maynard and C. Roux, *Forensic Science International*, 2006, **157**, 1–22.
54. W. A. Brand, *Isot. Environ. Health Stud.*, 2009, **45**, 135–149.
55. T. B. Coplen and H. Qi, *Isot. Environ. Health Stud.*, 2009, **45**, 126–134.
56. H. P. Qi, T. B. Coplen, H. Geilmann, W. A. Brand and J. K. Böhlke, *Rapid Commun. Mass Spectrom.*, 2003, **17**, 2483–2487.

57. D. Hofmann, K. Jung, J. Bender, M. Gehre and G. Schürmann, *J. Mass Spectrom.*, 1997, **32**, 855–863.
58. W. Meier-Augenstein, *J. Chromatogr. A*, 1999, **842**, 351–371.
59. W. Brand and A. R. Tegtmeyer, *Isot. Environ. Health Stud.*, 1992, **28**, 112.
60. D. A. Merritt and J. M. Hayes, *J. Am. Soc. Mass. Spectrom.*, 1994, **5**, 387–397.
61. W. A. Brand, A. R. Tegtmeyer and A. Hilkert, *Org. Geochem.*, 1994, **21**, 585–594.
62. C. C. Metges, K. J. Petzke and U. Hennig, *J. Mass Spectrom.*, 1996, **31**, 367–376.
63. S. A. Macko, M. E. Uhle, M. H. Engel and V. Andrusevich, *Anal. Chem.*, 1997, **69**, 926–929.
64. U. Flenker, M. Hebestreit, T. Piper, F. Hulsemann and W. Schanzer, *Anal. Chem.*, 2007, **79**, 4162–4168.
65. O. Sae-Khow and S. Mitra, *J. Chromatogr. A*, 2010, **1217**, 2736–2746.
66. L. Ellis and A. L. Fincannon, *Org. Geochem.*, 1998, **29**, 1101–1117.
67. W. A. Brand, *Finnigan MAT GmbH Patent application, GB-2270911-A (UK)*, 1993.
68. *GC Combustion Interface III; GC C III Operating manual; Thermo Finnigan*, 2002, **06**.
69. A. H. Meyer, H. Penning, H. Lowag and M. Elsner, *Environ. Sci. Technol.*, 2008, **42**, 7757–7763.
70. P. A. Eakin, A. E. Fallick and J. Gerc, *Chem. Geol.*, 1992, **101**, 71–79.
71. W. Meier-Augenstein, *Anal. Chim. Acta*, 2002, **465**, 63–79.
72. W. Meier-Augenstein, in *Handbook of Stable Isotope Analytical Techniques*, ed. P. A. De Groot, Elsevier, Amsterdam, 2004, vol. 1, pp. 153–176.
73. K. J. Goodman, *Anal. Chem.*, 1998, **70**, 833–837.
74. K. J. Leckrone and J. M. Hayes, *Anal. Chem.*, 1997, **69**, 911–918.
75. K. J. Leckrone and J. M. Hayes, *Anal. Chem.*, 1998, **70**, 2737–2744.
76. A. W. Hilkert, C. B. Douthitt, H. J. Schlüter and W. A. Brand, *Rapid Commun. Mass Spectrom.*, 1999, **13**, 1226–1230.
77. T. W. Burgoyne and J. M. Hayes, *Anal. Chem.*, 1998, **70**, 5136–5141.
78. H. J. Tobias, K. J. Goodman, C. E. Blacken and J. T. Brenna, *Anal. Chem.*, 1995, **67**, 2486–2492.
79. H. J. Tobias and J. T. Brenna, *Anal. Chem.*, 1996, **68**, 3002–3007.
80. Z. Sofer, *Anal. Chem.*, 1986, **58**, 2029–2032.

81. *ThermoFinnigan Training Folder*, 2003.
82. S. Bilke and A. Mosandl, *Rapid Commun. Mass Spectrom.*, 2002, **16**, 468–472.
83. A. L. Sessions, T. W. Burgoyne and J. M. Hayes, *Anal. Chem.*, 2001, **73**, 192–199.
84. A. L. Sessions and J. M. Hayes, *Geochim. Cosmochim. Acta*, 2005, **69**, 593–597.
85. A. L. Sessions, T. W. Burgoyne and J. M. Hayes, *Anal. Chem.*, 2001, **73**, 200–207.
86. S. J. Prosser and C. M. Scrimgeour, *Anal. Chem.*, 1995, **67**, 1992–1997.
87. http://www.sercongroup.com/20–20.htm.
88. *Micromass Application Note 300*.
89. G. Gremaud, C. Piguet, M. Baumgartner, E. Pouteau, B. Decarli, A. Berger and L. B. Fay, *Rapid Commun. Mass Spectrom.*, 2001, **15**, 1207–1213.
90. J. Jung, S. Sewenig, U. Hener and A. Mosandl, *Eur. Food Res. Tech.*, 2005, **220**, 232–237.
91. U. Hener, W. A. Brand, A. W. Hilkert, D. Juchelka, A. Mosandl and F. Podebrad, *Z. Lebensm. Unters. For*, 1998, **206**, 230–232.
92. Y. F. Huang and Y. H. Huang, *J. Hazard. Mater.*, 2009, **162**, 1211–1216.
93. C. F. Poole, *The Essence of Chromatography*, 1st edn., Elsevier, Amsterdam; Boston, 2003.
94. L. Zwank, M. Berg, T. C. Schmidt and S. B. Haderlein, *Anal. Chem.*, 2003, **75**, 5575–5583.
95. J. P. Foley and J. G. Dorsey, *Anal. Chem.*, 1983, **55**, 730–737.
96. H.-J. Kuss and S. Kromidas, *Quantification in LC and GC: a Practical Guide to Good Chromatographic Data*, Wiley-VCH, Weinheim, 2009.
97. M. Rautenschlein, K. Habfast and W. Brand, in *Stable Isotopes in Paediatric Nutritional and Metabolic Research*, ed. T. E. Chapman, R. Berger, D. J. Reijngoud and A. Okken, Intercept, Andover, Hampshire, 1990.
98. F. Bruner, G. P. Cartoni and A. Liberti, *Anal. Chem.*, 1966, **38**, 298–303.
99. W. A. Van Hook, *J. Phys. Chem.*, 1967, **71**, 3270–3275.
100. B. D. Gunter and J. D. Gleason, *J. Chromatogr. Sci.*, 1971, **9**, 191–192.
101. W. A. Van Hook, *J. Chromatogr. Sci.*, 1972, **10**, 191–192.
102. G. Jancso and W. A. Van Hook, *Chem. Rev.*, 1974, **74**, 689–750.
103. J. Bigeleisen, *J. Chem. Phys.*, 1961, **34**, 1485.

104. M. Wolfsberg, *J. Chim. Phys. Phys.-Chim. Biol.*, 1963, **60**, 15–22.
105. K. H. Freeman, J. M. Hayes, J. M. Trendel and P. Albrecht, *Nature*, 1990, **343**, 254–256.
106. F. P. Abramson, G. E. Black and P. Lecchi, *J. Chromatogr. A*, 2001, **913**, 269–273.
107. J. S. O. McCullagh, D. Juchelka and R. E. M. Hedges, *Rapid Commun. Mass Spectrom.*, 2006, **20**, 2761–2768.
108. *Org. Geochem.*, 1994, **21**, 561–827.
109. K. J. Goodman and J. T. Brenna, *Anal. Chem.*, 1994, **66**, 1294–1301.
110. H.-J. Hübschmann, *Handbook of GC/MS: Fundamentals and Applications*, 2nd, completely rev. and updated edn., Wiley-VCH, Weinheim, 2009.
111. W. Meier-Augenstein, W. Brand, G. F. Hoffmann and D. Rating, *Biol. Mass Spectrom.*, 1994, **23**, 376–378.
112. W. Meier-Augenstein, *J. High. Resolut. Chromatogr.*, 1995, **18**, 28–32.
113. J. A. Hall, J. A. C. Barth and R. M. Kalin, *Rapid Commun. Mass Spectrom.*, 1999, **13**, 1231–1236.
114. T. I. Taylor and H. C. Urey, *J. Chem. Phys.*, 1937, **5**, 597–598.
115. T. I. Taylor and H. C. Urey, *J. Chem. Phys.*, 1938, **6**, 429–438.
116. A. Braithwaite and F. J. Smith, *Chromatographic Methods*, 5th edn., Blackie Academic & Professional, London, 1996.
117. W. A. Brand and P. Dobberstein, *Isot. Environ. Health Stud.*, 1996, **32**, 275–283.
118. A. L. Sessions, S. P. Sylva and J. M. Hayes, *Anal. Chem.*, 2005, **77**, 6519–6527.
119. Y. Teffera, J. J. Kusmierz and F. P. Abramson, *Anal. Chem.*, 1996, **68**, 1888–1894.
120. R. J. Caimi and J. T. Brenna, *J. Chromatogr. A*, 1997, **757**, 307–310.
121. F. P. Abramson, B. L. Osborn and Y. Teffera, *Anal. Chem.*, 1996, **68**, 1971–1972.
122. M. Krummen, A. W. Hilkert, D. Juchelka, A. Duhr, H. J. Schlüter and R. Pesch, *Rapid Commun. Mass Spectrom.*, 2004, **18**, 2260–2266.
123. J.-P. Godin and J. S. O. McCullagh, *Rapid Commun. Mass Spectrom.*, 2011, **25**, 3019–3028.
124. S. P. Markey and F. P. Abramson, *Anal. Chem.*, 1982, **54**, 2375–2376.
125. M. McLean, M. L. Vestal, Y. Teffera and F. P. Abramson, *J. Chromatogr. A*, 1996, **732**, 189–199.

126. M. Moini and F. P. Abramson, *Biol. Mass Spectrom.*, 1991, **20**, 308–312.
127. Y. Teffera, F. P. Abramson, M. McLean and M. Vestal, *J. Chromatogr., Biomed. Appl.*, 1993, **620**, 89–96.
128. J. T. Brenna, *Prostaglandins Leukotrienes and Essential Fatty Acids*, 1997, **57**, 467–472.
129. K. Jorabchi, K. Kahen, P. Lecchi and A. Montaser, *Anal. Chem.*, 2005, **77**, 5402–5406.
130. F. P. Abramson, *Mass Spectrom. Rev.*, 1994, **13**, 341–356.
131. R. J. Caimi and J. T. Brenna, *J. Mass Spectrom.*, 1995, **30**, 466–472.
132. G. Jean, *Rapid Commun. Mass Spectrom.*, 2003, **17**, 419–428.
133. D. J. Morrison, K. Taylor and T. Preston, *Rapid Commun. Mass Spectrom.*, 2010, **24**, 1755–1762.
134. D. A. House, *Chem. Rev.*, 1962, **62**, 185–203.
135. A. Tsitonaki, B. Petri, M. Crimi, H. Mosbæk, R. L. Siegrist and P. L. Bjerg, *Crit. Rev. Environ. Sci. and Technol.*, 2010, **40**, 55–91.
136. C. von Sonntag, *Free-radical-induced DNA Damage and its Repair: a chemical perspective*, Springer, Berlin; London, 2005.
137. I. M. Kolthoff and I. K. Miller, *J. Am. Chem. Soc.*, 1951, **73**, 3055–3059.
138. C. E. H. Bawn and D. Margerison, *Transactions of the Faraday Society*, 1955, **51**, 925–934.
139. H. Galiba, L. J. Csanyi and Z. G. Szabo, *Z. Anorg. Allg. Chem.*, 1956, **287**, 152–168.
140. E. D. Katz and R. P. W. Scott, *J. Chromatogr.*, 1983, **268**, 169–175.
141. E. Hettmann, W. A. Brand and G. Gleixner, *Rapid Commun. Mass Spectrom.*, 2007, **21**, 4135–4141.
142. H. Penning and R. Conrad, *Geochim. Cosmochim. Acta*, 2006, **70**, 2283–2297.
143. J. P. Godin, G. Hopfgartner and L. Fay, *Anal. Chem.*, 2008, **80**, 7144–7152.
144. A. I. Cabañero, J. L. Recio and M. Rupérez, *J. Agric. Food. Chem.*, 2010, **58**, 722–728.
145. A. I. Cabañero, J. L. Recio and M. Rupérez, *J. Agric. Food. Chem.*, 2006, **54**, 9719–9727.
146. A. I. Cabañero, J. L. Recio and M. Rupérez, *Rapid Commun. Mass Spectrom.*, 2008, **22**, 3111–3118.
147. L. Elflein and K. P. Raezke, *Apidologie*, 2008, **39**, 574–587.
148. J. P. Godin, D. Breuille, C. Obled, I. Papet, H. Schierbeek, G. Hopfgartner and L. B. Fay, *J. Mass Spectrom.*, 2008, **43**, 1334–1343.

149. S. Bodé, K. Denef and P. Boeckx, *Rapid Commun. Mass Spectrom.*, 2009, **23**, 2519–2526.
150. H. T. S. Boschker, T. C. W. Moerdijk-Poortvliet, P. Van Breugel, M. Houtekamer and J. J. Middelburg, *Rapid Commun. Mass Spectrom.*, 2008, **22**, 3902–3908.
151. H. Schierbeek, D. Rook, F. Braake, K. Y. Dorst, G. Voortman, J. P. Godin, L. B. Fay and J. B. van Goudoever, *Rapid Commun. Mass Spectrom.*, 2009, **23**, 2897–2902.
152. X. L. Du, K. Ferguson, R. Gregory and S. R. Sprang, *Anal. Biochem.*, 2008, **372**, 213–221.
153. J. P. Godin, J. Hau, L. B. Fay and G. Hopfgartner, *Rapid Commun. Mass Spectrom.*, 2005, **19**, 2689–2698.
154. C. I. Smith, B. T. Fuller, K. Choy and M. P. Richards, *Anal. Biochem.*, 2009, **390**, 165–172.
155. S. Reinnicke, A. Bernstein and M. Elsner, *Anal. Chem.*, 2010, **82**, 2013–2019.
156. V. Heuer, M. Elvert, S. Tille, M. Krummen, X. P. Mollar, L. R. Hmelo and K. U. Hinrichs, *Limnology and Oceanography-Methods*, 2006, **4**, 346–357.
157. Y. Yang, *J. Sep. Sci.*, 2007, **30**, 1131–1140.
158. L. Zhang, D. M. Kujawinski, M. A. Jochmann and T. C. Schmidt, *Rapid Commun. Mass Spectrom.*, 2011, **25**, 2971–2980.
159. J. S. O. McCullagh, *Rapid Commun. Mass Spectrom.*, 2010, **24**, 483–494.

CHAPTER 4
Sample Preparation in Compound-specific Stable Isotope Analysis

4.1 SCOPE

This chapter covers the steps in the analytical process up to injection into one of the chromatographic systems described in Chapter 3. We start with a few general remarks and recommendations for sampling and sample preservation, with a focus on environmental samples. Further sample processing steps, including extraction, clean-up, preconcentration and injection will be discussed in two separate sections on volatile and semi-volatile compounds. We will focus on environmental samples for two reasons: first, many environmental samples are complex mixtures with target compounds being present only at trace concentrations thus requiring extended sample processing. Second, over the past years most developments and applications in the area of CSIA at trace concentration level have been in the environmental domain. Of course, many of the points discussed in these sections can be generalized to other application areas of CSIA. As a separate topic we cover derivatization procedures that are used to extend the application range of GC-based separations coupled to IRMS towards polar analytes.

4.2 SAMPLING, SAMPLE PRESERVATION AND STORAGE

In recent years there has been a renewed emphasis on the idea that sampling is part of the measurement process rather than a separate activity of little relevance to analysts.[1] This is also the case for sampling procedures of organic compounds prior to CSIA. Due to the huge field of applications that is covered by CSIA, the types of sample matrix vary from gaseous samples (e.g., VOCs in air) to liquids (e.g., contaminants in groundwater, surface water, drug or doping metabolites in urine) and solid samples (e.g., aminosugars in soil or fatty acids in ancient pottery).

Owing to the huge variety of sample matrices and the degree of heterogeneity, particularly in solid samples, the sampling strategy and procedure has to be adapted to the specific problem. Good sampling practice for subsequent isotope analysis does not differ substantially from that for concentration analysis of the same target compounds. However, mass discrimination during sampling procedures needs to be avoided and the sampling procedure has to be checked with regard to possible isotopic fractionation.

The sampling strategy must ensure that representative samples are obtained (e.g., by the duplicate method)[1] and the uncertainty of the isotopic abundance associated with or resulting from sampling should be reported. To that end, the Eurachem Guide to sampling uncertainty[2] advocates a logical sequence of events that will be familiar to analytical chemists, namely:

(i) validate a method by examining its uncertainty;
(ii) check that the uncertainty is consistent with fitness for purpose; and
(iii) check by quality control procedures that the uncertainty is not deteriorating during routine use of the method.[3]

All parameters potentially relevant for later data interpretation need to be recorded during sampling. To give an example, for sampling of VOCs in groundwater such parameters include:

(i) date and time of sampling;
(ii) geographical coordinates of sampling points or wells (using GPS, maps, GIS data);
(iii) weather conditions;
(iv) depth to the water table, sampling depth, pump type and flow, tubing material and length, pre-sampling purge volume, well condition, condition of rain gauge;
(v) chemical and physical data such as water temperature, pH, alkalinity, conductivity, and oxygen content measured on-site. Concentrations of other redox-sensitive constituents, such as Fe^{2+}, H_2S, etc., can be helpful during later data interpretation and need a separate sampling and determination;
(vi) sample preservation measures in the field (see below).

Sample storage vessels must be individually labelled (project code, location, date, sample number, collector's name, type of analysis required); the information has to be cross-linked with the field notebook and the sample collection sheets. The same type of sample storage vessels that are used for concentration analysis of target compounds in the same matrix are, in general, also suitable for isotope analysis. Since sample amount requirements might differ though, in many cases a number of identical vessels or larger ones need to be used for isotope analysis. This is particularly important when aiming for multiple isotope analysis since individual measurements are necessary for every element.

Samples for isotope analysis should always be processed and measured as soon as possible. If storage of samples is necessary it should always been done below 4 °C in the dark.[4] If possible, freezing of samples is recommended for long-term storage or very reactive environments. This has been demonstrated for TCE in aqueous suspensions containing zero-valent iron[5] but is also regularly used in storage of doping control urine samples.[6] Further sample preservation might be necessary to avoid microbial degradation. For carbon isotope analysis using pH adjustment by concentrated hydrogen chloride solution (to pH <2) or a 1% trisodium phosphate (TSP) solution (pH~10.5) is recommended for this purpose.[7] For many compounds, both methods are equally suited. For ethers, abiotic hydrolysis may occur at low pH, thus the TSP method should be favoured, whereas for chlorinated alkanes abiotic dehydrohalogenation is faster at high pH, thus acid preservation should be used. For a few compounds, including styrene, chloromethane and bromomethane, no preservation method was found to be suitable in control studies (for concentration analysis).[7] Validation of these or other preservation methods for isotope analysis of other elements has not been reported yet. The use of mercury salts or sodium azide is not generally recommended for preservation of environmental water samples since the samples need to be disposed of as hazardous waste after analysis. A more detailed account of sampling and sample preservation can be found in Hunkeler *et al.*[7]

4.3 SAMPLE PROCESSING

4.3.1 General Remarks

Sample preparation such as extraction, clean-up, preconcentration and, if necessary, derivatization is one of the most important aspects in the whole analytical process and often the most time-consuming step in CSIA. Sample preparation procedures are of the utmost importance because errors that occur in this step of the analytical process cannot be corrected subsequently.[8] Sample preparation is an individual procedure, depending on the target analytes as well as on the matrix involved. In the case of CSIA by GC- or LC-IRMS, a very good separation of analytes from matrix components (unresolved complex matter, UCM) is a prerequisite (see Chapter 3) and frequently requires clean-up.[9] Furthermore, the mass detector in IRMS instruments is inherently much less sensitive than in organic mass spectrometers in which electron multipliers are generally used for signal amplification (see Chapter 3). In the concentration domain, ng L^{-1} (or ppt) levels of target compounds can often be measured quite straightforwardly with modern MS-based equipment. Such concentration ranges are far below the lowest achievable levels in SIA. Thus, major emphasis is placed on improving and validating the best available preconcentration techniques in order to supply a sufficient amount of a compound to the separation column for subsequent isotope analysis.

The procedures involved in sample preparation can be diverse and some further examples for specific applications such as steroid isotope ratio determination will be presented in Chapter 6. In general, the following two rules should be obeyed:

(i) Every step of the sample preparation protocol (sampling, extraction, clean-up, preconcentration, derivatization, etc.) must be scrutinized for potential mass discriminating effects to avoid isotope discrimination of the analytes as far as possible.[9–11] The validation typically starts with clean matrix spiked with the isotopically characterized target compound as individual working standard but eventually should continue to a matrix as close as possible to the investigated one. A specific validation for each new target compound is necessary because no general explanation, or even prediction, of a potential isotope fractionation during sample preparation is possible yet.

(ii) A change in isotope composition can vary depending on analytical conditions such as the split ratio, extraction time, sorption or desorption conditions[4,7] but might also differ considerably among compounds measured under identical conditions.[12] Therefore, if the possibility of isotope fractionation cannot be ruled out conclusively, at least one internal standard, of a similar chemical nature and of known isotopic composition, should be added to the sample prior to sample preparation.[9]

Extraction and preconcentration methods to be used for subsequent GC-IRMS of target compounds primarily depend on their polarity, and hence volatility, as described by the magnitude of their partition constant between the sample matrix and air, and thus differ for volatile and semi-volatile compounds.

4.3.2 Sample Processing for Volatile Compounds

Extraction and Clean-up. For VOCs, i.e., those that show a sufficiently large air–water partitioning coefficient (or, in the case of dry samples, air–solid partitioning coefficient), the best clean-up step is to take an aliquot of air (typically 250 to 1000 µL) from the headspace above the sample. Thereby all non-volatile matrix components are excluded and frequently no further clean-up is necessary. Headspace (HS) analysis has been shown to produce no discernable carbon or hydrogen isotope effect in many studies in particular of BTEX compounds.[13–16] However, contrasting evidence has been reported with carbon isotope fractionation up to 1.3‰ for volatile chlorinated hydrocarbons[17] and up to 4‰ and poor precision for MTBE (methyl tert-butyl ether).[18] Thus, even with HS analysis, a thorough method validation is required. For analysis of volatiles in gas samples, active or passive collection of analytes on suitable sorbents, mostly followed by thermodesorption as a means of sample introduction is utilized. Recently, this has been thoroughly studied for benzene and its alkylated homologues using Tenax TA as sorbent.[19]

Preconcentration. If concentration levels of the target compounds are too small for direct HS analysis, preconcentration from the headspace is necessary. Over the past few years several studies have investigated the use of sorptive extraction techniques, in particular solid-phase microextraction (SPME) and purge and trap (P&T) for rather volatile compounds. For guidance in selection of an appropriate preconcentration method, Figure 4.1 gives ranges of achievable limits of detection (LOD) using these approaches in comparison to standard liquid or HS injection.

Note, however, that it is difficult to compare reported LOD among different studies since the numbers given depend substantially on the procedure used to generate these values. Thus, we refrain here from providing comprehensive lists of LOD and, rather, refer to previous overviews for VOCs.[7,20] Furthermore, in Chapter 5 we present an approach to determine LODs in CSIA that we have recently introduced.[21]

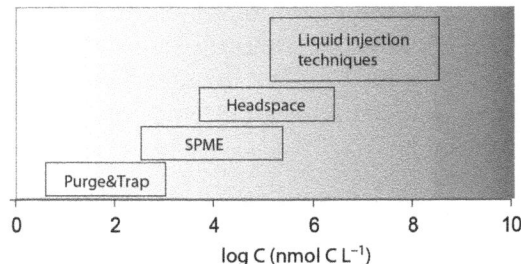

Figure 4.1 Overview of achievable limits of detection for carbon isotope analysis using various preconcentration techniques for VOC.

In the following, we will briefly discuss only investigations that have addressed analytical development issues beyond a mere application. In most cases, only a negligible isotope fractionation within the standard error has been found in the use of both headspace-SPME[17,22] and direct immersion-SPME.[12,17,23–27] Nevertheless, for carbon tetrachloride Zwank et al. found an isotope offset from the reference value as high as −7.1‰ together with a poor precision (SD 2.2‰), which could not be explained.[12] A rather strong isotope shift for carbon tetrachloride was confirmed later (but still not explained) by Palau et al.[28] Smaller, but significant deviations have been found for ethanol and acetaldehyde,[29] tert-butanol[24] and some of the investigated nitroaromatic compounds and anilines.[26] In most but not all cases deviations were negative, i.e., isotope values determined by SPME were more depleted in ^{13}C than the reference values. Stronger and positive deviations from reference values have been found for $\delta^{15}N$ of anilines.[26] In contrast, in a later study by the same group good agreement of SPME and EA data was observed under slightly different extraction conditions and a large equilibrium isotope effect in $^{15}N/^{14}N$ associated with the protonation of substituted anilines was shown. This work demonstrated the potential relevance of pH during extraction on resulting isotope data. However, so far it is not possible to derive any general conclusions or even predictions of such isotope effects based on molecular structure of target compounds, and previous attempts to do so have failed.[28] All discussed findings underline the need for method validation with isotopically characterized working standards mentioned above. SPME fibres used comprise both purely absorptive phases such as polydimethylsiloxane (PDMS) and mixed mode phases such as carboxen-PDMS without notable difference with regard to isotope integrity. Note, however, that in the case of mixed mode fibres competition in sorption with matrix components may influence extraction efficiency and isotope fractionation.[12] Thus, a careful validation is particularly important when investigating complex mixtures of analytes at largely varying concentration levels as is common for oil- or fuel-related contaminations.

Despite the many phase transfer processes involved, P&T has been found in several studies to yield reliable isotope values even at the trace concentration level.[12,18,21,30–32] This is partly due to the almost background-free analysis resulting in flat baselines (except for inevitable column bleed). With commercial P&T methods, the lowest LODs reported so far in CSIA have been achieved, in some cases below 1 µg L^{-1}.[12,21] Still lower LODs have been reported with home-made systems. A continuous-flow P&T system allows the extraction of large and adjustable water samples. In order to facilitate exchange between water and the purge gas helium, water is passed through an ultrasonic nebulizer to form an aerosol.[30] Another approach used a vacuum extraction as first preconcentration step prior to commercial P&T.[31] However, such systems have not yet found widespread application and are not designed to be fully automated, in contrast to commercial P&T systems.

In carbon isotope analysis of volatile halogenated and oxygenated compounds in the atmosphere, LODs in the ppt$_v$ range are required if natural background concentrations are to be measured. Such low LODs have been achieved by multi-bed adsorbent tubes[33] or combinations of cryotraps and adsorbent tubes.[34,35]

Only a few studies address low concentration stable isotope measurements of VOC for hydrogen, in particular of fuel oxygenates with P&T.[32,36] In this case, Kujawinski et al. emphasized the need for a good separation of residual water released from the trap during thermodesorption despite previous dry purging. Water entering the pyrolysis unit led to irreproducible results due to partial consumption of the reducing capacity and generation of additional hydrogen falsifying the hydrogen signal from co-eluting target compounds.[32] Water management was also a critical issue in thermodesorption of alkylated benzenes from atmospheric samples for subsequent hydrogen (and carbon) isotope analysis.[9]

4.3.3 Sample Processing for Semi-volatile Compounds

Extraction. For semi-volatile compounds typically an extraction step by an organic solvent (liquid and solid samples) or a solid phase (liquid and gas samples) is used as the primary step.

For liquid (typically aqueous) samples liquid-liquid extraction (LLE) and solid-phase extraction (SPE) followed by elution with an organic solvent are both rather common. By deploying semi-permeable membrane devices (SPMDs) filled with, for example, triolein directly in an aqueous compartment, sampling and sample extraction can be combined.[37] Moreover, it allows for measuring a time-averaged isotope composition of target compounds. No isotope fractionation by this sampling procedure, even followed by extensive clean-up has been found for carbon and hydrogen isotope analysis of *n*-alkanes and polycyclic aromatic hydrocarbons (PAHs). For solid samples (including particulate matter from the air phase), extraction by the traditional Soxhlet method is still common. Newer methods include ultrasonic-assisted extraction, microwave-assisted extraction and accelerated solvent extraction (ASE). In a comprehensive study of various extraction methods, Graham *et al.* found ASE with dichloromethane best suited for extraction of PAHs due to a low amount of co-extracted, interfering material.[38] ASE was also successfully used for *n*-alkane extraction from sediments.[39]

Clean-up. Regardless of the extraction method, a clean-up of the primary extract will frequently be required since samples contain complex mixtures of analytes or strongly interfering matrix components. In general, UCM that is co-eluting with target compounds causes a problem with background correction if signal sizes of target compounds become small or large differences in isotope composition of UCM and target analytes occurs.[9,40] This has been studied in detail for isotope analysis of PAHs in soil, sediment and aerosol samples, thus PAH analysis will be discussed in more detail here.

Typical clean-up steps comprise several further extraction steps by solid phases, often followed by fractionated elution with various organic solvents, including *n*-pentane, *n*-hexane, toluene and dichloromethane. The most commonly used stationary phases for clean-up are alumina and silica, sometimes both in mixed columns or applied in series. Activated copper on top of the columns is used in many cases to remove residual elemental sulfur. HPLC separation using aminopropyl and/or cyanopropyl functionalized silica with off-line fraction collection,[41–43] gel permeation chromatography to remove large interfering compounds[44] and preparative thin-layer chromatography[44] have also been used. However, Yan *et al.* questioned the use of further HPLC fractionation since incomplete recovery can lead to significant shifts in the observed isotope composition. This was observed particularly for low molecular weight compounds.[45] Before storing sample extracts one needs to validate the materials in contact with the sample for potential of contamination. In particular, phthalates from inappropriate polymers (e.g., septa) can pose a considerable problem as illustrated in Figure 4.2.

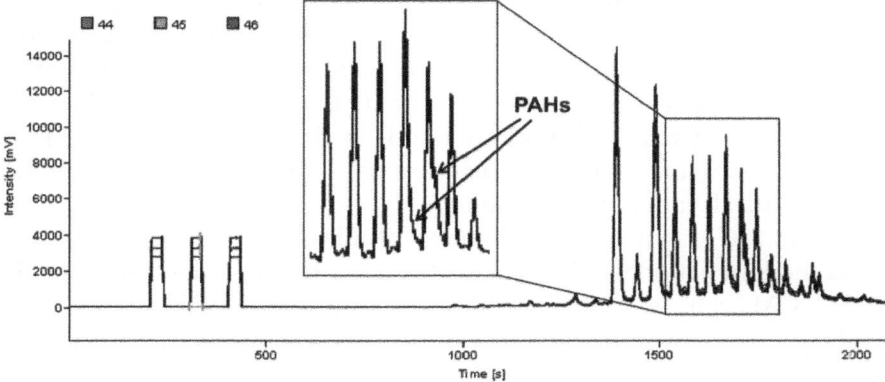

Figure 4.2 GC/IRMS chromatogram of an *n*-pentane extract containing phthalates (main peaks) leached out of septum material. Target peaks from co-eluted PAHs could not be resolved, which inhibited an isotope analysis. (Graphic from Blessing *et al.*,[4] reprinted with permission from Springer.)

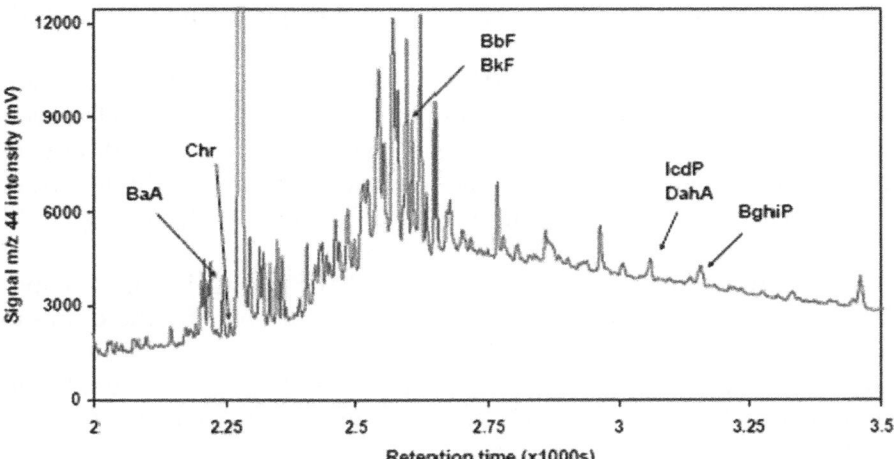

Figure 4.3 Example of a chromatogram obtained for PAH isotope analysis in urban road dust after extraction and two-step clean-up, LVI (large volume injection) of 100 μL.
(Graphic from ref. 46, reprinted with permission from Taylor & Francis.)

Even after extensive clean-up, UCM and non-target individual peaks may still be present in final extracts to a large extent and might even dominate the chromatograms. In these cases, derived δ-values need to be extremely critically evaluated. To give a recent example, Figure 4.3

shows a part of a chromatogram for higher molecular weight PAHs from urban road dust.[46] Obviously, UCM is still present and the authors point out that, in particular, small peaks will be considerably affected by the background isotope composition and thus no isotope values for Benzo[g,h,i]perylene (indicated as BghiP in the figure) are given. Furthermore, baseline separation cannot be achieved in such complex samples for all target compounds.

In the case of source apportionment studies where often rather small differences in isotope signatures have been utilized, co-elution of peaks and UCM can blur the sought-after information. Thus, we warn explicitly against an overinterpretation of small differences in isotope abundance values (less than 1–2‰ even in best case scenarios) in such studies. This is of particular importance when studies rely on just a few samples and compounds at small concentrations and thus small signal sizes.

Validation of Sample Preparation Procedures. As mentioned above, for all extraction and clean-up procedures, isotope integrity, i.e., the absence of isotope fractionation in the sample processing steps, needs to be confirmed.[9–11] Best suited for this purpose would be the use of certified reference materials (CRMs) of the matrix under investigation with the isotope composition of target compounds known. However, no such materials are available so far. Therefore, in most method validations a comparison of isotope composition with and without further clean-up steps is done for compounds not (strongly) affected by UCM co-elution (e.g., refs 38, 40, 42).

Alternatively, standard mixtures of PAHs of known isotope composition can be processed in order to check for any bias by the clean-up step.[44] However, the latter approach cannot capture any effects from the matrix and thus seems to be adequate only as an additional validation step. In the absence of CRMs, a superior alternative is to spike real samples prior to the extraction and clean-up procedure with PAHs of known isotopic composition, which are not present in the real sample and chromatographically well resolved from other compounds. This approach has been described in detail by Abrajano and coworkers.[40,47,48] Typically, they used one aromatic compound (acenaphthylene or acenaphthene) after checking its absence from their samples. In addition they used two *n*-alkanes (n-C_{25} and n-C_{30}) to monitor the reliability of background correction. Walker *et al.* used eicosane (n-C_{20}) and showed rather good between-sample agreement of its measured $\delta^{13}C$-value (standard error ±0.2‰) as well as good agreement with its off-line determined reference $\delta^{13}C$-value (+0.5‰ of the mean).[49] Okuda et al. used acenaphthene-d_{10}, p-terphenyl-d_{14}, and n-C_{34} for the same purpose.[50] Alternatively, it might also be possible to use a standard addition method of the target compound. In this case, various amounts of the isotopically characterized target compound are added to different aliquots of the sample that are measured independently. The isotope composition of the target compound in the original sample is then calculated by an isotope mass balance or by plotting data in a modified Keeling plot.[6] A disadvantage of this approach is the huge measuring effort for each sample.

Similar approaches are also used for clean-up and fractionation of alkanes and biomarkers such as hopanoids and steroids. It was shown that sample clean-up by shape-selective chromatographic separation on a zeolite molecular sieve did not induce carbon isotope fractionation for n-alkanes[51] and biomarkers.[52] Recently, Li et al. suggested the use of hollow fibre liquid-phase microextraction as a one-step sample extraction and clean-up for isotope analysis of n-alkanes from crude oil in water, for example, after a spill event.[53] To date, no application to real spills has been reported, but the use of membrane-based extraction techniques surely will gain importance for the investigation of aqueous samples since interferences by very large organic components can be efficiently and conveniently removed. A further improvement of separation of target compounds in complex mixtures from co-eluting compounds and UCM may be achieved by two-dimensional GC (see Chapter 7).

No similar validation of sample preparation procedures has been reported yet for compound-specific hydrogen (or nitrogen for N-containing compounds) isotope analysis.

Liquid Injection. The knowledge of proper injection techniques in GC-IRMS is necessary because preservation of sample integrity is extremely important. In other words, the transfer from the syringe into the GC-column must be carried out without isotope fractionation.[9,11,54] This means that changes in the sample composition by

(i) mass discrimination of sample compounds,
(ii) thermal decomposition,
(iii) selective adsorption or catalytic decomposition at active sites and
(iv) sample contamination through the injection process by inadequate injection syringe cleaning, residues in the liner (e.g., septum rubber, precipitates) and leaching rubber septa have to be avoided.

As well as these requirements, small solute bands should be obtained during sample introduction to prevent unnecessary peak broadening.

Within the scope of this book, it is impossible to cover the whole field of liquid injection techniques in detail. For an in-depth discussion we refer to the monographs by Konrad Grob.[55,56]

Split/splitless inlet systems are the most widespread type of injectors for liquid injection and can be operated in one of these two modes. A) In splitless injection, 1–2 µL of the organic solvent containing the analytes are injected as a liquid by a microlitre syringe through a septum into a heated vaporization chamber (liner) of the injection port. The injection temperature depends on the chemical nature of the compounds but a fast evaporation requires temperatures at least 50 °C above the boiling point of the solvent and the major high boiling components. Higher temperatures lead to faster evaporation but if the temperature chosen is too high, thermal decomposition of target compounds can occur. Ideal vaporization temperatures and possible compound degradation are discussed in the related GC method development literature. B) In split injection, only a defined fraction of the vaporized sample is transferred onto the head of the column while most of the vaporized sample is removed from the injection port via a split-vent line.

Two construction principles of split/splitless injectors are common. The first is based on backpressure regulation, the second on head-pressure regulation. Instruments interfaced to Agilent™ or Varian™ GCs are fitted with the former type of injector while TraceGCs made by Thermo are fitted with the latter type. Modern injectors also incorporate a septum purge, which is a small flow that prevents septum bleed entering the column. For a fast non-discriminating evaporation, the needle tip position relative to the analytical column is important and depends on the individual type of injector. Furthermore, the column installation depth has to be correct depending on the type of injector and the mode of injection. Proper installation of the column with regard to its position in the injector as well as setting needle penetration depth according to manufacturers' recommendation is mandatory. It is important that the solvent vapour volume does not exceed the capacity of the liner volume. Otherwise, due to a phenomenon called 'flashback', contamination of the septum by condensation and of the carrier gas inlet and septum purge line by back-diffusion can occur. Possible consequences are ghost peaks or high carbon background values, which will compromise accuracy and precision of isotope ratio measurements.

Seemingly contradictory statements exist with regard to isotope fractionation associated with split injection.[54] It has been reported that split injection of gaseous and liquid samples may cause isotope fractionation effects.[57–60] In a dedicated study of injection conditions, Schmitt *et al.* have shown substantial dependency of isotope fractionation on injected analyte mass for injection of pure CO_2 or *n*-hexane but not toluene at split ratios below 1:120.[60] Deviations from the reference value for CO_2 were up to 6‰ at a split ratio of 1:12. In contrast, other authors found no significant fractionation at any split ratio.[61,62] In a study by Smallwood *et al.* four different injection techniques for analysing MTBE samples with GC-IRMS were compared with dual-inlet isotope measurement.[18] They concluded that both split and headspace injection of neat MTBE into a split/splitless injector with large split ratios (100:1) provided accurate results for determining $\delta^{13}C$ values. Other users, including the authors of this book, also found that problems associated with isotope fractionation can occur at split ratios smaller than 20:1.[54] A possible explanation for this could be that at high split ratios the flow control is carried out more precisely and the flow conditions in the injector may cause a mass-dependent isotope fractionation at low split ratios.

Preconcentration. Preconcentration for semi-volatiles typically involves gentle evaporation of the solvent prior to injection. If the solvent–air partition constants are rather small, target compounds are lost, and if evaporative loss exceeds 90% substantial isotope fractionation might occur (see also discussion of isotope fractionation in physical processes in Section 6.6.4). So far, there are hardly any reports on the use of microextraction techniques for semi-volatile compounds and subsequent isotope analysis although the literature is abundant on the use of such techniques for concentration analysis. Remarkable exceptions are the use of SPME for extraction of nitroaromatic compounds[26] and anilines[27] from aqueous samples for carbon and nitrogen isotope analysis. Certainly, the potential of SPME and related techniques has not yet been sufficiently explored in isotope analysis of semi-volatiles.

Sensitivity of isotope analysis of semi-volatiles can also be increased significantly by large volume injection (LVI) of the final solvent extract (50 to 150 µL). Compared with splitless injection of typically 1 µL, this allows a maximum improvement of a factor 50 to 150 (assuming identical peak shapes). Indeed, Mikolajczuk *et al.* confirmed these values in an experimental study for various PAHs,[63] achieving LODs of 0.1 to 0.3 mg L^{-1} in the injected solvent extract. A prerequisite for LVI is to ensure a sufficiently high purity of the extractant to permit its use without compromising the accuracy of the determination. LVI has been successfully applied to the investigation of aerosols[63] and soil samples.[4,64] In these studies it has been demonstrated that careful selection of appropriate solvents and injection conditions is mandatory to avoid

(i) partial or complete loss of more volatile analytes and (ii) severe isotope fractionation. For example, naphthalene and acenaphthene cannot be analysed with LVI when using cyclohexane as solvent due to the insufficient difference in their respective boiling points. Here, n-pentane is better suited as solvent but a cooled PTV (programmed temperature vaporizer) must be used. Blessing reported good recoveries and accurate isotope values for 2- and 3-ring PAHs at 20 °C injector temperature.[64] PAHs of four and more rings resulted in unreproducible isotope signatures under the same injection conditions and cyclohexane was found to be a better suited solvent for these less volatile target compounds.[64] However, at a PTV temperature of 45 °C, Mikolajczuk et al. successfully used n-pentane for LVI of PAHs less volatile than anthracene.[63] Although applications of LVI are still scarce, they likely will gain importance for the isotope analysis of semi-volatile compounds to overcome the limited sensitivity in CSIA.

4.4 DERIVATIZATION

4.4.1 General Considerations and Corrections

GC measurements are limited to compounds with boiling points that are low enough to permit their vaporization without thermal decomposition within the injector and the analytical column. However, high molecular weight substances such as proteins, carbohydrates or lipids as well as substances with various polar chemical functionalities such as pesticides, pharmaceuticals and doping agents have increasingly become the target of CSIA. To make biochemically relevant macromolecules amenable to GC-IRMS they have to be transformed into lower molecular weight subunits such as amino acids, monosaccharides and fatty acids. This cleavage in subunits is often carried out by hydrolysis. However, the monomeric subunits generally possess one or more bipolar functional groups with H-donor and H-acceptor properties such as hydroxyl (-OH), amino ($-NH_2$), thiol (-SH) or organic acid (-COOH) groups. Such compounds need to be derivatized prior to GC-IRMS. A considerable number of applications in which derivatization is a main part of the analytical procedure prior to isotope ratio determination are presented in Chapter 6. Prominent examples are the measurement of fatty acid methyl esters (FAMEs) and steroids.

As discussed in Chapter 3, GC columns with highly polar stationary phases such as nitroterephthalic acid modified polyethyleneglycole, so-called 'free fatty acid phases' (FFAP) and polyethyleneglycole based (WAX) coatings for the separation of underivatized organic acids and alcohols are commercially available. However, these phases have maximum operating temperatures around 250 °C, which can lead to high levels of column bleed, long analysis times resulting in peak broadening and reduced resolution, and thus limit the molecular size of the target compounds.[11] This was observed particularly for compounds with carbon numbers exceeding C_{24}.[65] In such cases, chemical modification by derivatization of the polar compounds into less polar and more volatile derivatives can improve chromatographic separation but has to be handled with care if it is applied for compound-specific isotope ratio measurements by GC-IRMS. Although a huge number of derivatization methods exist for GC analysis, only a few of these have been utilized in GC-IRMS.[66]

Derivatization reactions follow the general equation:

$$\text{Compound} + \text{Derivative} \rightarrow \text{Derivatized compound} \qquad (4.1)$$

where the compound, for example an amino acid, is derivatized by adding a derivatization reagent or derivative. The result of this reaction is the derivatized compound. To simplify the following discussion, in this section the compound will be abbreviated by the index c, the derivative group by d and the derivatized compound by dc. Prerequisites for the derivatization are that (i) the isotope composition of the introduced derivative group must be constant and reproducible, (ii) the derivatization reagent should not have an adverse effect on the conversion process into measurable gas species,[54] and (iii) as discussed in Chapter 3 complete chromatographic baseline resolution of the compounds to be detected is necessary, a condition that must also be met by the derivatized compounds.[67]

Derivatization can introduce a variety of uncertainties when used in CSIA. Therefore, the applied derivatization reaction has to be chosen carefully, taking into account various potential scenarios that are depicted in Figure 4.4 and explained below.

Scenario 1. If a derivatization reagent is used that does not contain atoms of the element measured and no bonds of this element are involved in the rate-limiting step of the derivatization reaction, no *KIE* will occur, and the reaction can be used without correction since the isotope signature of the compound will not be altered by the introduction of the derivative.

Scenario 2. If the derivatization reagent contains atoms of the element measured but no bonds of this element are involved in the reaction so that no *KIE* occurs, a correction of the compound's isotope value is necessary. The change of the isotope value is caused only by the dilution

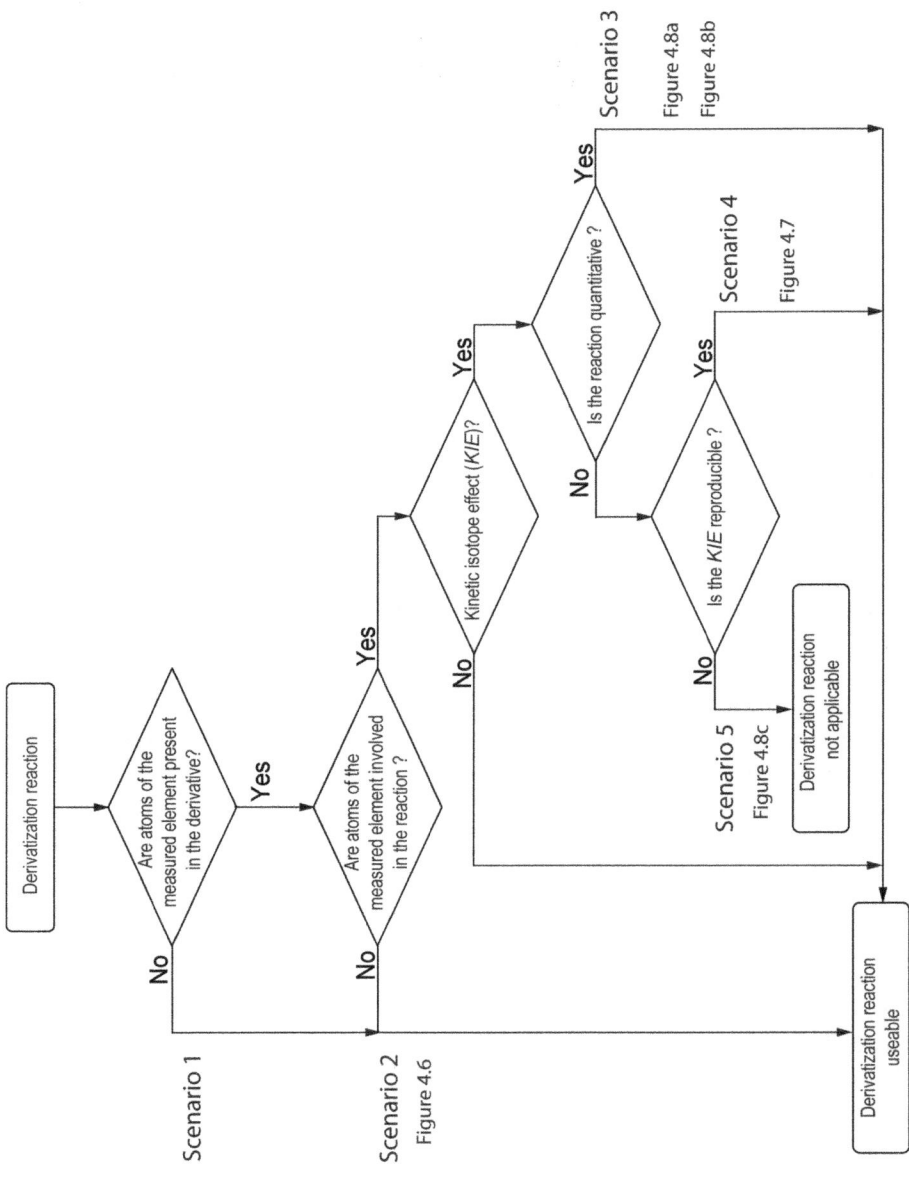

Figure 4.4 Decision scheme for the applicability of a derivatization method.

of the original isotope value by addition of isotopes of the same element. This situation comprises, for example, silylation of -NH and -OH groups,[65] in which the silicon atom reacts with the oxygen or nitrogen atom of the compound's functional group.

The introduction of additional atoms of the determined element can be corrected for, to obtain the original isotope signature of the target compound. For the case where only one kind of derivative group is added to one functional group in the compound a simple mass balance such as given by eqn (4.2) can be employed to correct the compound's δ-value,

$$\delta^h E_c = \frac{n_{dc}\delta^h E_{dc} - n_d \delta^h E_d}{n_c} \quad (4.2)$$

in which n is the number of atoms of the measured element in the compound, derivative and derivatized compound. Note that if more than one identical functional group is present in the compound, n_d refers to the total number of atoms of the measured element in the introduced derivatives. For the correction, the isotope ratios of the derivatized compound and of the derivatization reagent have to be measured independently. The uncertainty associated with the derivatization can be expressed by Gaussian error propagation:

$$\sigma^2_{\delta^h E_c} = \sigma^2_{\delta^h E_{dc}} \left(\frac{n_c + n_d}{n_c}\right)^2 + \sigma^2_{\delta^h E_d} \left(\frac{n_d}{n_c}\right)^2 \quad (4.3)$$

The uncertainty induced in the compound's δ-value depends on the compound-to-derivative element abundance ratio.

Using carbon as the most important element in CSIA, Figure 4.5 shows the uncertainty introduced by addition of various numbers of derivative carbon atoms n_d by the derivatization agent. The obtained curves are calculated by using eqn (4.3), where the uncertainty of the derivatized compound $\sigma_{\delta^h E_{dc}}$ measured by GC-IRMS is assumed to be ± 0.3‰ and the uncertainty of the derivatization reagent $\sigma_{\delta^h E_d}$ measured by EA-IRMS or dual-inlet is assumed to be ± 0.1‰. To give an example, we consider the common derivatization of fatty acids using BF_3-methanol for carbon isotope measurement. For palmitic acid ($C_{16:0}$), ($n_c = 16$, $n_d = 1$) application of eqn (4.3) and assuming the abovementioned uncertainties, an uncertainty in the compound's δ-value $\sigma_{\delta^h E_c}$ of 0.32‰ results, i.e., the measurement uncertainty is hardly altered by the introduced derivative carbon. When the number of carbon atoms introduced by derivatization rises, meaning that the compound-to-derivative element abundance ratio becomes smaller, the introduced

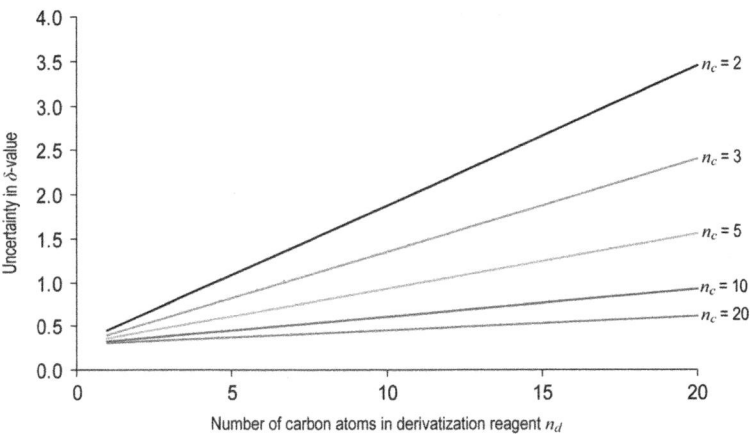

Figure 4.5 Uncertainty induced in the measured δ-value of a compound through the addition of derivative carbon.

uncertainty increases. It is therefore recommended to keep the number of additionally introduced carbon atoms as low as possible.

Sometimes it is necessary for more than one functional group to be derivatized. In such situations, a more general case can be given. A compound possesses z different functional groups that are derivatized with z different derivatives.[65] Then,

$$\delta^h E_c = \frac{\left(n_c + \sum_{i=1}^{z} n_{d_i}\right) \delta^h E_{dc} - \sum_{i=1}^{z} n_{d_i} \delta^h E_{d_i}}{n_c} \quad (4.4)$$

is valid and the associated contribution to the uncertainty is given by:[65]

$$\sigma^2_{\delta^h E_c} = \sum_{i=1}^{z} \frac{\sigma^2_{\delta^h E_{d_i}} n_{d_i}^2}{n_c^2} + \sigma^2_{\delta^h E_{dc}} \left(\frac{n_c + n_d}{n_c}\right)^2 \quad (4.5)$$

Scenario 3. Derivatization reactions can be associated with *KIE*s. As pointed out in Chapter 2 such effects can cause a measurable isotope fractionation, altering the original compound's isotope ratio during derivatization. For derivatization reactions only primary *KIE*s are significant, which are associated with bonds containing the element measured being involved in the rate-determining step.[65,67] If a reaction is accompanied by a *KIE* but the reaction is quantitative as far as the target compound is concerned so that the isotopic difference $\Delta\delta^h E_{c/dc} = 0$ then no isotope fractionation with regard to the target compound will occur. In such cases, the mass balance considerations in Scenario 2 apply, and eqns. (4.2) to (4.5) can be used for incorporation of extra atoms of the measured element.

Scenario 4. In cases where the reaction is not quantitative, the *KIE* for the derivatization reaction can be expressed as described in Chapter 2. These *KIEs* have to be corrected for; factors for *KIE* correction can be defined as the apparent isotope composition of the derivative element, introduced through derivatization, that takes into account the isotope fractionation that is associated with the reaction.[67]

If a reproducible *KIE* is involved in the derivatization reaction, the correction factor can be determined indirectly, by measuring the δ-value of an underivatized standard of the target molecule (e.g., by closed-tube combustion and dual-inlet-IRMS or EA-IRMS) as well as the δ-value of the standard after derivatization (by GC-IRMS) and using a modified mass balance equation. If the indirect method to determine the δ-value of the derivative groups is employed, the following equation can be used:

$$\delta^h E_d = \frac{n_{ds}\delta^h E_{ds} - n_s \delta^h E_s}{n_d} \qquad (4.6)$$

where n_d is the total number of atoms of the measured element in the added derivatives, n_{ds} and n_s are the number of atoms of the element of interest in the derivatized standard and in the standard, respectively and $\delta^h E_{ds}$ and $\delta^h E_s$ are the δ-values of the derivatized standard and the standard, respectively. Subsequently, the δ-value of individual compounds from natural materials $\delta^h E_c$ can be determined by introducing the value for $\delta^h E_d$ obtained by eqn (4.6) into eqn (4.2). The uncertainty associated with the indirect determination of $\delta^h E_c$ in the presence of a reproducible *KIE* is given by:[67]

$$\sigma^2_{\delta^h E_c} = \sigma^2_{\delta^h E_s}\left(\frac{n_s}{n_c}\right)^2 + \sigma^2_{\delta^h E_{ds}}\left(\frac{n_s + n_d}{n_c}\right)^2 + \sigma^2_{\delta^h E_{dc}}\left(\frac{n_c + n_d}{n_c}\right)^2 \qquad (4.7)$$

In many cases, n_s and n_c will be equivalent, i.e., the same compound is used as standard to derive the correction factor. To give an example, we consider a deoxy sugar such as rhamnose, for which $n_s = n_c = 6$ carbon atoms. Rhamnose contains 5 OH- groups which, upon the frequently used acetylation, will be derivatized with $n_d = 10$ carbon atoms. Assuming again the above-mentioned uncertainties error propagation using eqn (4.7) yields an overall uncertainty in the compound's δ-value $\sigma_{\delta^h E_c}$ of 1.3‰. A good example of the indirect approach with the necessary corrections is given by Sauvinet *et al.* for plasma glucose.[68]

For the general case, when several derivatization reagents are involved that have been determined by the indirect method, the uncertainty can be written as:

$$\sigma^2_{\delta^h E_c} = \sum_{i=1}^{z} \frac{\sigma^2_{\delta^h E_{s_i}} n_s^2}{n_c^2} + \sum_{i=1}^{z} \frac{\sigma^2_{\delta^h E_{sd_i}} (n_s + n_{d_i})^2}{n_c^2} + \sigma^2_{\delta^h E_{dc}} \left(\frac{n_c + \sum_{i=1}^{z} n_{d_i}}{n_c}\right)^2 \quad (4.8)$$

Giving again a common example, we consider the derivatization of the amino acid alanine ($n_s = n_c = 3$ carbon atoms) by trifluoroacetate/isopropanol ($n_{d_1} = 2$, $n_{d_2} = 3$). Error propagation using eqn (4.8) now yields an overall uncertainty in the compound's δ-value $\sigma_{\delta^h E_c}$ of 1.3‰.

Scenario 5. Finally, if the *KIE* is not reproducible the method cannot be used for derivatization and another derivatization strategy has to be found.

4.4.2 Overview of Derivatization Reactions

Volatilization by Reduction. The formation of hydrides by reduction is used in the analysis of organometal(loid) compounds by GC-ICPMS.[69] To give an example, eqn (4.9) shows the reduction of trimethylarsine oxide (TMAsO) by sodium borohydride to the volatile trimethylarsine (TMAs). TMAs can be purged via a helium gas stream and enriched in a cryo trap before it is separated from other volatile organic compounds by GC.

$$TMAsO + NaBH_4 + HCl + 2H_2O \rightarrow TMAs \uparrow + B(OH)_3 + NaCl + 3H_2 \uparrow \quad (4.9)$$

The method can be applied to volatilize compounds for carbon isotope ratio analysis because no additional carbon is introduced into the molecule; a quantitative derivatization can be achieved, even in complex matrices, and the derivatization reagent reacts directly with the metal(loid); thus, no isotope fractionation is expected for carbon. Because of the volatility of the derivatized compound a simple separation from the matrix is possible.[69]

Corso *et al.* reported the reduction of FAMEs by lithium aluminium hydride (LiAlH$_4$) to fatty alcohols for chromatographic separation and subsequent pyrolytic position-specific isotope analysis (PSIA).[70]

Volatilization by Oxidation. In some cases, an oxidation rather than reduction of compounds may generate derivatized compounds more suitable for GC-IRMS analysis. In steroid analysis, oxidation of hydroxyl to ketone functional groups by potassium dichromate has been used to convert corticosteroids to their corresponding triones with simultaneous loss of the 17β side chain.[71,72] An advantage of the method is that no extraneous carbon is introduced by this derivatization step. An isotope fractionation by breakage of the carbon-carbon bond in the loss of the side chain has not been observed.

Decarboxylation. Zaideh et al. used enzymatic decarboxylation of tyrosine and phenylalanine to enhance the volatility of these compounds for $\delta^{15}N$ analysis by GC-IRMS.[73] Such enzyme-catalysed reactions are highly specific and less prone to side reactions.[73]

Silylation. Silylation is a very common derivatization technique in GC-MS analysis.[74] An advantage of silylation is that the reaction can be carried out quantitatively in a rapid one-step reaction without the necessity for product clean-up.[66] Silylation reagents will derivatize almost all OH- and NH- groups, thus it is important that they do not come into contact with stationary phases containing these functional groups, such as polyethylene glycole phases. Trimethylsilyl (TMS) chloride, tert-butyldimethylsilyl (*t*BDMS) chloride or BSTFA (N,O-bis(trimethylsilyl)trifluoroacetamide) are widely applied to derivatize amino acids in carbon and nitrogen isotope ratio analysis by GC-IMRS[75,76] but also for the derivatization of fatty acids.[77] In Figure 4.6, the silylation of an alcohol by trimethylsilyl chloride is shown as an example.

The addition of a base such as pyridine or trimethylamine promotes the silylation reaction and a quantitative reaction can be achieved.[77] For carbon isotope ratio analysis the method has the advantage that no carbon bonds are affected during the silylation and thus no carbon isotope fractionation occurs.[8,65] Silylation is generally carried out with a huge excess of derivatization reagent, so that it is not likely to cause nitrogen and oxygen isotopic fractionation because of the complete reaction of the target compounds.[77] Nevertheless, silylation involves the addition of excessive numbers of carbon atoms, three for TMS and six for *t*BDMS per derivatized functional group, thus reducing the precision of carbon isotope determination[78] and can lead to incomplete compound conversion,[9] especially when carbohydrates with several hydroxyl groups are measured.[78] It is also suspected that silicon is an oxidation catalyst poison due to formation of silicon oxide or, more likely, silicon carbide in the combustion oven leading to additional uncertainties in isotope values.[79] The derivatization of amino acids and α-ketocarboxylic acids can result in the formation of several different compounds such as mono-, di-, tri-TMS esters and *cis/trans*-isomers, when TMS chloride is

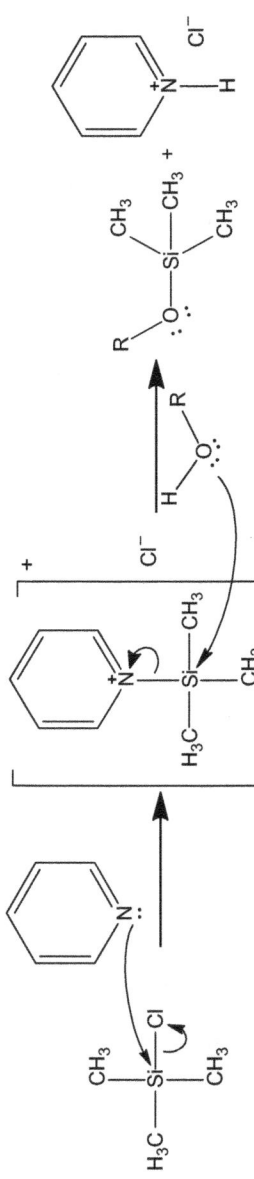

Figure 4.6 Reaction mechanism for silylation of an alcohol. In the first step of the reaction a cation with the base pyridine is formed under abstraction of chloride from the trimethylsilyl chloride. Then a nucleophilic attack of a free electron pair of the oxygen atom of an alcohol occurs at the silicon atom.

used.[8] Through the derivatization with a silylation reagent the choice of analytical columns is limited to apolar phases (see Chapter 3), which can lead to separation problems for complex mixtures.[8] Another disadvantage of this method is the instability of the derivatization reagents, which decompose through humidity and have to be stored in septum-sealed bottles under nitrogen atmosphere. Moreover, the derivatized samples have only a limited storage lifetime (a few days even if stored in the dark at subzero temperatures).[8]

Esterification. Reactions such as methylation, acetylation, isopropylation and transesterification can be summarized as esterification. All these reactions involve a carbon centre, which is located in the final derivatized compound so that a *KIE* can possibly affect determination of carbon isotope ratios.[65] Two cases have to be distinguished:

(i) In most esterification reactions the derivatization reagent, often an alcohol with a catalyst, is added in excess. The atom at which the reaction occurs is in the compound and the reaction is fast and quantitative so that no *KIE* affecting the compound per se will be observed and only corrections according to eqns (4.2) and (4.3) are necessary. For methylation reactions see separate section below.

(ii) In acetylation reactions the situation is different. Here, the atom involved in the rate-determining step, is located in the derivatization reagent. Due to this, even if the compound of interest reacts quantitatively, a *KIE* at the acetyl carboxy carbon is possible.[65] Acetylation is frequently used for carbon isotope ratio analysis of steroids (see Chapter 6) and carbohydrates.[80] An acetylation such as shown in Figure 4.7 is usually carried out with an excess of acid (e.g., acetic acid) or an anhydride and a base such as pyridine[80] or *n*-methylimidazole [67] for mediating the reaction.

In this type of reaction the *KIE* is constant[80] but a correction for the additional carbon is necessary. However, if alditols are measured, generally a reduction of keto-groups by, for example, sodium borohydrate is used, so that in the case of acetylation, a loss of information occurs. Two different pentoses or hexoses can result in the same product. This information loss can be prevented by methyl-boroacetylation and adjacent silylation with TMS chloride. This reaction is quantitative and no measurable *KIE* occurs.[78]

For the derivatization of amino acids trifluoroacetate/isopropanol (TFA/IP) has been established as the most common two-step derivatization reagent for carbon isotope analysis.[81] However, a correction of the *KIE* is necessary. Furthermore, during the combustion/thermolysis of TFA derivatized compounds HF is formed which yields stable CuF_2

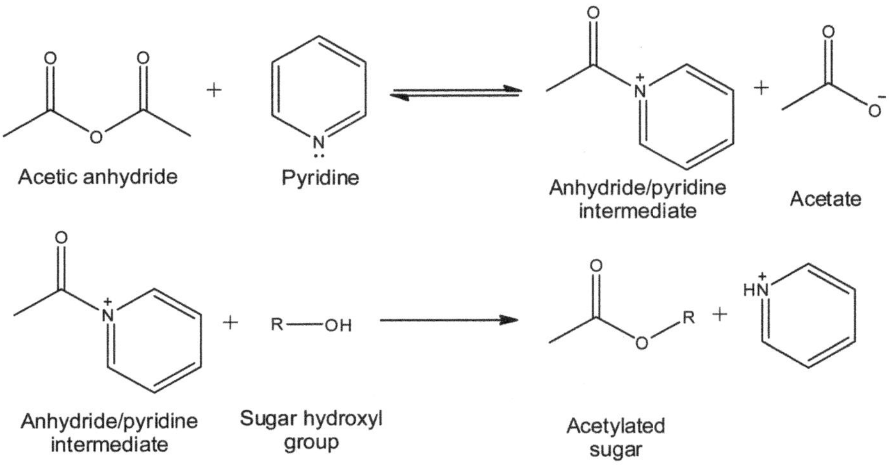

Figure 4.7 Mechanism of an acetylation reaction. In the first step of the reaction an anhydride/pyridine intermediate and acetate are formed from the acetic anhydride and pyridine. In a second step the anhydride/pyridine intermediate acetylates hydroxyl groups (e.g., of sugars).

and NiF_2 with the CuO and NiO wires in the combustion oven and irreversibly poisons platinum.[8]

Corr et al. investigated the derivatization of 15 amino acids into N-acetylmethyl (NACME) esters, N-acetyl n-propyl (NANP) esters, N-acetyl i-propyl (NAIP) esters, N-trifluoroacetyl-i-propyl (TFA-IP) esters, N-pivaloyl methyl (NPME) esters, N-pivaloyl n-propyl (NPNP) esters and N-pivaloyl i-propyl (NPIP) esters.[82] They concluded that the NACME derivative was best suited, mainly because it has the highest compound-to-derivative carbon abundance ratio, resulting in the lowest analytical errors for amino acid $\delta^{13}C$-value determinations, ranging from ±0.6‰ for phenylalanine, leucine and isoleucine to ±1.1‰ for serine and glycine.

Saudan et al. reported a method for GC-IRMS of γ-hydroxybutyric acid (GHB) after intramolecular esterification to the corresponding γ-butyrolactone (GBL) to avoid introduction of extraneous carbon by a derivatizing reagent.[83]

Methylation. In GC-MS analysis the methylation by BF_3-methanol and diazomethane is often used.[74] Figure 4.8 shows different methylation methods. Methylation has the advantage that carbon measurements are affected by the introduction of only one carbon atom with the derivative group.[76,77] The methylation shown in Figure 4.8a has been widely used in GC-IRMS.[8] Generally, commercially available

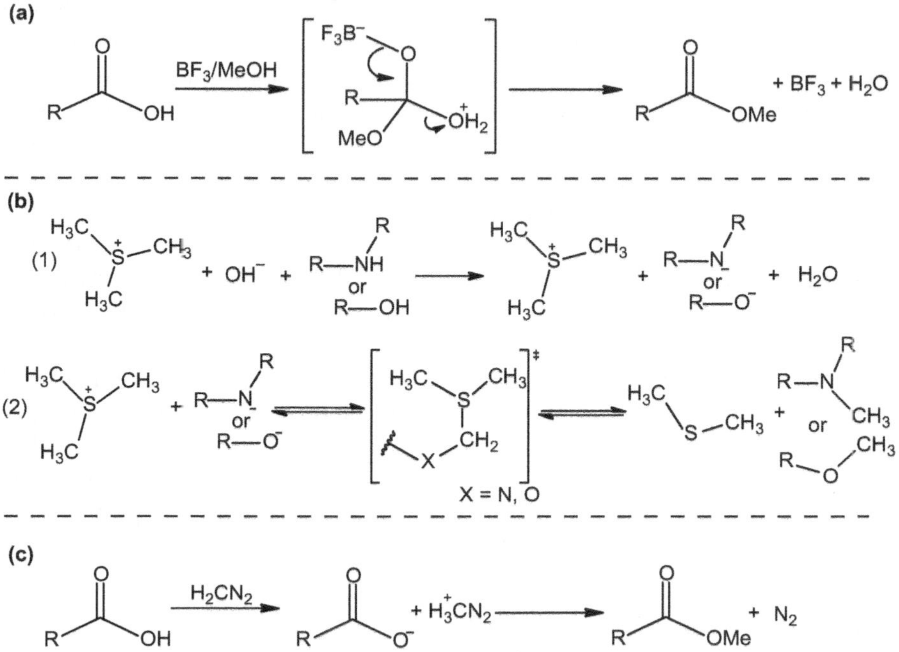

Figure 4.8 Different methylation reactions used for derivatization. a) Methylation of a carboxylic acid with BF$_3$-methanol. b) Methylation of H-acidic groups such as amino or hydroxyl groups with trimethylsulfonium hydroxide (TMSH). c) Methylation of a carboxylic acid with diazomethane.

BF$_3$-methanol reagent has been employed to transfer fatty acids into their corresponding methyl esters (FAMEs).[8]

The reaction takes place at the carbon centre of the fatty acid so that in the case of carbon measurements a *KIE* can be expected in principle. However, in this reaction an excess of the alcohol is employed and the reaction is generally complete, so that no isotope fractionation is observed at the carbon centre. Other advantages of this method are that the BF$_3$-methanol complex is commercially available (14% (w/v),[77] is easy to handle and the Lewis acid BF$_3$ that catalyses the reaction is non-corrosive and very volatile, thus it can be removed by purging before analysis of the derivatized compound. The use of BF$_3$ as catalyst instead of HCl or HBr avoids problems with corrosion in the analytical system.[77] However, artefacts and side reactions when using BF$_3$-methanol have been reported so that this reaction, too, has to be used carefully.[76]

Recently, Reinnicke *et al.* used trimethylsulfonium hydroxide (TMSH) for the methylation of the herbicides 3-isopropyl-1H-2.1.3-benzothiadia-zone-4(3H)-one (bentazone) and (4-chloro-2-methylphenoxy) acetic acid

(MCPA) (see Figure 4.8b).[76] An interesting aspect of this methylation reaction is that it was carried out automated in the packed glass bead liner of a PTV injector using LVI. To that end, the target compounds were pre-mixed with derivatization reagent and solvent (methanol) before injection. After injection, the solvent was removed through the split-vent. During a subsequent flash heating of the injector, the derivatization reaction took place, after which the derivatized molecules were transferred onto the analytical column.[76] Due to the fact that no nitrogen is introduced during derivatization, the stoichiometric addition of TMSH gave accurate and reproducible nitrogen isotope values. In contrast, reproducible carbon isotope values needed a $\geq$250-fold excess of TMSH.[76]

Methylation with diazomethane, as shown in Figure 4.8c, is fast and irreversible but the reagent has to be used in excess and in the rate-limiting step the reaction occurs in contrast to the methylation with BF_3-methanol on the carbon of the reagent diazomethane, which leads to a non-reproducible *KIE*.[65,76] Therefore, the method is not applicable for CSIA measurements of carbon. Furthermore, diazomethane is explosive and highly toxic.[76]

Concluding Remarks on Derivatization. In summary, it can be said that derivatization extends the field of GC-IRMS but a rigorous control of the derivatization reaction with regard to potential isotope fractionation is necessary. A few isotopically characterized derivatizing reagents, such as phthalic acid and acetic anhydride (normalized against two internationally accepted standards) can be obtained from Indiana University. It was pointed out by Sessions that there is a clear need for further types of such standards.[54] To date, most derivatization reagents still have to be isotopically characterized by the user as discussed above or as mentioned in more general terms in Chapter 5. Further application examples of derivatization reactions are given in Chapter 6.

REFERENCES

1. M. Thompson, ed., *The Duplicate Method for the Estimation of Measurement Uncertainty Arising from Sampling, Royal Society of Chemistry*, 2009.
2. M. H. Ramsey and S. L. R. Ellison, eds., *Eurachem/EUROLAB/ CITAC/Nordtest/AMC Guide-Measurement Uncertainty Arising from Sampling—A Guide to Methods and Approaches http://www.eurachem.org/guides/UfS_2007.pdf*, Eurachem, 2007.

3. M. Thompson, ed., *Measurement Uncertainty Arising from Sampling: The New Eurachem Guide*, Royal Society of Chemistry, London, 2008.
4. M. Blessing, M. A. Jochmann and T. C. Schmidt, *Anal. Bioanal. Chem.*, 2008, **390**, 591–603.
5. M. Elsner, G. L. Couloume and B. Sherwood Lollar, *Anal. Chem.*, 2006, **78**, 7528–7534.
6. T. Piper, U. Mareck, H. Geyer, U. Flenker, M. Thevis, P. Platen and W. Schanzer, *Rapid Commun. Mass Spectrom.*, 2008, **22**, 2161–2175.
7. D. Hunkeler, R. U. Meckenstock, T. C. Schmidt, B. Sherwood Lollar and J. T. Wilson, *A Guide for Assessing Biodegradation and Source Identification of Organic Ground Water Contaminants using Compound Specific Isotope Analysis (CSIA)* EPA 600/R-08/148, U.S. Environmental Protection Agency, Office of Research and Development, Ada, OK, 2008.
8. W. Meier-Augenstein, in *Handbook of Stable Isotope Analytical Techniques*, ed. P. A. de Groot, Elsevier B.V., Amsterdam, 2004, vol. I, pp. 153–176.
9. W. Meier-Augenstein, *J. Chromatogr. A*, 1999, **842**, 351–371.
10. J. T. Brenna, T. N. Corso, H. J. Tobias and R. J. Caimi, *Mass Spectrom. Rev.*, 1997, **16**, 227–258.
11. W. Meier-Augenstein, *LC-GC Europe*, 1997, **10**, 17–25.
12. L. Zwank, M. Berg, T. C. Schmidt and S. B. Haderlein, *Anal. Chem.*, 2003, **75**, 5575–5583.
13. J. R. Gray, G. Lacrampe-Couloume, D. Gandhi, K. M. Scow, R. D. Wilson, D. M. Mackay and B. Sherwood Lollar, *Environ. Sci. Technol.*, 2002, **36**, 1931–1938.
14. S. A. Mancini, A. C. Ulrich, G. Lacrampe-Couloume, B. E. Sleep, E. A. Edwards and B. Sherwood Lollar, *Appl. Environ. Microbiol.*, 2003, **69**, 191–198.
15. G. F. Slater, H. S. Dempster, B. Sherwood Lollar and J. Ahad, *Environ. Sci. Technol.*, 1999, **33**, 190–194.
16. J. A. M. Ward, J. M. E. Ahad, G. Lacrampe-Couloume, G. F. Slater, E. A. Edwards and B. Sherwood Lollar, *Environ. Sci. Technol.*, 2000, **34**, 4577–4581.
17. D. Hunkeler and R. Aravena, *Environ. Sci. Technol.*, 2000, **34**, 2839–2844.
18. B. J. Smallwood, R. P. Philp, T. W. Burgoyne and J. D. Allen, *Environ. Forensics*, 2001, **2**, 215–221.
19. C. Vitzthum von Eckstaedt, K. Grice, M. Ioppolo-Armanios, G. Chidlow and M. Jones, *J. Chromatogr. A*, 2011, **1218**, 6511–6517.

20. T. C. Schmidt, L. Zwank, M. Elsner, M. Berg, R. U. Meckenstock and S. B. Haderlein, *Anal. Bioanal. Chem.*, 2004, **378**, 283–300.
21. M. A. Jochmann, M. Blessing, S. B. Haderlein and T. C. Schmidt, *Rapid Commun. Mass Spectrom.*, 2006, **20**, 3639–3648.
22. H. Dayan, T. Abrajano, N. C. Sturchio and L. Winsor, *Org. Geochem.*, 1999, **30**, 755–763.
23. R. F. Dias and K. H. Freeman, *Anal. Chem.*, 1997, **69**, 944–950.
24. D. Hunkeler, B. J. Butler, R. Aravena and J. F. Barker, *Environ. Sci. Technol.*, 2001, **35**, 676–681.
25. L. Zwank, M. Berg, M. Elsner, T. C. Schmidt, S. B. Haderlein and R. P. Schwarzenbach, *Environ. Sci. Technol.*, 2005, **39**, 1018–1029.
26. M. Berg, J. Bolotin and T. B. Hofstetter, *Anal. Chem.*, 2007, **79**, 2386–2393.
27. M. Skarpeli-Liati, A. Turgeon, A. N. Garr, W. A. Arnold, C. J. Cramer and T. B. Hofstetter, *Anal. Chem.*, 2011, **83**, 1641–1648.
28. J. Palau, A. Soler, P. Teixidor and R. Aravena, *J. Chromatogr. A*, 2007, **1163**, 260–268.
29. K. Yamada, R. Hattori, Y. Ito, H. Shibata and N. Yoshida, *Isot. Environ. Health Stud.*, 2010, **46**, 392–399.
30. N. R. Auer, B. U. Manzke and D. E. Schulz-Bull, *J. Chromatogr. A*, 2006, **1131**, 24–36.
31. H. I. F. Amaral, M. Berg, M. S. Brennwald, M. Hofer and R. Kipfer, *Environ. Sci. Technol.*, 2010, **44**, 1023–1029.
32. D. M. Kujawinski, M. Stephan, M. A. Jochmann, K. Krajenke, J. Haas and T. C. Schmidt, *J. Environ. Monit.*, 2010, **12**, 347–354.
33. M. I. Mead, M. A. H. Khan, I. D. Bull, I. R. White, G. Nickless and D. E. Shallcross, *Environ. Chem.*, 2008, **5**, 340–346.
34. M. E. Archbold, K. R. Redeker, S. Davis, T. Elliot and R. M. Kalin, *Rapid Commun. Mass Spectrom.*, 2005, **19**, 337–342.
35. B. M. Giebel, P. K. Swart and D. D. Riemer, *Anal. Chem.*, 2010, **82**, 6797–6806.
36. T. Kuder, J. T. Wilson, P. Kaiser, R. Kolhatkar, P. Philp and J. Allen, *Environ. Sci. Technol.*, 2005, **39**, 213–220.
37. Y. Wang, Y. Huang, J. N. Huckins and J. D. Petty, *Environ. Sci. Technol.*, 2004, **38**, 3689–3697.
38. M. C. Graham, R. Allan, A. E. Fallick and J. G. Farmer, *Sci. Total Environ.*, 2006, **360**, 81–89.
39. M. Zech and B. Glaser, *Rapid Commun. Mass Spectrom.*, 2008, **22**, 135–142.
40. V. P. O'Malley, T. A. Abrajano and J. Hellou, *Org. Geochem.*, 1994, **21**, 809–822.

41. T. Okuda, H. Kumata, H. Naraoka and H. Takada, *Org. Geochem.*, 2002, **33**, 1737–1745.
42. L. Mazeas and H. Budzinski, *Org. Geochem.*, 2002, **33**, 1253–1258.
43. P. J. Yanik, T. H. O'Donnell, S. A. Macko, Y. Qian and M. C. Kennicutt II, *Org. Geochem.*, 2003, **34**, 291–304.
44. M. K. Kim, M. C. Kennicutt and Y. R. Qian, *Environ. Sci. Technol.*, 2005, **39**, 6770–6776.
45. B. Yan, T. A. Abrajano, R. F. Bopp, L. A. Benedict, D. A. Chaky, E. Perry, J. Song and D. P. Keane, *Org. Geochem.*, 2006, **37**, 674–687.
46. A. Mikolajczuk, E. P. Przyk, B. Geypens, M. Berglund and P. Taylor, *Isot. Environ. Health Stud.*, 2010, **46**, 2–12.
47. A. Smirnov, T. A. Abrajano, A. Smirnov and A. Stark, *Org. Geochem.*, 1998, **29**, 1813–1828.
48. A. Stark, T. Abrajano, J. Hellou and J. L. Metcalf-Smith, *Org. Geochem.*, 2003, **34**, 225–237.
49. S. E. Walker, R. M. Dickhut, C. Chisholm-Brause, S. Sylva and C. M. Reddy, *Org. Geochem.*, 2005, **36**, 619–632.
50. T. Okuda, H. Kumata, H. Naraoka and H. Takada, *Geochem. J.*, 2004, **38**, 89–100.
51. K. Grice, R. de Mesmay, A. Glucina and S. Wang, *Org. Geochem.*, 2008, **39**, 284–288.
52. F. Kenig, B. N. Popp and R. E. Summons, *Org. Geochem.*, 2000, **31**, 1087–1094.
53. Y. Li, Y. Xiong, J. Fang, L. Wang and Q. Liang, *J. Chromatogr. A*, 2009, **1216**, 6155–6161.
54. A. L. Sessions, *J. Sep. Sci.*, 2006, **29**, 1946–1961.
55. K. Grob, *Classical Split and Splitless Injection in Capillary Gas Chromatography: with some remarks on PTV injection*, 2nd edn., Huethig, Heidelberg, 1988.
56. K. Grob, *Split and Splitless Injection for Quantitative Gas Chromatography: Concepts, Processes, Practical Guidelines, Sources of Error*, 4th completely rev. edn., Wiley-VCH, Weinheim; Chichester, 2001.
57. S. A. Baylis, K. Hall and E. J. Jumeau, *Org. Geochem.*, 1994, **21**, 777–785.
58. B. Glaser and W. Amelung, *Rapid Commun. Mass Spectrom.*, 2002, **16**, 891–898.
59. W. Meier-Augenstein, P. W. Watt and C. D. Langhans, *J. Chromatogr. A*, 1996, **752**, 233–241.
60. J. Schmitt, B. Glaser and W. Zech, *Rapid Commun. Mass Spectrom.*, 2003, **17**, 970–977.
61. Y. Wang and Y. Huang, *Org. Geochem.*, 2001, **32**, 991–998.

62. M. Li, Y. Huang, M. Obermajer, C. Jiang, L. R. Snowdon and M. G. Fowler, *Org. Geochem.*, 2001, **32**, 1387–1399.
63. A. Mikolajczuk, B. Geypens, M. Berglund and P. Taylor, *Rapid Commun. Mass Spectrom.*, 2009, **23**, 2421–2427.
64. M. Blessing, Eberhard-Karls-University, 2008.
65. G. Rieley, *Analyst*, 1994, **119**, 915–919.
66. L. T. Corr, R. Berstan and R. P. Evershed, *Rapid Commun. Mass Spectrom.*, 2007, **21**, 3759–3771.
67. G. Docherty, V. Jones and R. P. Evershed, *Rapid Commun. Mass Spectrom.*, 2001, **15**, 730–738.
68. V. Sauvinet, L. Gabert, D. Qin, C. Louche-Pélissier, M. Laville and M. Désage, *Rapid Commun. Mass Spectrom.*, 2009, **23**, 3855–3867.
69. O. Würfel, R. A. Diaz-Bone, M. Stephan and M. A. Jochmann, *Anal. Chem.*, 2009, **81**, 4312–4319.
70. T. N. Corso, B. A. Lewis and J. T. Brenna, *Anal. Chem.*, 1998, **70**, 3752–3756.
71. E. Bourgogne, V. Herrou, J. C. Mathurin, M. Becchi and J. De Ceaurriz, *Rapid Commun. Mass Spectrom.*, 2000, **14**, 2343–2347.
72. C. Buisson, C. Mongongu, C. Frelat, M. Jean-Baptiste and J. de Ceaurriz, *Steroids*, 2009, **74**, 393–397.
73. B. I. Zaideh, N. M. R. Saad, B. A. Lewis and J. T. Brenna, *Anal. Chem.*, 2001, **73**, 799–802.
74. T. Toyo'oka, *Modern Derivatization Methods for Separation Sciences*, Wiley, Chichester; New York, 1999.
75. D. Hofmann, M. Gehre and K. Jung, *Isot. Environ. Health Stud.*, 2003, **39**, 233–244.
76. S. Reinnicke, A. Bernstein and M. Elsner, *Anal. Chem.*, 2010, **82**, 2013–2019.
77. W. Meier-Augenstein, *Anal. Chim. Acta*, 2002, **465**, 63–79.
78. S. Gross and B. Glaser, *Rapid Commun. Mass Spectrom.*, 2004, **18**, 2753–2764.
79. S. R. Shinebarger, M. Haisch and D. E. Matthews, *Anal. Chem.*, 2002, **74**, 6244–6251.
80. S. A. Macko, M. Ryan and M. H. Engel, *Chem. Geol.*, 1998, **152**, 205–210.
81. J. A. Silfer, M. H. Engel, S. A. Macko and E. J. Jumeau, *Anal. Chem.*, 1991, **63**, 370–374.
82. L. T. Corr, R. Berstan and R. P. Evershed, *Anal. Chem.*, 2007, **79**, 9082–9090.
83. C. Saudan, M. Augsburger, P. Mangin and M. Saugy, *Rapid Commun. Mass Spectrom.*, 2007, **21**, 3956–3962.

CHAPTER 5
Referencing Strategies and Quality Assurance for Compound-specific Stable Isotope Analysis

5.1 ACCURACY, UNCERTAINTY, PRECISION AND ERROR

The measurements of all physical quantities, such as the isotope ratios and δ-values discussed here, are subject to uncertainty. A prerequisite in order to draw valid conclusions and interpret data correctly is a proper indication of these uncertainties. A measurement therefore consists of two essential components: (i) the best estimate numerical value of the measured quantity, and (ii) the degree of uncertainty associated with the estimated value. The uncertainty thereby characterizes the range of values in which the true value is asserted to lie with a defined probability (e.g., 2-sigma level, 2σ). Additionally, it is useful to state how many measurements n have been carried out to obtain the result.

An example from the determination of carbon isotope values is $\delta^{13}C = -29.4‰ \pm 0.3‰$ (2σ, $n = 6$) or $\delta^{13}C = -0.0294 \pm 0.0003$, ($2\sigma$, $n = 6$) which indicates that the measurement is believed to be closest to $-29.4‰$ but it could have been $-29.7‰$ or $-29.1‰$ with a probability of 95%. The estimation of the (combined) uncertainty should address errors from all possible effects whether systematic or random. Only if all sources of error are included the uncertainty is the most appropriate means for result accuracy. When only the random error is included in the estimation of uncertainty, only the precision of the measurement is reflected. The source for random errors can be manifold, for example,

---

Compound-specific Stable Isotope Analysis
By Maik A. Jochmann and Torsten C. Schmidt
© The Royal Society of Chemistry 2012
Published by the Royal Society of Chemistry, www.rsc.org

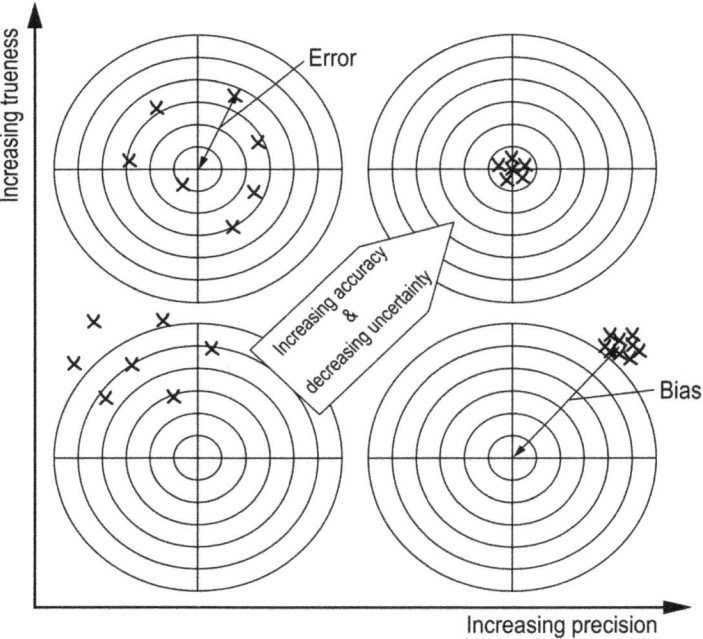

Figure 5.1 Illustration of the difference in the widely used terms precision, trueness, accuracy and uncertainty. For description see text.

preparation of standards by different operators, noise in the measurement and others. It is not possible to correct for random errors and only repeated observations can be used to determine the magnitude of its contribution to uncertainty. The type of contribution that can only be determined from repeated observations is also called 'type A' uncertainty. The so-called 'type B' uncertainties are contributed by systematic errors and cannot be determined by repeated measurements. A systematic error, as shown in the lower right bull's eye in Figure 5.1 tends to shift all measurements in a systematic way so that the mean value of a number of measurements constantly differs in a predictable way. The total systematic error is called the bias of the method or, vice versa, the absence of bias is defined as the trueness. In CSIA, a systematic error is comparable with a reproducible offset caused by a reproducible isotopic fractionation in the course of the analytical procedure. If known, these errors should always be indicated and corrected or included in the uncertainty estimation. Combined standard uncertainties u_c are then determined through:

$$u_c = \sqrt{u_{typeA}^2 + u_{typeB}^2} \qquad (5.1)$$

The formal process of determining the uncertainty of measurement is beyond the scope of this book but is explained in detail in the ISO Guide for the expression of uncertainty.[1] For a quick introduction into the terminology we refer to a technical brief of the Royal Society of Chemistry.[2]

5.2 INTERNATIONAL PRIMARY REFERENCE MATERIALS AND CERTIFIED REFERENCE MATERIALS

5.2.1 Overview

The availability, as well as the proper use of suitable internationally accepted primary reference materials or international measurement standards, is a basic precondition to ensure the compatibility[3] of stable isotope data by different laboratories in different disciplines worldwide at different times.[4–6] Nowadays, stable isotope laboratories often are service facilities, and customers tend to blindly trust the analytical results they receive. For example, they might compare with other data sets (e.g., the same samples measured by another laboratory or comparison with older data). Thus, such laboratories have to provide accurate data with a low degree of uncertainty that can be traced back to an internationally used reference scale.[7] This is of special importance when results are used for assessing the behaviour of a compound in the environment, such as carbon dioxide and its role in climate change[8,9] or even more for measurements relevant in legal cases, for example, doping control or arson investigations.[5] Therefore, professional laboratories usually welcome inter-laboratory tests and want to learn about deficiencies of analytical results if they arise.[10] Additionally, such laboratories always include blind reference materials in order to test laboratory performance independently.[10]

As pointed out in Chapter 2, in general it is more important to quantify the differences in isotopic ratios between samples or compounds than to know their absolute abundances, which are important for correction of isobaric interferences (see Chapter 3). Thus, the δ-value of a sample, which is defined as the relative difference between the isotope ratio of a sample or compound $R(^hE/^lE)_c$ and an internationally accepted reference material $R(^hE/^lE)_{ref}$ (see Chapter 2) is widely applied at the natural abundance level. The δ-scale is based on the relative stable isotope ratio in an agreed-upon international reference material.[4] The stable isotope ratios of these primary reference materials serve as the end members of a traceability chain and define a conventional scale based on numbers of isotopes N. Such reference materials have to fulfil special requirements. One requirement is that the material should be in a stable chemical form so that its isotopic composition will remain constant during storage and distribution. Another is that the material should be fully homogeneous so that aliquots of the material are equal with regard to its chemical and isotopic composition. Material homogeneity is of emerging importance because the applied analytical methods can be carried out with smaller and smaller amounts of material. The material should also be available for a long period or, at best, it should be non-exhaustive, not hygroscopic and should not exchange with atmospheric gases such as carbon dioxide. A major demand on such a material is that it should have a sufficiently small combined uncertainty u_c for the measurement purpose.[4,5]

In the past, some of the arbitrary primary reference materials did not fulfil these demands. It was splendidly pointed out by Taylor et al.[5] that many of these materials 'started their life at the spur of the moment ... in response to a direct and immediate need'. These materials were chosen to represent a major homogeneous pool of an element such as ocean water for hydrogen and oxygen, ocean chloride for chlorine, atmospheric nitrogen for nitrogen and marine carbonates for carbon and oxygen. In several cases the selected material is even virtual; it does not exist in physical form.[4] In such cases the isotopic composition is defined in terms of another real existing material (reference material) as anchor point with a defined isotopic distance to the scale origin. Although the US National Bureau of Standards (NBS; now National Institute of Standards and Technology (NIST) in Gaithersburg, Maryland, US) distributed isotopic reference materials in the 1950s, the International Atomic Energy Agency (IAEA) in Vienna, Austria, has been the primary agency promoting isotopic reference materials and encouraging improvements in isotopic measurements that have led to improved comparability of analytical results.[6] Internationally distributed isotope reference materials that serve as anchors for stable isotope scales are compiled in Table 5.1. The standard reference material defining the scale for hydrogen and oxygen is Vienna Standard Mean Ocean Water (VSMOW), where the 'V' indicates the role of the IAEA based in Vienna during the definition of this scale, the preparation of the reference materials, as well as a demarcation from the formerly used scale Standard Mean Ocean Water (SMOW). To correct for scale compression or expansion, a second precisely known anchor point has been introduced for several of the scales. In the case of the SMOW scale, Standard Light Antarctic Precipitation (SLAP) was defined as second anchor. It is recommended to express δ^2H-values of water and all H-bearing materials to the $\delta^2H_{VSMOW-SLAP}$ scale.[11,12]

Table 5.1 Compilation of the international stable isotope ratio scales. Data are adapted from Coplen et al.[6]

Element	$^hE/^1E$	Scale	Anchor1 Material	$10^3 \delta^h E$	Anchor2 Material	$10^3 \delta^h E$
Hydrogen	$^2H/^1H$	$\delta^2H_{VSMOW-SLAP}$	VSMOW[a]	0	SLAP[b]	−428
		or δ^2H_{VSMOW}	VSMOW2	0 ± 0.3	SLAP2	−427.5 ± 0.3
Carbon	$^{13}C/^{12}C$	$\delta^{13}C_{VPDB-LSVEC}$	NBS 19	+1.95	L-SVEC	−46.6
Nitrogen	$^{15}N/^{14}N$	$\delta^{15}N_{AIR}$	N$_2$ in air	0	USGS32[c] (KNO$_3$)	+180[c]
Oxygen	$^{17}O/^{16}O^d$	$\delta^{17}O_{VSMOW}{}^d$	VSMOW	0		
			VSMOW2	0 ± 0.03		
	$^{17}O/^{16}O^d$	$\delta^{17}O_{AIR}{}^d$	O$_2$ in air	0		
	$^{18}O/^{16}O$	$\delta^{18}O_{VSMOW-SLAP}$	VSMOW	0	SLAP[b]	−55.5
		or $\delta^{18}O_{VSMOW}$	VSMOW2	0 ± 0.03	SLAP2	−55.5 ± 0.02
	$^{18}O/^{16}O$	$\delta^{18}O_{VPDB}$	NBS 19	−2.2		
	$^{18}O/^{16}O$	$\delta^{18}O_{AIR}$	O$_2$ in air	0		
Sulfur	$^{34}S/^{32}S$	$\delta^{34}S_{VCDT}$	IAEA-S1	−0.3		

[a]Supply of VSMOW reference water is exhausted and has been replaced by VSMOW2 reference water.
[b]Supply of SLAP reference water is exhausted and has been replaced by SLAP2 reference water.
[c]Assignment of +180‰ for the δ^{15}N-value of USGS32 relative to N$_2$ in air was adopted by Böhlke and Coplen[16] but has not been officially recommended by IUPAC or IAEA.
[d]There is no second anchor for δ^{17}O-values; therefore, authors should provide δ^{17}O-values of SLAP2 and VSMOW2 so that data can be normalized by readers if necessary.

For carbon, Vienna Pee Dee Belemnite (VPDB) defines the zero point of the δ-scale (note that VPDB does not materially exist and that NBS 19 and L-SVEC carbonates actually define the scale).[13,14]

For oxygen VPDB and AIR-O$_2$ are accepted references for the δ^{18}O-scale.[10,15]

For nitrogen isotope ratio measurements AIR-N$_2$, for sulfur Vienna Cañon Diabolo Troilite (VCDT) and for chlorine Standard Mean Ocean Chloride (SMOC) define the international scales. The δ-values of these materials possess no uncertainty (uncertainty values are zero).[6] Only replacement materials, such as SLAP2 reference water, have an uncertainty greater than zero (see Table 5.1), which is determined during their production and calibration.[6]

The origins of the δ-scales are set to zero by definition, which is comparable with a tare-setting on a balance.[7]

Up-to-date revisions of reference values can be found on the website of the CIAAW (see http://ciaaw.org/).[6] In the following, we will look at the international stable isotope ratio scales and available calibration and reference materials for hydrogen, carbon, nitrogen, oxygen, sulfur and chlorine in more detail.

5.2.2 Hydrogen and Oxygen Calibration and Reference Materials

At the historical beginning of hydrogen isotope ratio determinations, Lake Michigan water was used as a reference material. This water would have a δ^2H-value of $-42‰$ on the VSMOW scale.[17] In 1953, the average ^{18}O content of ocean water was determined through measurements by Epstein and Mayeda.[18] In 1961, Harmon Craig recommended defining SMOW, to express the relative variations of ^{2}H/^{1}H and ^{18}O/^{16}O ratios in natural waters.[19,20] At first, there was no material called SMOW. Instead, the isotopic composition of SMOW was defined with respect to the existing water calibration standard NBS 1, which consisted of Potomac River steam condensate water and was provided by the National Bureau of Standards.[21,22] The relationships between SMOW and NBS 1 for hydrogen and oxygen were given by:[19]

$$R(^2H/^1H)_{SMOW} = 1.050 R(^2H/^1H)_{NBS\ 1} \qquad (5.2)$$

$$R(^{18}O/^{16}O)_{SMOW} = 1.008 R(^{18}O/^{16}O)_{NBS\ 1} \qquad (5.3)$$

Another calibration standard called NBS 1a was obtained from melted snow from Yellowstone Park and showed a lower abundance of the heavier isotopes. In 1965, an IAEA initiated inter-laboratory comparison showed that these materials underwent changes in isotopic composition over time.[4] Therefore, in 1966 it was recommended that two new primary water reference materials should be established.[23] To obtain water with an isotopic composition close to the defined SMOW, Harmon Craig and Ray Weiss from the Scripps Institution of Oceanography in La Jolla, USA, produced a standard of around 70 litres by mixing distilled ocean water (collected in the Pacific Ocean in July 1967 at latitude 0° and longitude 180°) with other waters. This water was introduced by the IAEA as SMOW to define the international scales for hydrogen and oxygen isotope ratio measurements. The ^{18}O/^{16}O ratio remained the same as for SMOW within the limits of analytical uncertainty.[4] In order to avoid confusion between the virtual SMOW scale defined in terms of NBS 1 reference water and the scale defined in terms of the water prepared by H. Craig, at the IAEA's Consultants' Meeting in 1976 it was recommended that the water prepared by H. Craig be renamed VSMOW, after Vienna-SMOW.[6]

Because the δ^2H-value of NBS 1a reference water ($\sim -183‰$) was not sufficiently negative to encompass terrestrial water, a second calibration material originating from a firn ice sample from Antarctica (Vostok) was obtained by Picciotto and was introduced as SNOW and later renamed Standard Light Antarctic Precipitation (SLAP)[24] to improve the coherence between δ-values of different laboratories and to correct scale compression (due to insufficient correction for memory)[6] in case of large measured isotopic differences. The isotopic composition of SLAP against VSMOW has been fixed at δ^2H = $-428‰$ and δ^{18}O = -55.5. SLAP was accepted as second anchor point, completing the definition of the VSMOW scale (see Table 5.1).[25]

Willi Dansgaard from the University of Copenhagen, Denmark obtained another standard, which was called GISP (Greenland Ice Sheet Precipitation).[26] The GISP standard is often used as quality assurance (QA) material when comparing measurements in inter-laboratory studies.

In 2007, VSMOW2 and SLAP2 were introduced to replace the calibration materials VSMOW and SLAP, which were exhausted by that time. The isotopic compositions for both δ^2H and δ^{18}O were adjusted to be as close as possible to the predecessor materials.[27]

Lin et al. showed in a calibration of the δ^{17}O- and δ^{18}O-values of VSMOW, VSMOW2, SLAP and SLAP2 that the new international measurement standards, VSMOW2 and SLAP2, were indistinguishable from their precursors considering measurement uncertainties.[28]

Table 5.2 gives a compilation of calibration and reference materials for hydrogen and oxygen.

It has to be remembered that the VPDB primary reference material is a solid carbonate material and not CO_2 obtained by the reaction with phosphoric acid.[38] However, most reported values are given relative to the solid carbonate and not the gas liberated by the acidic reaction. On the VSMOW scale, VPDB has a δ^{18}O-value of $+30.92‰$.[26,34] Recently, the variability of δ^{13}C and δ^{18}O in CO_2 generated from NBS19-calcite was investigated by Brand et al.[9]

5.2.3 Carbon Calibration and Reference Materials

As the largest homogeneous carbon reservoir on earth, marine bicarbonate dissolved in the ocean controls the composition of atmospheric carbon dioxide. Atmospheric carbon dioxide and marine bicarbonate provide the starting material and the sink for the majority of processes in the biogeochemical carbon cycle.[39] Therefore, to define a δ-scale a carbon material that stems from marine carbonate material was preferred.

Table 5.2 δ^2H of hydrogen isotope reference materials vs. VSMOW and $\delta^{18}O$-values and oxygen vs. VSMOW and VPDB. Data compiled from CIAAW list of isotopic reference materials.[29]

Reference materials (distributor)	Material description	$10^3\delta^2H_{VSMOW}$	$10^3\delta^{18}O_{VSMOW}$	$10^3\delta^{18}O_{VPDB}$
SLAP – Standard Light Antarctic Precipitation* (IAEA, NIST R 8537)	water	−428[25]	−55.5[25]	
SLAP2 – Standard Light Antarctic Precipitation2[a] (IAEA, NIST RM 8536)	water	−427.5 ± 0.3[27]	−55.5 ± 0.02[27]	
GISP – Greenland Ice Sheet Precipitation (IAEA, NIST RM 8536)	water	−189.73[19]	−24.78[19]	
IAEA-CH-7 (IAEA, NIST RM 8540)	Polyethylene foil		−100[19]	
IAEA-CO-9 (IAEA)	barium carbonate		−15.04[b]	−15.40[30]
IAEA-NO-3 (IAEA, NIST RM 8549)	potassium nitrate		+25.32[31]	
IAEA-SO-5 (IAEA)	barium sulfate		+12.13[31]	
IAEA-SO-6 (IAEA)	barium sulfate		−11.35[31]	
IAEA-600 (IAEA)	caffeine		−3.48[31]	
IAEA-601 (IAEA)	benzoic acid		+23.14[31]	
IAEA-602 (IAEA)	benzoic acid		+71.28[31]	
NBS 18 (IAEA, NIST RM 8543)	carbonatite		+7.20[b]	−23.01[30]
NBS 19*	limestone		+28.65[b]	−2.2[32]
NBS 22 (IAEA, NIST RM 8544)	oil	−117[33]		
NBS 28 (IAEA, NIST RM 8539)	silica sand (optical)		+9.58[19]	

Reference	Material	Value 1	Value 2
(IAEA, NIST RM 8546) NBS 30	biotite	-66^{19}	$+5.1^{34}$
(IAEA, NIST RM 8538) NBS 127	barium sulfate		$+8.59^{31}$
(IAEA, NIST RM 8557)			
CO_2-Heavy Palaeomarine Origin (IAEA, NIST RM 8562)	carbon dioxide (palaeomarine origin)		$+11.86^{b,30}$ -18.49^{30}
CO_2-Light Petrochemical Origin (IAEA, NIST RM 8563)	carbon dioxide (petrochemical origin)		$-3.64^{b,30}$ -33.52^{30}
CO_2-Biogenic Modern Biomass Origin (IAEA, NIST RM 8564)	carbon dioxide modern biomass origin		$+20.52^{b,30}$ -10.09^{30}
USGS32 (IAEA, NIST RM 8558)	potassium nitrate		$+25.4^c$
USGS34 (IAEA, NIST RM 8568)	potassium nitrate		-27.78^{31}
USGS35 (IAEA, NIST RM 8569)	sodium nitrate		$+56.81^{35}$
USGS42	human hair (Tibetan)	-78.5^{35}	$+8.56^{35}$
USGS43	human hair (Indian)	-50.3^{35}	$+14.11^{35}$
NGS1 (IAEA, NIST RM 8559)	natural gas (coal origin)	$-138(CH_4)$	
NGS2 (IAEA, NIST RM 8560)	natural gas (petroleum origin)	$-173(CH_4)$	
NGS2 (IAEA, NIST RM 8560)	natural gas (petroleum origin)	$-121(C_2H_6)^{d,26}$	
NGS3 (IAEA, NIST RM 8561)	natural gas (biogenic)	$-176(CH_4)^{d,26}$	

*Exact values defining the δ-scale.
$^a 10^3 \delta^2 H$ vs. VSMOW/SLAP and $10^3 \delta^{18}O$ vs. VSMOW/SLAP.
^{b} Calculated from data in Verkouteren et al.[30] and the relation between VPDB and VSMOW by Coplen et al.[36]
^{c} Calculated from data by Böhlke et al.[37] and Brand et al.[31]
^{d} Value for the corresponding gas species.

The carbon standard PDB had its origin as a working standard in the Chicago group of Harold Urey.[40–42] Urey and the geologist Heinz Lowenstam sampled belemnite guards of *Belemnitella americana* from the cretaceous Pee Dee formation at the Pee Dee river in South Carolina, USA, for their paleo-thermometric studies with $^{18}O/^{16}O$ ratios. The CO_2 obtained from PDB by treatment with 100% phosphoric acid at 25 °C was initially adopted as a reference standard in these investigations.[19,43–47] PDB has carbon and oxygen isotope ratios close to those of limestone of marine origin, and is considerably enriched in ^{13}C with respect to organic carbon compounds.[39] In 1957, Harmon Craig evaluated the isotope ratios of the PDB-derived CO_2 in order to establish the correction equations for the interference of ions of the same mass in mass spectrometric determinations (see Chapter 3).[48] The PDB isotope ratio values $R(^{13}C/^{12}C) = 11237.2 \times 10^{-6}$ and $R(^{17}O/^{16}O) = 380 \times 10^{-6}$ were derived indirectly by Craig[48] from measurements performed by Nier.[49] PDB was proposed as the international reference material for carbon fixing the $\delta^{13}C_{PDB}$ scale.[42,48] In this case the scale origin $\delta^{13}C_{PDB} = 0$ and the anchor point are the same (see Figure 5.2).

By early 1980, the original material was exhausted and in 1983, Gonfiantini[32] reported that $\delta^{13}C$-values of the same homogeneous material analysed by 31 laboratories in 19 countries ranged between +1.75‰ and +2.33‰ relative to PDB.[6] One of the major reasons for replacing the PDB scale was the inhomogeneous character at grain-size level.[10] Thus, NBS 19 was prepared by I. Friedman, J. R. O'Neil and G. Cebula of the US Geological Survey. The reference material NBS 19 was obtained from a single slab of white marble of an unknown origin (in literature also referred to as (Toilet Seat) TS-Limestone). It was crushed to a grain size ranging from 200–300 μm.[19] It is much better defined and its homogeneity has been proven to be better than 0.003‰ for ^{13}C.[9] NBS 19 has been indirectly calibrated versus PDB, and consensus values were found and agreed to as $\delta^{13}C = +1.95‰$ and $\delta^{18}O = -2.2‰$ exactly.[50] In order to make sure that this is recognized as a new, more precise scale it got the name VPDB. Thus, the origin of PDB was not changed but the associated ambiguities were removed by fixing it with the value for NBS 19 as anchor point, that is $\delta^{13}C_{VPDB} = +1.95‰$ for carbon (see Figure 5.2).

A fundamental problem remained: materials with $\delta^{13}C$-values far from 0‰, such as NBS 22 oil, had much poorer uncertainties. Stalker *et al.* reported that $\delta^{13}C_{VPDB}$-values of NBS 22 and other organic reference materials are too positive.[51]

In 2004, the IAEA convened a panel to review carbon reference materials and to recommend a second material for two-point normalization of the $\delta^{13}C_{VPDB}$ scale. Therefore, four international laboratories

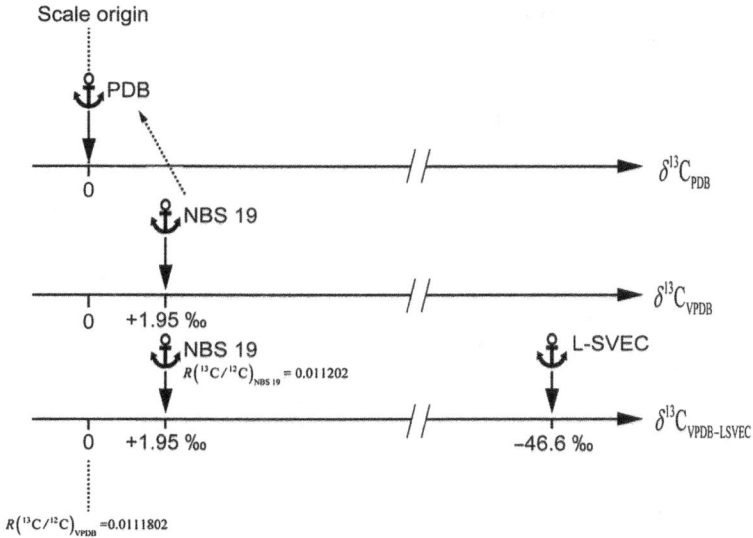

Figure 5.2 Evolution of the δ-scale for carbon. PDB was proposed as international reference material for carbon, fixing the $\delta^{13}C_{PDB}$ scale.[42,48] NBS 19 has been indirectly calibrated versus PDB (see hatched arrow), and consensus values were found and agreed to as $\delta^{13}C = +1.95‰$ and $\delta^{18}O = -2.2‰$ exactly.[50] In order to make sure that this is recognized as a new, more precise scale it got the name VPDB. Thus, the origin of PDB was not changed but the associated ambiguities were removed by fixing it with the value for NBS 19 as anchor points. Lithium carbonate L-SVEC was chosen as second anchor for carbon isotope ratio measurements with a $\delta^{13}C_{VPDB} = -46.6‰$.

performed measurements on 1055 samples of 13 reference materials using state-of-the art continuous-flow EA mass spectrometry with uncertainty analysis by NIST using ISO-GUM with multivariate Bayesian determination of consensus means.[6,13,14]

Lithium carbonate L-SVEC which was introduced by H. Svec from Iowa State University with a $\delta^{13}C_{VPDB} = -46.6‰$ was chosen as second anchor for carbon isotope ratio measurements (see Figure 5.2).[13] The consistency of $\delta^{13}C$ measurements can be improved by ~40% by anchoring the scale with two isotopic reference materials, $\delta^{13}C_{VPDB-LSVEC}$, differing substantially in $^{13}C/^{12}C$.[14]

Zhang et al. reported a $R(^{13}C/^{12}C)$ value of $(11\,202 \pm 28) \times 10^{-6}$ for NBS 19,[52] and thus a $^{13}C/^{12}C$ ratio of $11\,180.2 \times 10^{-6}$ for VPDB is agreed upon.[53] Previously, the PDB isotope ratio value determined by Craig,[48] which was 5‰ higher at $11\,237.2 \times 10^{-6}$, was accepted.[47] This value is now obsolete.

For organic compounds, organic carbon isotope reference materials have been developed and it is recommended that these materials are used according to the principle of identical treatment when measuring organic materials. A prominent and widely used EA-IRMS reference

Table 5.3 δ^{13}C-values of carbon isotope reference materials vs. VPDB and absolute carbon isotope ratios. Data compiled from CIAAW list of isotopic reference materials[29] and IRMM.[54]

Reference materials (distributor)	Material description	$10^3 \delta^{13}C_{VPDB}$
NBS 19* (IAEA, NIST RM 8544)	calcium carbonate (toilet seat limestone)	+1.95[32]
L-SVEC* (IAEA, NIST RM 8545)	lithium carbonate	−46.6[14]
IAEA-CO-1 (IAEA, NIST)	calcium carbonate (Carrara marble)	+2.49[14]
IAEA-CO-8 (IAEA, NIST)	calcium carbonate	−5.76[14]
IAEA-CO-9 (IAEA, NIST)	barium carbonate	−47.32[14]
IAEA-CH-3 (IAEA, NIST)	cellulose	−24.72[14]
IAEA-CH-6 (IAEA, NIST RM 8542)	sucrose	−10.45[14]
IAEA-CH-7 (IAEA, NIST RM 8540)	polyethylene foil	−32.15[14]
IAEA-600 (IAEA, NIST)	caffeine	−27.77[14]
IAEA-601 (IAEA, NIST)	benzoic acid	−28.81[14]
IAEA-602 (IAEA, NIST)	benzoic acid	−28.85[14]
NBS 18 (IAEA, NIST RM 8543)	carbonatite, (Fen, Norway)	−5.01[14]
NBS 22 (IAEA, NIST RM 8539)	oil	−30.03[14]
CO$_2$-Heavy Paleomarine Origin (IAEA, NIST RM 8562)	carbon dioxide (paleomarine origin)	−3.72[14]
CO$_2$-Light Petrochemical Origin (IAEA, NIST RM 8563)	carbon dioxide (petrochemical origin)	−41.59[14]
CO$_2$-Biogenic Modern Biomass Origin (IAEA, NIST RM 8564)	carbon dioxide modern biomass origin	−10.45[14]
USGS24 (IAEA, NIST RM 8541)	graphite	−16.05[14]
USGS40 (IAEA, NIST RM 8573)	L-glutamic acid	−26.39[14]
USGS41 (IAEA, NIST RM 8574)	L-glutamic acid	+37.63[14]
NGS1 (IAEA, NIST RM 8559)	natural gas (coal origin)	−29.0(CH$_4$)[b,26]
NGS2 (IAEA, NIST RM 8560)	natural gas (petroleum origin)	−44.5(CH$_4$)[b,26]
NGS3 (IAEA, NIST RM 8560)	natural gas (biogenic)	−72.8(CH$_4$)[b,26]
BCR-656 (IRMM)	Ethanol from wine	−26.91 ± 0.07[c,55]
BCR-657 (IRMM)	sugar	−10.76 ± 0.04[c,55]
BCR-660 (IRMM)	Ethanol in water	−26.72 ± 0.09[c,55]

*Exact values defining the δ-scale.
[a]Calculated from data in ref. 14 and the relation between VPDB and VSMOW given on page 36 of ref. 12.
[b]Value for the corresponding gas species.
[c](1σ-level).

material is NBS 22, a mineral oil which was introduced by Silverman of Chevron Oil Company, La Habra, US.[56] Because of its liquid nature it can be assumed to be homogeneous at all amount levels.[4] One problem is that it was distributed in polyethylene bottles, and it was mentioned at an IAEA meeting that the big supply bottle has a dark residue sitting at the bottom of the oil. This means that the NBS 22 oil has changed compositionally over time.[57]

Further organic reference materials provided by IAEA and NIST are listed in Table 5.3. For direct normalization in GC-IRMS and LC-IRMS, ethanol from wine obtained as BCR-656 as well as the benzoic acids (IAEA-601 and IAEA-602) are also interesting. Furthermore, for LC-IRMS, all sugars and the IAEA-600 caffeine as well as the two L-glutamic acids (USG40 and USG41) prepared by Qi and Coplen are useful.[58]

5.2.4 Nitrogen Calibration and Reference Materials

At 79% (v/v) nitrogen is the most abundant gas in the atmosphere, which is by far the largest nitrogen pool on earth. Hence it cannot be altered easily by exchange with other compartments.[10] Its isotope ratio does not change within measurement precision over time and is not exhaustible.[47,59,60] The reason for its stable value is that it is involved in natural processes as source and sink so that a net change close to zero is established.[47] Because of this it was a likely choice to define ubiquitously available atmospheric nitrogen as the international standard. Therefore, AIR-N_2 has been adopted as primary reference material for nitrogen isotope ratio analysis and is set to 0‰ by definition.[60] The standard is not sold, it must be prepared by each lab itself. For the preparation, CO_2 as well as water have to be removed by cryogenic trapping and oxygen is removed by reaction with copper oxide.[17] Around 1% argon remains and can influence the isotope ratio measurement.[10] NSVEC was prepared as a nitrogen gas standard by Junk and Svec,[59] and was later split into aliquots and sealed in glass tubes by the USGS.[4,61] However, it has to be mentioned that NSVEC is not regarded to be a primary reference material. An assignment of +180‰ for the $\delta^{15}N$-value of USGS32 relative to AIR-N_2 was adopted by Böhlke and Coplen[16] but has not been officially recommended by IUPAC or IAEA.[6]

Additionally, the solid reference materials ammonium sulfate and potassium nitrate salts as well as urea and the already mentioned L-glutamic acids USGS 40 and USGS 41, shown in Table 5.4, are available.[6] These solid nitrogen-bearing standards are easier to handle in

Table 5.4 δ^{15}N-values of nitrogen isotope reference materials vs. AIR-N$_2$. Data compiled from CIAAW list of isotopic reference materials.[29]

Reference materials (distributor)	Material description	$10^3\delta^{15}N_{AIR-N2}$
NSVEC (IAEA, NIST RM 8552)	nitrogen gas	-2.78[62]
IAEA-N-1 (IAEA, NIST RM 8547)	ammonium sulfate	$+0.43$[62]
IAEA-N-2 (IAEA, NIST RM 8548)	ammonium sulfate	$+20.41$[62]
IAEA-NO-3; formerly IAEA-N-3 (IAEA, NIST RM 8549)	potassium nitrate	$+4.72$[62]
USGS 25 (IAEA, NIST RM 8550)	ammonium sulfate	-30.41[62]
USGS 26 (IAEA, NIST RM 8551)	ammonium sulfate	$+53.70$[62]
USGS 32* (IAEA, NIST RM 8558)	potassium nitrate	$+180$[62]
USGS 34 (IAEA, NIST RM 8568)	potassium nitrate	-1.8[37]
USGS 35 (IAEA, NIST RM 8569)	sodium nitrate	$+2.7$[37]
USGS 40 (IAEA, NIST RM 8573)	L-glutamic acid	-4.52[58]
USGS 41 (IAEA, NIST RM 8574)	L-glutamic acid	$+47.57$[58]

*Interim consensus values used for scale normalization.

the laboratory and referencing with continuous-flow EA-IRMS systems makes vacuum line separation of atmospheric nitrogen superfluous.

The stable nitrogen isotope abundance ratio $R(^{15}N/^{14}N)$ is calculated to be $(3676 \pm 4) \times 10^{-6}$ from the IUPAC recommended value of 272.0 ± 0.3 for $n(^{14}N)/n(^{15}N)$ in atmospheric nitrogen.[63,64]

5.2.5 Sulfur Calibration and Reference Materials

Since 1997, the international standard for sulfur isotopic measurements is VCDT.[6,65,66] The original standard Cañon Diablo Troilite (CDT)[67,68] was prepared from the FeS (troilite) phase of the iron meteorite found at the Barringer meteor crater in Arizona. CDT was chosen because of its expected primordial isotopic composition in the meteorite sulfur,[4] but it turned out that CDT was not sufficiently homogeneous to be used as primary reference material. The variability in its $^{34}S/^{32}S$ isotope ratio is at least ± 0.4‰ and greatly exceeds the achievable analytical uncertainty of ± 0.05‰.[66,69] Several attempts to introduce a natural sulfur-bearing material such as sphalerite or precipitated BaSO$_4$ from sea water failed because of material inhomogeneities.[4] To overcome this problem, an agreed-upon δ^{34}S-value of -0.3‰ of silver sulfide IAEA-S-1 against VCDT was defined.[66] In the case of SO$_2$, isobaric interferences at ion current m/z 66 caused by the contribution of $^{32}S^{16}O^{18}O^+$ have to be taken into account.[47] Due to the isobaric dilution effect, δ^{66}S-values in general are smaller numbers than the corresponding δ^{34}S-values. Adopting the $^{18}O/^{34}S$ ratio from the VSMOW/VCDT pair, δ^{66}S-values must be multiplied by 1.091 for obtaining δ^{34}S. The procedure requires

Table 5.5 δ^{34}S-values of sulfur isotope reference materials vs. VCDT. Data adapted from CIAAW list of isotopic reference materials.[29]

Reference materials (distributor)	Material description	$10^3\delta^{34}S_{VCDT}$
IAEA-S-1; formally NZ-1[a] (IAEA, NIST RM 8554)	silver sulfide	−0.30[71]
IAEA-S-2; formally NZ-2 (IAEA, NIST RM 8555)	silver sulfide	+22.67[36]
IAEA-S-3 (IAEA, NIST)	silver sulfide	−32.55[36]
IAEA-S-4 (Soufre de lacq) (IAEA, NIST RM 8553)	sulfur (elemental)	+16.90[72]
IAEA-SO-5 (IAEA)	barium sulfate	+0.49[36]
IAEA-SO-6 (IAEA)	barium sulfate	−34.05[36]
NBS 123 (IAEA, NIST RM 8556)	sphalerite	+17.44[73]
NBS 127 (IAEA, NIST) (IAEA, NIST RM 8557)	barium sulfate	+21.1[36]

[a]Calibration material.

the oxygen isotopic composition of the sample and reference in question to be identical.[10] So far, a second anchor to improve comparability of δ^{34}S data has not been selected (see Table 5.1).[6]

Compound-specific isotope measurements of sulfur so far have been carried out only by coupling GC with a MC-ICP-MS with SF_6 as reference gas (see Chapter 7).[70] Available IAEA and NIST reference materials for sulfur are given in Table 5.5.

5.2.6 Chlorine and Bromine Calibration and Reference Materials

For chlorine isotope measurements, SMOC was proposed as primary reference material. It was supposed that the δ^{37}Cl-value of seawater is homogeneous within a range of ±0.15‰ and therefore each lab used individual seawater. However, it turned out in 2002 that a substantial variability exists.[74] The NIST reference material SRM 975 has been the basis for absolute isotope abundance measurements of chlorine. This material was exhausted in the 1960s and has been replaced by SRM 975a. Table 5.6 gives chlorine δ-values for chlorine reference materials.

There is no formal international standard for bromine isotope ratios δ^{79}Br and so far, there have been no tests to check variations of ocean water bromine composition. Due to the fact that the residence time of bromine in ocean water is larger than that of chlorine the variation in isotope composition is assumed to be even smaller than that for chlorine. Therefore, it is used as a reference material, which is called Standard Mean Ocean Bromide (SMOB).[76]

Table 5.6 δ^{37}Cl-values of chlorine isotope reference materials vs. SMOC. Data compiled from CIAAW list of isotopic reference materials.[29]

Reference materials (distributor)	Material description	$10^3\delta^{37}Cl_{SMOC}$
SRM 975 (exhausted since 1960s)	NaCl	+0.43[75]
SRM 975a (NIST)	NaCl	+0.2[75]
ISL 354 (IAEA)	NaCl	+0.05[75]

5.3 NORMALIZATION OF STABLE ISOTOPE DATA

In the last section, the available international reference scales, calibration and reference materials were discussed. It was pointed out that these materials are necessary to express measured isotope data in terms of the internationally used scales. The procedure to convert measured data into internationally comparable data is commonly called normalization. There are several normalization algorithms to convert the raw δ-value of a sample measured in the laboratory against the local working reference into a value expressed on an international scale $\delta^h E_{c,i-ref}$.[77] The normalization algorithms differ in the number of anchoring standards used as well as in the respective normalization errors associated with each method. In general, it is necessary that the sample is measured in the same way as the reference materials by 'identical treatment'.[47]

Paul *et al.* reviewed six existing and generally applied normalization methods:[77]

(i) normalization of an isotope ratio in a compound relative to the normalized isotopic composition of a working or reference gas;
(ii) normalization versus a reference material;
(iii) modified normalization versus a reference material;
(iv) normalization by linear shift calculated from the difference between a measured and internationally accepted value of two (certified) reference materials;
(v) two-point normalization;
(vi) multi-point normalization.

Here, we will focus on methods (i), (ii) and (v) as representatives of the most widely used methods. These algorithms are either applied directly by the instrument software or can be applied externally to the obtained data. Generally, they can be applied independent of the involved instruments.

(i) Normalization Relative to the True Isotopic Composition of a Working or Reference Gas. A widely applied method is normalization

relative to the isotopic composition of a working or reference gas. This method is normally carried out automatically by the IRMS software.

For this kind of normalization procedure the δ-value of the used working or reference gas relative to an international reference material $\delta^h E_{rg,i-ref}$ has to be determined, for example, by dual-inlet.

For the calculation of the internationally accepted value $\delta^h E_{c,i-ref}$ we can rearrange the following equations for the involved δ-values:

$$\delta^h E_{c,i-ref} = \frac{R(^h E/^l E)_c}{R(^h E/^l E)_{i-ref}} - 1 \quad (5.4)$$

$$\delta^h E_c = \frac{R(^h E/^l E)_c}{R(^h E/^l E)_{rg}} - 1 \Rightarrow R(^h E/^l E)_c = R(^h E/^l E)_{rg} \times (\delta^h E_c + 1) \quad (5.5)$$

$$\delta^h E_{rg,i-ref} = \frac{R(^h E/^l E)_{rg}}{R(^h E/^l E)_{i-ref}} - 1 \Rightarrow R(^h E/^l E)_{i-ref} = \frac{R(^h E/^l E)_{rg}}{(\delta^h E_{rg,i-ref} + 1)} \quad (5.6)$$

By substituting the values for $R(^h E/^l E)_c$ from eqn (5.5) and $R(^h E/^l E)_{i-ref}$ from eqn (5.6) in eqn (5.4) we can derive the following expression for the normalization:

$$\delta^h E_{c,i-ref} = \delta^h E_c + \delta^h E_{rg,i-ref} + \delta^h E_c \times \delta^h E_{rg,i-ref} \quad (5.7)$$

This expression is also known as 'standard conversion identity' and can also be used for conversion of δ-values between various standards.[48] If we suppose a normalized CO_2 reference gas value of $\delta^{13}C_{rg,i-ref} = -0.0375$ as can be found for pressurized gas cylinders, and a measured compound value $\delta^{13}C_c = 0.0126$ against this reference gas we can calculate an internationally accepted δ-value of $0.0126 + (-0.0375) + (0.0126 \times (-0.0375)) = 0.0254$ by using eqn (5.7). If no value for $\delta^i E_{rg,i-ref}$ is entered in the gas configuration of the mass spectrometer software and the value is set to zero, the second and third terms will be zero and only $\delta^h E_c$ is calculated. In the traditional form of eqn (5.7) a factor of 1000 can be found in the denominator of the third or so-called 'scaling term'. As pointed out in Chapter 2, the approach used here is recommended in order to obtain clarity and to avoid confusion and errors.[78] There are generally two problems that can occur by using this normalization method. The first one is that the error for the normalization becomes large if $|\delta^h E_{c,i-ref} - \delta^h E_{rg,i-ref}|$ increases. Therefore, the reference gas value should be similar to that of the sample. A problem when using gas cylinders containing CO_2 or SO_2 is that a mass-dependent isotopic fractionation can occur in gas cylinders.[79] The reason for this fractionation is the coexistence of a liquid and a gaseous phase in the cylinder. This fractionation can be enormous and variable when the cylinders stay outside and are exposed to temperature variations by sun or frost. It is therefore recommended that the gas cylinders and the gas removal station are placed within the same lab, where the temperature is held constant by air-conditioning. The best way, proposed by Willi Brand of the Max Planck Institute for Biogeochemistry in Jena, is to keep the CO_2 gas cylinders at 40 °C above the critical temperature of CO_2 gas (31 °C).[10] The isotopic composition of the reference gas against an international reference material should therefore be determined regularly by carbonate reference material.

However, users are strongly recommended not to rely on this normalization strategy. Users should seldom, if ever, need to measure the δ-value of a reference gas used for injection into, for example, EA-IRMS or GC-IRMS.[6] What they ought to be doing is interspersing isotopic reference materials among their unknowns and normalizing the δ-values of the unknowns relative to the δ-values of the isotopic reference materials as described in the following.[10,80]

(ii) Normalization Versus a Reference Material. By using normalization of a sample or a compound versus a reference material (single-point anchoring) an internationally accepted δ-value ($\delta^h E_{i-ref}$) is analysed alongside the unknown sample by identical treatment. Both are measured with respect to the working gas so that a measured δ-value of the well-known reference material on the individual instrument is obtained ($\delta^h E_{m-ref}$). The mathematical description of this normalization procedure is given by eqn (5.8):[81]

$$\delta^h E_{c,i-ref} = \left[\frac{(\delta^h E_c + 1)(\delta^h E_{i-ref} + 1)}{(\delta^h E_{m-ref} + 1)}\right] - 1 \quad (5.8)$$

As in the case of normalization versus a working or reference gas, the normalization error is higher if the isotopic composition of the sample is significantly different from that of the used reference material. The error also increases if the principle of identical treatment is not strictly followed.[77] A detailed derivation of eqn (5.8) has been given by Paul *et al.*[77]

(iii) Two-point Normalization. In the case of two-point normalization, a linear regression of measured ($\delta^h E_{m-ref1}$ and $\delta^h E_{m-ref2}$) and internationally accepted δ-values of two reference materials ($\delta^h E_{i-ref1}$

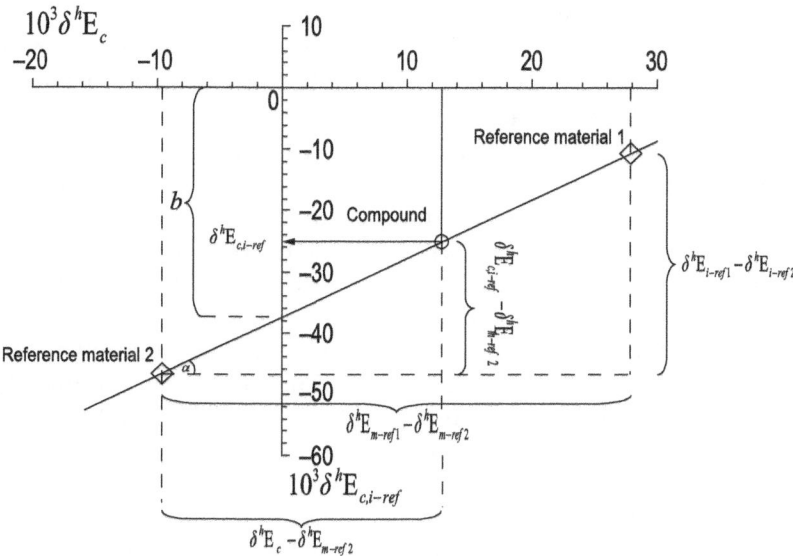

Figure 5.3 Illustration of two-point normalization. The measured δ-values are plotted on the abscissa and the related internationally accepted or normalized δ-values are plotted on the ordinate.

and $\delta^h E_{i-ref2}$) is used to normalize a sample $\delta^i E_c$ as illustrated in Figure 5.3.

The slope m of the linear relationship in Figure 5.3 can be expressed by the two-point form of a straight line such as is given in the following equation:

$$m = \tan \alpha = \frac{\delta^h E_{c,i-ref} - \delta^h E_{i-ref2}}{\delta^h E_c - \delta^h E_{m-ref2}} = \frac{\delta^h E_{i-ref1} - \delta^h E_{i-ref2}}{\delta^h E_{m-ref1} - \delta^h E_{m-ref2}} \quad (5.9)$$

Rearranging eqn (5.9) leads to eqn (5.10) which can be used to calculate the normalized value of the sample.

$$\delta^h E_{c,i-ref} = \frac{\delta^h E_{i-ref1} - \delta^h E_{i-ref2}}{\delta^h E_{m-ref1} - \delta^h E_{m-ref2}} \times (\delta^h E_c - \delta^h E_{m-ref2}) + \delta^h E_{i-ref2} \quad (5.10)$$

A prerequisite assumption that has to be made is that systematic error introduced by the mass spectrometric analysis is linear within a dynamic range. The factor m is also referred to as the 'expansion factor' and the y-intercept b as the 'additive correction factor', representing the normalized δ-value of the working or reference gas.

Applying two-point normalization by using a second anchor point has the advantage that it compensates for differences in δ-scaling because δ-values are often compressed, but rarely expanded.[6,10] To this end, the two-point normalization results in a better accuracy than the single-point anchoring methods. Thus, whenever possible a two-point calibration should be carried out.[82,83]

The use of multi-point normalization is an extension of the two-point calibration. Paul *et al.* pointed out that in the case of random error associated with the analysis of a standard used to anchor the linear scale, the slope of the regression line will be different from that expected. Such random error can be caused by incomplete combustion of sample in, for example, the elemental analyser. This will introduce error in the computed normalized isotopic composition of an unknown sample depending on the position of the sample along the regression line. Thus, the error will be large for the sample that plots near the standard that has been affected by the random error. In such cases the error can be reduced by using a best-fit regression line passing through more than two anchor points.[77] However, a word of caution is necessary. Not all values of the references in use are known with the same precision and accuracy. Hence, multi-point referencing might actually worsen the situation. Instead, a two-point normalization with the values for the two points clearly stated opens the possibility to recalculate values later, should the reference value of one of the scaling reference materials be adjusted.[10] In case of a multi-point calibration with fitted regression, this can be rarely done.

5.4 REFERENCING IN CSIA

Isotope ratio analysis incorporates two main steps: (i) the measurement of ion current ratios and (ii) the conversion of these ratios into isotopic abundances.[84] For reproducible measurements, referencing by expression of isotopic abundances relative to the internationally accepted primary reference materials introduced in Section 5.2 is necessary.

However, the standardization in continuous-flow analysis is not as straightforward as in isotope ratio analysis by dual-inlet measurements. In dual-inlet, referencing is achieved by alternating introduction of the sample and an isotopically characterized reference gas. The obtained ion currents can be compared and relative isotopic abundances can be calculated as described in Chapter 3 of this book.

In contrast, in continuous-flow-IRMS it is not possible to reference the sample to a standard of precisely known isotopic composition introduced in exactly the same manner.[85,86] To obtain the highest possible accuracy, it is preferable to compare sample and standard within each chromatogram and the sample and standard should be handled as similarly as possible according to the 'principle of identical treatment'.[47] According to this principle, to minimize errors affecting the accuracy of the results the sample and standard should undergo the same steps in the analytical procedure, such as injection, chromatographic separation (chromatographic isotope effect, peak distortion, integration), conversion into a gas and transfer via the interface into the IRMS source.

Various procedures to introduce standards for CSIA into the IRMS are possible and have been investigated. These include:[84–87]

(i) introduction of reference gas pulses from a reference gas volume directly into the ion source;
(ii) introduction of reference gas pulses into the continuous carrier gas stream;
(iii) introduction of a gaseous reference compound between chromatographic column and conversion reactor;
(iv) addition of reference compounds into the mixture injected into the gas chromatograph;
(v) 'full chromatogram' normalization.

In case (i) a normalized reference gas (CO_2, H_2, N_2, CO) is introduced from a bellows and changeover valve into the ion source of the mass spectrometer.[85]

In the case (ii) a switching mechanism such as a rotary sample valve[88] or, more commonly, a moving capillary interface is used,[84,85] with which the reference gas is introduced via a small helium stream ($\sim 30\,\mu L\,min^{-1}$) into the ion source.[89] The technical aspects of such moving capillary interfaces as they are used in commercially available instruments are described in Chapter 3. For direct injection of a reference gas via a helium stream into the ion source, there are essentially two methods. The first case, shown in Figure 5.4a, is that the reference gas peaks are set before the chromatographic system is connected to the IRMS via setting, for example, 'open-split in' and 'backflush out', so that the gas peaks are measured without the chromatographic background.

This procedure is very often used but here the contribution of the chromatographic background is not included in referencing. Sometimes three peaks are set at the end of a chromatographic run. This kind of procedure can be used to check the individual conditions of the mass spectrometer before and after a chromatographic run.[87] However, closer to the principle of identical treatment is to set the reference gas pulses on the chromatographic background at the beginning of the chromatogram or, even better, in direct vicinity to the measured compound (see Figure

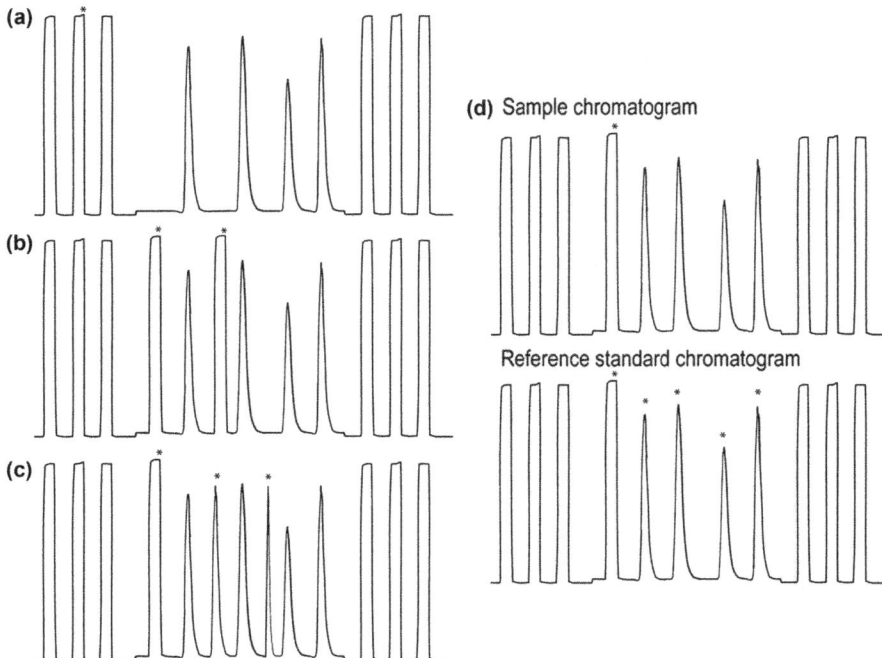

Figure 5.4 Referencing strategies for CSIA. a) Reference gas peak pulses (flat-topped peaks marked with asterisks) are set before the chromatographic system is connected to the IRMS. b) Reference gas pulses are set on the chromatographic background at the beginning of the chromatogram or, even better, in direct vicinity to the measured compound within a free slot in the chromatogram. c) Co-injection of multiple organic standards (Gaussian shaped peaks marked with asterisks) with well-known isotopic composition and similar chemical composition in a slot in direct vicinity to the target compounds. The flat-topped reference gas pulse marked with an asterisk is used for referencing. d) 'Full chromatogram' normalization. In this method a sample chromatogram and a reference standard chromatogram are always measured alternately, resembling the dual-inlet procedure for normalization.

5.4b). By using this method, the chromatographic background is taken into account. Both principles suffer from the fundamental drawback that the reference gas does not experience the same conditions as analytes with respect to injection, separation and the chemical conversion into a gas[89] and, thus, the principle of identical treatment is not fulfilled completely. Whenever possible, calibration by this method should be avoided.

Case (iii) is an interesting approach, so far not implemented in commercially available instruments, that has been demonstrated by Meier-Augenstein.[87] This approach includes the chemical conversion step;

therefore, an isotopically well-known organic reference standard is introduced as gas (e.g. butane) at a point between the chromatographic column and the combustion reactor in which the chemical conversion takes place. Meier-Augenstein showed the applicability of such an interface for FAMEs. Advantages of such a system are that the reference gas peaks behave in the same way as analyte peaks (Gaussian peak shape) and are treated as such by the data evaluation software. Additionally the size of every peak can be adjusted to match the size of a compound peak and the performance of the conversion reactor can be tested without injection into the gas chromatograph.[87]

However, to fulfil the principle of identical treatment including potential sources of mass discrimination and hence isotopic fractionation by the analyte injection and chromatographic separation, it is necessary to co-inject multiple organic standards with well-known isotopic composition and similar chemical composition (case (iv)) (see Figure 5.4c). A prerequisite for this method are free regions or slots in the chromatogram, which are in direct vicinity to the compound of interest. For simple mixtures such as fatty acids from vegetable oils[89] or n-alkanes[84] this approach works well. In some cases deuterated standards can be used for carbon isotope ratio measurements or ^{13}C-labelled standards for hydrogen measurements. An example for such a procedure is shown in Figure 5.5.

However, often deuterated or ^{13}C-labelled standards are not commercially available or are very expensive and the use of deuterated

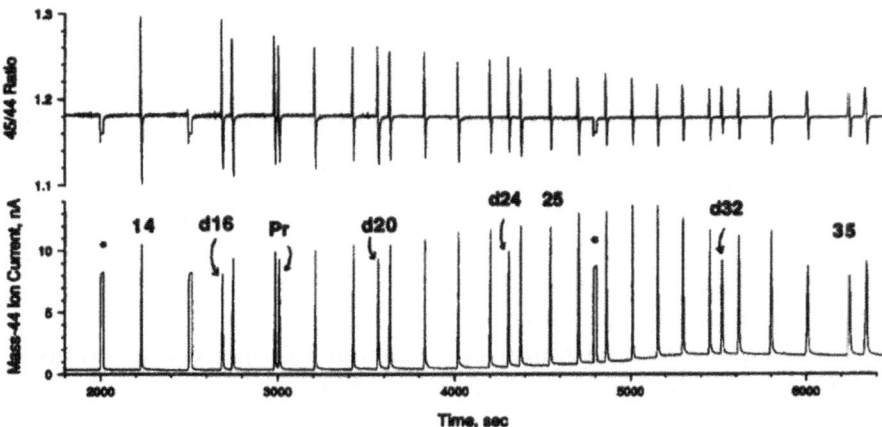

Figure 5.5 Chromatogram of homologous n-alkanes with four fully deuterated alkane standards (d_{16}, d_{20}, d_{24}, d_{32}) in free slots of the chromatogram.[84] The asterisks designate CO_2 reference gas standards and (Pr) pristine. (Graphic reprinted with permission from Elsevier.)

compounds permanently ruins the GC column for any hydrogen isotopic work. Another problem is samples with complex matrices such as are often observed for environmental samples. In such samples free slots in the chromatogram are rare to non-existent and can vary from chromatogram to chromatogram,[89] and co-elution of standards, matrix and target analytes on the effluent stream leads to serious systematic errors for all isotopic values referenced by the co-eluted standard.[84]

In case (v), the 'full chromatogram' normalization method, a sample chromatogram and a reference standard chromatogram are measured alternately, resembling the dual-inlet procedure for normalization (see Figure 5.4d). In general, conditions of the chromatographic separations are very reproducible with regard to injection and retention times. If the sample mixture is sufficiently similar to the reference standard mixture, this should be the preferred method for EA-IRMS and GC- or LC-IRMS methods, following the identical treatment principle. Additionally, the method can be combined with the gas injection methods (i)–(iii).

In an extensive investigation, Merritt and co-workers compared the use of reference gas and co-injected organic standards for isotopic referencing. A main conclusion of this investigation is that both methods can provide identical results for carbon isotopes within the experimental error ($<0.1‰$) in the absence of fractionation during combustion of analytes in the conversion reactor, and that in case of inadequate combustion performance caution is imperative. In such cases they observed for carbon isotope measurements systematic offsets $>2‰$ when using only reference gas peaks for calibration.[84] With regard to the problems occurring by complex chromatograms an optimal solution is thus to introduce both reference gas and co-injected reference standards. The former is used for isotopic calibration, while comparison with the latter allows for assessment of systematic bias due to analyte processing.[89] Otherwise, if this is not possible because of the lack of free spots in the chromatogram for reference gas pulses or multicomponent reference working standards, at a minimum, the CO_2 working standard should be analysed on the chromatographic background at the beginning of each sample run or, better, the 'full chromatogram' normalization method should be used.

The need for normalization of measured isotopic compositions to primary reference materials and to compensate for differences in δ-scaling in IRMS instruments, means that the use of two different reference materials of CSIA-compatible organic compounds is mandatory. Unfortunately, the lack of widely available, GC- and LC-amenable organic standards of known isotopic composition remains a significant problem for practitioners of CSIA. The development of suitable organic stable isotope reference standards for CSIA has not kept pace with emerging needs of users.[90] An example of this lack is that NIST does not offer ^{15}N-containing organic reference materials for GC-IRMS although the method is used since 1988.[90] Also, no organic O-isotopic standards are available, and standardization for ^{18}O remains particularly problematic.[89,91] Recently, an attempt was made to develop an isotopic Grob-like test as a multi-isotopic calibration tool for GC-IRMS of hydrogen-, carbon-, nitrogen- and oxygen-containing compounds.[92,93] However, Grob-like mixtures are chemically complex and have failed spectacularly as isotope reference materials because the ingredients tend to react with each other, thus introducing isotopic fractionation. The mixtures have very short shelf life.[57]

The normalization procedure in CSIA is not as straightforward as in dual-inlet or bulk isotope ratio analysis (BSIA) with EA-IRMS. In case of CSIA often more than one normalization step is necessary to trace the measured isotopic signature back to a calibration material and thus to a primary reference material.[94] Possible normalization pathways for CSIA with LC- and GC-IRMS are illustrated in Figure 5.6.

Generally, reference materials are normalized against the available calibration materials. This procedure is promoted by the IAEA and carried out by a few selected laboratories worldwide. The reference

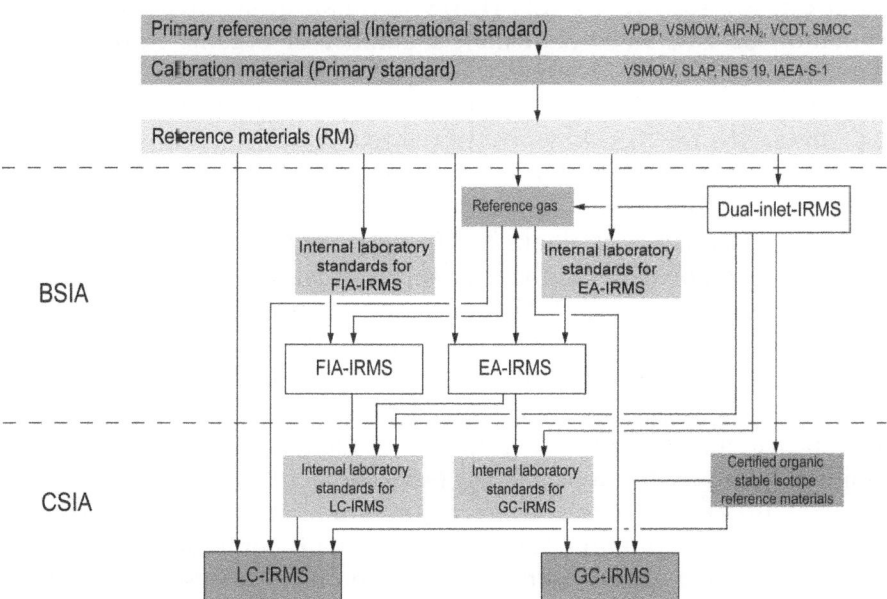

Figure 5.6 Possible standardization and normalization pathways in CSIA.

materials can be purchased to calibrate internal laboratory (local) standards. These internal laboratory standards can be used to normalize samples for bulk analysis by dual-inlet-, EA-, flow injection analysis (FIA)-IRMS or if possible directly for CSIA by LC- and GC-IRMS.

The easiest and most direct, but not recommended, way for normalization in CSIA is by using a reference gas. As mentioned above this method can lead to inaccurate results, for example when chemical conversion is incomplete. However, for referencing, each laboratory has to cross-calibrate their in-house reference gas against international standard materials. This cross-calibration can be done by conventional dual-inlet or EA-IRMS. The dual-inlet approach often involves a sample preparation, which tends to be forgotten in the error assessment. As an example, the EA-IRMS measurement of NBS 22 can provide superior, i.e. less biased results. So, instead of the dual-inlet, using the EA, NBS 22 can be combusted and the reference gas can be calibrated 'by hindsight'. Any error associated with this would cancel, provided the principal conditions remain identical.[10] For CSIA, individual materials (compounds), which have been separately calibrated on the EA-system, can be injected and used as the 'reference standard mixture' for identical treatment. (See normalization methods (iv) and (v) and Figure 5.4 above.) Thus, the precise value of the reference gas is not urgently required.

Apart from a few reference materials (see Section 5.2.3), which can be used directly for GC- or LC-IRMS normalization, such as BCR-656 (ethanol) or IAEA-600 (caffeine), a number of organic stable isotope reference materials with known isotope ratios are available, which are normalized against international reference materials. The most useful set of standards for CSIA work is offered by Arndt Schimmelmann, Department of Geological Sciences, Indiana University in Bloomington, Indiana, US, for research applications. The list of available compounds includes *n*-alkanes with chain length between C-1 to C-50, pure fatty acid esters and mixtures of them as well as other widely used compounds for EA-, GC- and LC-IRMS, such as nicotine, acetanilide, *cis*-1,2-dichloroethylene, dibenzothiophene, phenanthrene, methanol, ethanol and others (for contact address see Section 5.8.1). The determination of δ^2H- and δ^{13}C-values of these compounds is carried out by multiple offline measurements using the 'principle of identical treatment' and two-point isotopic calibrations against NBS 19, L-SVEC, VSMOW, SLAP, etc. The multiple analyses for each compound are performed via combustion of milligram-amounts of individual compounds in quartz ampoules and cryogenic purification of combustion gases in a vacuum line.[90] Isotopically characterized *n*-alkanes are also offered by Chiron AS, Trondheim, Norway. However, it has to be mentioned that the procedures for δ-value assignment are not always transparent. The documentation of a full traceability chain remains with the users of the materials, not the supplier.[57]

Many target compounds with previously determined isotopic signatures for normalization and data validity checking are not available from these sources. Therefore, most labs generate their own isotopic working standards by calibration using other forms of isotopic analysis such as dual-inlet-IRMS, EA-IRMS, FIA-IRMS[95] (see Figure 5.6), which leads to concerns about inter-laboratory data comparability.[89] For closed tube combustion and dual-inlet, protocols developed by Schimmelmann et al.[90] and Caimi et al.[96] (Fatty acids and their corresponding methyl esters) can be followed.

As standards, substances with the highest grade of purity should be purchased and additionally the purity of the internal laboratory and daily working standards should be tested by GC-MS/LC-MS because even small amounts of contaminants can affect the ability to accurately characterize the compound of interest. When the compounds are measured by EA-IRMS application of the procedures outlined by Qi et al.[58] and Werner et al.[47] is recommended. As mentioned before, these standards should be checked regularly to obtain consistent isotope ratio data. It is necessary to prepare sufficient amounts of these standards. The USEPA guideline EPA 600/R-08/148 recommends storing two dozen sealed glass ampoules for use when needed to cross-check the isotopic value in the future, or for inter-laboratory comparisons. In addition, a second set of 30–40 sealed glass ampoules to be used as daily working standards for daily standardization, controls on experiments, and for correcting problems with the performance of the instrument are recommended. If the standards are stored in screw cap bottles or vials these bottles should be stored without headspace to minimize possible fractionation through volatilization.[97] However, this is obviously not valid for all types of compounds. Some can evaporate through the septa or become contaminated by contact with the septa. Hence, the right method of storage has to be evaluated on an individual basis. If in doubt, flame-sealed tubes are the best bet.[10]

5.5 REPEATABILITY, REPRODUCIBILITY, LINEARITY, STABILITY, DETECTION LIMITS AND 'TOTAL UNCERTAINTY'

The repeatability of an analytical method is the precision, which is estimated under repeatability conditions. These conditions are characterized by using the same method on identical test items (same batch)

Table 5.7 Limits on the precision of isotopic measurements as well as typically obtained precisions. Data according to Sessions.[89]

Isotopes	Conversion gas	Ionization efficiencies estimated (10^{-3} ions/ molecule)	Theoretical sensitivity (nmol)	Typical sensitivity (nmol)	Typical precision (‰)
^{2}H/^{1}H	H$_2$	0.1^a	21	10–50	2–5
^{13}C/^{12}C	CO$_2$	1.2	0.024	0.1–5	0.1–0.3
^{15}N/^{14}N	N$_2$	0.9^a	0.11	1–10	0.3–0.7
^{18}O/^{16}O	CO	0.9	0.19	4–14	0.3–0.6
^{34}S/^{32}S	SO$_2$	1.7	0.0048	n.a.	n.a.
^{37}Cl/^{35}Cl	CH$_3$Cl	1.2	0.00066	n.a.	n.a.

^{a}The calculation accounts for the fact that two moles of H and N are required for each mole of sample gas.

in the same laboratory by the same operator using the same equipment within short time intervals. When a CSIA is run in duplicate or triplicate under identical operating conditions, the standard deviation of the mean of the replicate measurements is typically 0.1 to 0.3‰ for δ^{13}C-values (see Table 5.7), depending on the sample.

Reproducibility is defined as the precision obtained under conditions where test results are obtained with the same method on identical test items such as samples or reference materials in different laboratories with different operators using different GC- or LC-IRMS instruments for CSIA.[1,2] It has already been pointed out that in these cases the same normalization procedures have to be used. However, in contrast to concentration analysis, there are hardly any inter-laboratory comparisons or round-robin studies reported for CSIA and often only instrumental precision or repeatability are determined.

Stability relating to instrumental drift is of importance because in contrast to dual-inlet measurements it is not possible to compare reference gas and sample within a few seconds. In CSIA it is not always possible to use a reference gas within a chromatographic run, therefore the instrumental drift should be as low as possible. The instrumental drift, for example, for carbon isotope ratio measurements of modern IRMS instruments is in the order of ± 0.2‰ h^{-1}.[98] However, this is only a guiding value, which cannot be achieved for all measurements. An example is tree ring analysis with GC runs exceeding 10 000 seconds. There is no way that the instrument is stable to ± 0.2‰ over such a long time period. Hence, one needs to follow the state of the instrument by repeated reference injections and local standards.[10] The most important measure in such demanding analyses, however, is to keep moisture out of the ion source, which can be done routinely by immersing the transfer line between the Nafion™ dryer and the open-split into a dry-ice trap.

This avoids variable conditions for ion-molecule reactions in the ion source.[10]

The term linearity as it is used in analytical chemistry usually refers to a linear increase of the signal with increasing amount of analyte. In contrast, the term linearity as used in isotope ratio analysis indicates that (within an acceptable range) the obtained isotope ratio is independent of the amount of compound injected. This so-called ratio linearity L_R in nA^{-1} is defined for viscous flow as well as for continuous-flow instruments as:

$$L_R = \frac{R_2 - R_1}{i_2 - i_1} \qquad (5.11)$$

whereby isotope ratios R_1 and R_2 with the corresponding ion currents i_1 and i_2 are measured. In continuous-flow instruments L_R can be obtained by introducing two different amounts of each target compound and application of eqn (5.11). However, in order to validate the ratio linearity it would be better to use more than two different amounts of each target compound for a regression analysis. In CSIA, the knowledge of the linear range or amount independence is of importance because target compounds will elute with different intensities from the column depending on their concentration in the sample and also with different retention times as peaks tend to broaden during long separation times. In contrast to dual-inlet an equilibration of peak heights of sample and reference gas pulses is not possible. Therefore it is necessary to evaluate the linear range within a method development extensively for all target compounds. Prior to isotope analysis of real samples, concentrations of target analytes need to be known and, if necessary, adapted to the linear range either by dilution or by use of improved extraction methods. A typical linearity evaluation of a laboratory working standard run by CSIA over a wide range of different peak sizes (or signal sizes) by varying the amount of analyte introduced is shown in Figure 5.7.[99] The linear range is limited on the lower concentration side by the detection limit and on the upper concentration side by column and conversion interface capacity, space charge effects and isobaric interferences in the ion source.

There is an increasing interest in CSIA in studying compounds at low concentrations in a sample. (see Chapters 4 and 6).[88] In such studies it is necessary to know the lowest measurable concentrations that allow for an accurate isotope analysis or, in other words, the detection limits.

In first instance, the precision of an isotope ratio measurement depends on the counting statistic and instrumental noise.[89] In modern

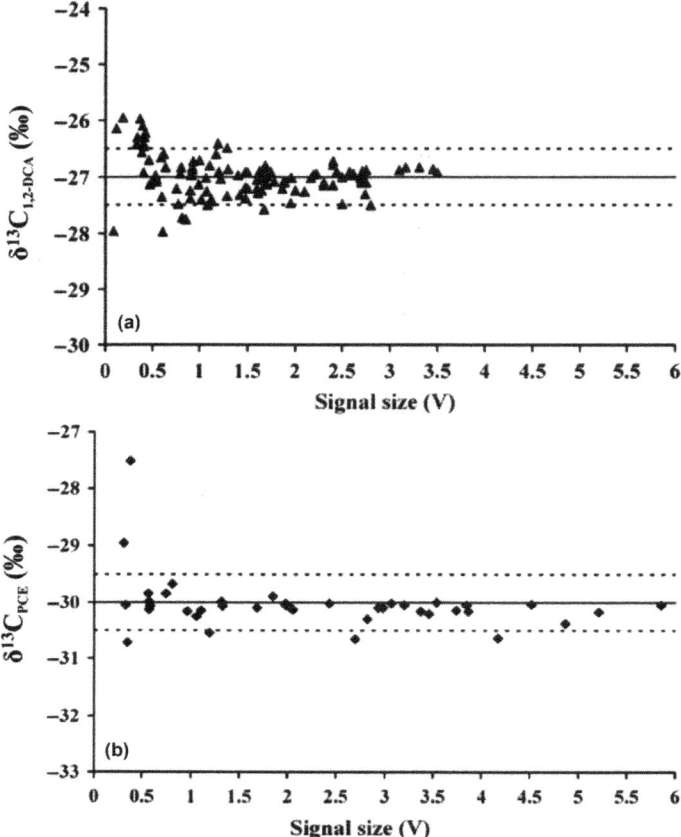

Figure 5.7 Typical linearity test of a laboratory working standard run by CSIA over a wide range of different peak sizes (or signal sizes) by varying the amount of analyte introduced. a) Results from experiments measuring $\delta^{13}C$-values of isotopically characterized laboratory working standard 1,2-dichloroethane (1,2-DCA) by CF-IRMS. CO_2 isotopic reference standard gas is set at 1–2 V, while the signal size for 1,2-DCA was varied in the range from 0.2 to 3.5 V. The mean $\delta^{13}C$-value of all analyses is -27.0 ± 0.4‰ (indicated by the solid black line through the data) and by the hatched lines above and below. Variance in $\delta^{13}C$-values for different signal size intervals is < 0.5 (0.27), 0.5–1 (0.14), 1–2 (0.070), and > 2 V (0.036). b) Results from experiments measuring $\delta^{13}C$-values of isotopically characterized laboratory working standard perchloroethylene (PCE) by CF-IRMS. CO_2 isotopic reference standard gas is set at 1–2 V, while the signal size for PCE was varied in the range from 0.3 to 6 V. The mean value of all analyses is -30.0 ± 0.5‰ (indicated by the solid black line through the data) and by the hatched lines above and below. Variance in $\delta^{13}C$-values for different signal size intervals is < 0.5 (1.9), 0.5–1 (0.031), 1–2 (0.036), and > 2 V (0.035).
(Graphic from Sherwood Lollar et al.[99] printed with permission of American Chemical Society.)

IRMS the precision obtained in ion beam measurements is the only significant source of noise. This is called the 'shot-noise limit'.

The precision with which a δ-value can be determined in terms of its standard deviation, can be expressed by a Gaussian propagation of errors.[100] If we assume that the associated standard deviations are independent of each other we can write,

$$\sigma^2_{\delta^h E_{c,ref}} = \left(\frac{\partial \delta^h E_{c,ref}}{\partial R(^h E/^l E)_c}\right) \sigma^2_{R(^h E/^l E)_c} + \left(\frac{\partial \delta^h E_{c,ref}}{\partial R(^h E/^l E)_{ref}}\right) \sigma^2_{R(^h E/^l E)_{ref}} \quad (5.12)$$

where $\partial \delta^h E_{c,ref}/\partial R(^h E/^l E)_c$ and $\partial \delta^h E_{c,ref}/\partial R(^h E/^l E)_{ref}$ are the derivation of the sample δ-value to the abundance ratio of the sample and the reference material and $\sigma^2_{R(^h E/^l E)_c}$ and $\sigma^2_{R(^h E/^l E)_{ref}}$ are standard deviations of the sample and reference abundance ratios, respectively.

The calculation of the shot-noise limit is based entirely on the ion counting statistics. As we have seen in Chapter 3, the abundance ratio $R(^h E/^l E)$ of two isotopic species is given by the ion currents of the minor and major ion beam $^h i$ and $^l i$, respectively, and we can write,

$$\sigma^2_{R(^h E/^l E)} = \left(\frac{\partial R(^h E/^l E)}{\partial ^h i}\right) \sigma^2_{^h i} + \left(\frac{\partial R(^h E/^l E)}{\partial ^l i}\right) \sigma^2_{^l i} \quad (5.13)$$

where $\sigma_{^h i}$ and $\sigma_{^l i}$ denote the standard deviations of the minor and major ion currents, respectively. After evaluation of the derivatives and substitution of these results in eqn (5.13) we get

$$\sigma^2_{R(^h E/^l E)} = R(^h E/^l E)^2 \left[\left(\frac{\sigma_{^h i}}{^h i}\right)^2 + \left(\frac{\sigma_{^l i}}{^l i}\right)^2\right] \quad (5.14)$$

Any ion current is given by $i = N q_e / t$, where N is the number of ions collected in a time interval t in seconds and q_e is the charge in Coulombs carried by an ion. The variance in N at the shot-noise limit is N and we can note that:

$$\sigma^2_i = \left(\frac{\partial i}{\partial N}\right) \sigma^2_N = \left(\frac{q_e}{t}\right)^2 N \quad (5.15)$$

Thus, the relative standard deviation of an ion current is given by

$$\left(\frac{\sigma_i}{i}\right)^2 = \frac{1}{N} \quad (5.16)$$

which can be written in terms of the number of ions collected from the minor and major ion beam currents:

$$\left(\frac{\sigma_{R(^hE/^lE)}}{R(^hE/^lE)}\right)^2 = \left(\frac{1}{^hN}\right) + \left(\frac{1}{^lN}\right) \tag{5.17}$$

If we recall $R(^hE/^lE) = {}^hN/{}^lN = {}^hi/{}^li$ from Chapter 3, eqn (5.17) can be rewritten in terms of the number of ions of the major ion beam.

$$\left(\frac{\sigma_{R(^hE/^lE)}}{R(^hE/^lE)}\right)^2 = \frac{1}{{}^lN}\left(\frac{1 + R(^hE/^lE)}{R(^hE/^lE)}\right) \tag{5.18}$$

As discussed in Chapter 3, the molar sensitivity E_M of the IRMS can be described in terms of the number of molecular ions received at the collector per molecule introduced into the ion source. This term is often called efficiency of an IRMS. To a very good approximation, the efficiency is independent of the isotopic composition. With the number of moles n of sample gas molecules introduced and the Avogadro's constant N_A it follows that

$$^lN = \frac{E_M n N_A}{(1 + R(^hE/^lE))} \tag{5.19}$$

By incorporation of eqn (5.19) in eqn (5.18) we obtain

$$\left(\frac{\sigma_{R(^hE/^lE)}}{R(^hE/^lE)}\right)^2 = \frac{(1 + R(^hE/^lE))^2}{E_M n N_A R(^hE/^lE)} \tag{5.20}$$

Together with the definition of the standard deviation for the δ-value and the fact that in practice the values of $R(^hE/^lE)_c$ and $R(^hE/^lE)_{ref}$ are numerically almost equal and differ only in the third or fourth significant figure, we can rewrite eqn (5.12) in terms of a single $R(^hE/^lE)$. When we obtain $\partial\delta/\partial R(^hE/^lE)_{ref} = -R(^hE/^lE)_c/R(^hE/^lE)_{ref}^2$ we can recognize that this quotient differs very little from $-1/R(^hE/^lE)$ and we get an expression for the variance of the δ-value as a function of $R(^hE/^lE)$.

$$\sigma^2_{\delta^hE} = 2\left(\frac{\sigma_{R(^hE/^lE)}}{R(^hE/^lE)}\right)^2 = 2\left(\frac{(1 + R(^hE/^lE))^2}{E_M n N_A R(^hE/^lE)}\right) \tag{5.21}$$

In this equation $E_M n N_A$ is the number of major ion beam ions, for example ions with the m/z 44 in case of carbon measurements. Expressed in terms of the molar amount n of an element required to obtain a certain precision we can rearrange eqn (5.21) to

$$n = 2 \frac{(1 + R(^hE/^lE))^2}{\sigma^2_{\delta^hE} E_M N_A R(^hE/^lE)} \quad (5.22)$$

Sessions *et al.* calculated theoretical sensitivities in terms of nanomoles of gas required by the IRMS to achieve a standard deviation of 0.1‰ when operating at the shot-noise limit for a sample containing a compound at natural isotope abundance. Table 5.7 gives theoretically calculated limits on the precision on isotopic measurements.

For a first rough estimation of the mass of an individual compound in ng that must be delivered on column $m_{E_{OC}}$ to reach the theoretical precision of 0.1‰ the theoretical sensitivity S_{theo} from Table 5.7 can be used:

$$m_{E_{OC}} = \frac{S_{theo} \times M_c}{N_{E_c}} \quad (5.23)$$

where M_c is the molecular weight of the sample (compound) in $g\,mol^{-1}$ and N_{E_c} is the number of atoms of the measured element in the compound. As an example, for the carbon and hydrogen isotope ratio determination of benzene (C_6H_6, $M_c = 78.18\,g\,mol^{-1}$) theoretically ~ 0.3 ng C and ~ 273 ng H are required on the column to achieve a standard deviation of 0.1‰. However, when using the typical sensitivities and precisions given in Table 5.7 the required masses are often higher.

Even in concentration analysis there is no generally accepted or regulated definition of an expression for the lowest analyte concentration to be measured with a given accuracy. Thus, many different methods, for example, based on signal to noise ratios, calibration data or precision at low concentrations are used. In isotope ratio analysis, the detection limit is, rather, given by a minimal amount of element required to yield a threshold precision of the isotope ratios determined. This amount is then often given on a per mol or per gram element basis (see Table 5.7 and previous example).[88] Apart from these theoretical limits given by shot-noise, precision of measurements varies from compound to compound and is also affected by the analytical separation techniques as well as the sample preparation.[101] It is therefore favourable to determine the total or combined uncertainty of a method that is necessary to resolve investigated isotopic fractionation processes or to distinguish between samples based on differences in isotopic composition.[99] As an example, in contaminant hydrology a combined uncertainty of ± 0.5‰ for $\delta^{13}C$ is required to resolve biodegradation processes.[97] In contrast to this, the determination of changing isotopic CO_2 signatures in the atmosphere require combined uncertainties in the order of ± 0.01‰.[6]

In order to transfer these demands into measuring practice, often conservative amplitude offset values are defined, below which uncertainties become unacceptable for a certain compound, method and research question.

Another approach is the determination of the method detection limit (MDL) by a moving mean method as illustrated in Figure 5.8 for carbon isotope analysis. In a first step, the mean δ^{13}C-values of the three highest concentration levels are determined. A predefined combined uncertainty interval is set around the calculated mean value or the EA or dual-inlet value. This interval incorporates the analytical uncertainty including the internal repeatability on triplicate measurements as well as the trueness of the measurement with respect to reference material.[102,103] The moving mean procedure is repeated consecutively by including the δ^{13}C-value of the next lower concentration level into the mean value calculation. The last concentration for which the δ^{13}C-value falls within this iterative interval and for which the standard deviation is lower than the predefined uncertainty (e.g., ± 0.5‰ in contaminant hydrology) for triplicate measurements is defined as the MDL. In case of known EA- or dual-inlet-derived isotope values of the pure target compound this value can be used instead of the moving mean.

5.6 QUALITY ASSURANCE (QA) AND QUALITY CONTROL (QC) IN CSIA

To obtain accurate results for CSIA measurements the sensitivity should be <2000 molecules/ion and the instrumental linearity should be better than 0.1‰ nA^{-1}.[104] For isotope ratio determination of individual compounds by CSIA several procedures are necessary prior, during and after measurements to ensure reliable data. The measures that should be taken to maintain quality for CSIA measurements in the indicated regular intervals are indicated in Table 5.8.

It is useful to keep control charts for these parameters to understand possible failures and changes in accuracy and precision, to take steps to restore optimal conditions as fast as possible and allow for continuous monitoring of QA/QC over the long term. Spreadsheets with automatic generation of the diagrams are advantageous. Before starting a sample measurement campaign the following steps are recommended to be applied for QA and realization of accurate measurements as introduced in the *EPA Guideline for assessing biodegradation and source*

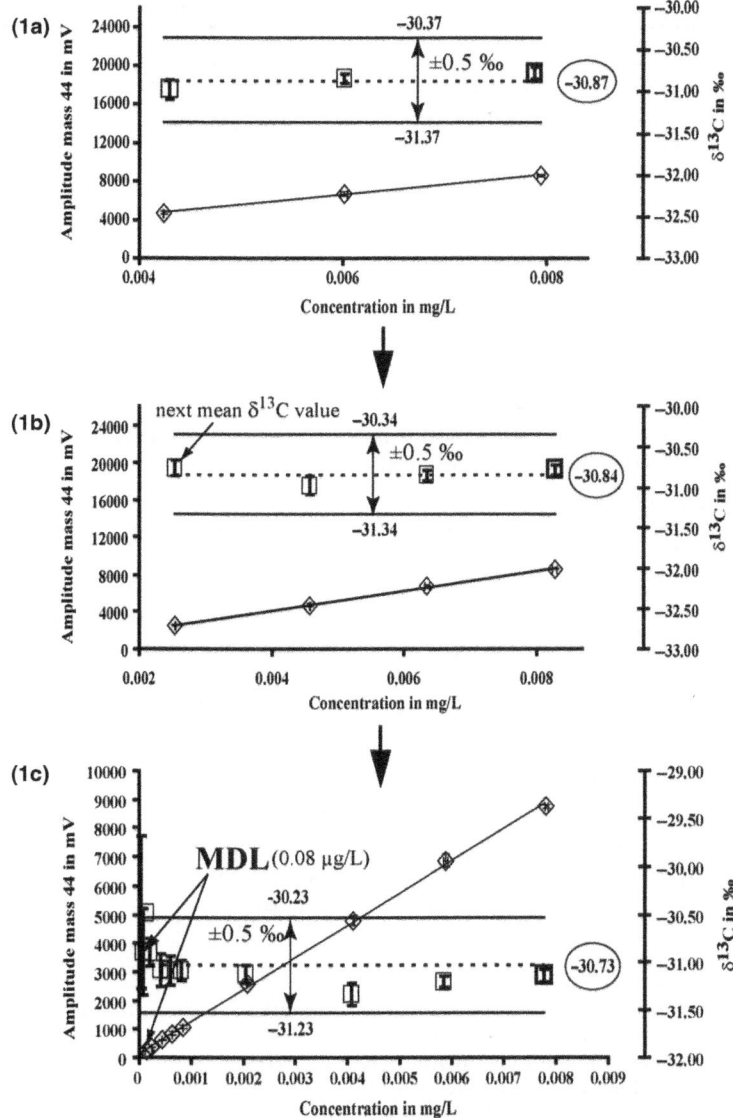

Figure 5.8 Schematic illustration of MDL determination for fluorobenzene. 1a) the determination of the moving mean of mean δ^{13}C-values (each n = 3) for the first three concentration levels. A ±0.5‰ interval was set around this moving mean value. All δ^{13}C mean values are within this interval. Because of this, in 1b), the next moving mean value incorporating the mean δ^{13}C-value at the next lower concentration was calculated. This procedure was iterated as long as either a δ^{13}C-value was outside the ±0.5‰ interval around the moving mean or the standard deviation of a mean δ^{13}C-value was higher than ±0.5‰. In 1c), both abort criteria can be observed. The MDL is then defined as the last concentration with a δ^{13}C mean value that fulfils both criteria.

Table 5.8 Maintenance and tests in CSIA quality assurance.

Maintenance and tests	Interval
Checking background values by magnetic field scan (background scan) and/or by monitoring single m/z: Water (m/z 18) Nitrogen (m/z 28) Oxygen (m/z 32) (These background values depend on the kind of separation technique (LC, GC) and the used interface, according to manufacturer specifications.)	daily
Check vacuum	daily
Leak check with argon (m/z 40), IRMS only, straight mode (open-split in, backflush off)	daily
Standard on/off test with reference gas pulses to check stability, precision and performance of the IRMS	daily
Linearity test with reference gas pulses and standards	daily
H_3^+ factor in case of hydrogen measurements (typical value is 10 ppm nA^{-1} with a stability of 0.02 ppm nA^{-1} h^{-1})[105]	daily
5 min CO_2 background stability check in case of LC-IRMS	daily
Peak shape test	weekly to monthly
Open the gas ballast of rotary vane pump for around 15 min.	monthly
Amplifier test Informs about the background noise of electronic devices, therefore the source has to be switched off. For specifications of each single cup see manufacturer manuals.	
Signal stability test Describes the stability of the signal intensity, therefore the intensity on the top of a mass peak is measured for a certain period of time. For specifications see manufacturer manuals.	
System stability test Describes the high voltage and magnetic field stability, therefore the slope of intensity at 50% peak flank is measured because high voltage fluctuations can be better observed. The slope of the relative mass drift versus time in min^{-1} and the scatter of the mass in min^{-1} is determined by the standard deviation of the slope. For specifications see manufacturer manuals.	monthly to annually
Changing oil of the rotary vane pump and lubricant reservoir of the turbo pumps	annually
Cleaning the ion source	annually (depending on the number and kind of samples measured)

identification of organic groundwater contaminants using CSIA.[97] For additional steps that are used by WADA in doping control analysis we refer to Chapter 6 and specialized literature.[106]

(i) The samples should be screened by LC-/GC-MS or -FID prior to CSIA measurements to avoid overloading of the GC- and LC-IRMS system with non-target analytes and to adapt concentrations to the linear range of the instrument and exclude contamination and diminished reactor capacity by main sample components.[107]

(ii) A chromatographic blank needs to be measured before sample runs to ensure no high column bleeding or ghost peaks from the system.

(iii) A solvent blank or a blank of an enrichment device (Purge&Trap, SPME fibre, etc.) should be measured prior to sample run to exclude the presence of interfering compounds.

(iv) The linearity of the instrument should be tested with the compound-specific working standards over the typical range of operating conditions that will be used. The operating conditions include the range of concentrations, split or flow settings, and the technique used to prepare the samples, such as a headspace sampler, SPME or Purge&Trap etc.

(v) Ideally, samples should be measured in triplicate and an isotopically characterized multicomponent working standard should be included each ninth measurement to control accuracy. However, this procedure is applicable for scientific investigations but is often unacceptable for commercial labs because of the small sample throughput. Even in commercial environments, at least every fifth sample should be a replicate and at least every tenth sample should be the compound-specific working standard.

(vi) All samples should stay within the previously established range of acceptable linearity and above the established detection limit. If a sample falls outside the acceptable range, the concentrations of the analytes should be adjusted, if possible, to bring the sample within the established range, and the sample analysed a second time.[97]

5.7 WEB FINDINGS

Caltech Oral Histories – The Caltech Institute Archives
Interview with Heinz A. Lowenstam from 1991 by the oral history project of the California Institute of Technology Archives. Lowenstam

joined the Urey's Chicago group in 1948 and worked there on the field of palaeothermometry. The interview is available as a pdf file.
http://oralhistories.library.caltech.edu/66/

LIMS for Light Stable Isotopes is a Laboratory Information Management System

Tylor Coplen provides a Laboratory Information Management System LIMS for light stable isotopes based on Microsoft Access for managing samples, analyses, normalization, and other data in a stable isotope laboratory.[108]
http://isotopes.usgs.gov/research/topics/lims.html

Commission on Isotopic Abundances and Atomic Weights (CIAAW)
Up-to-date revisions of reference values.
http://ciaaw.org/index.htm

Vendors and Sources of Isotopic Reference Material

International Atomic Energy Agency (IAEA) Section of Isotope Hydrology
P.O. Box 100
1400 Vienna, Austria
A catalogue of stable isotope IAEA standard reference materials is available at http://curem.iaea.org/catalogue/SI/index.html

National Institute of Standards and Technology (NIST) Standard Reference Material Program
Room 204, Building 202 Gaithersburg, MD 20899–0001 USA
NIST standard reference materials are available at
http://www.nist.gov/ts/msd/srm/index.cfm

Institute for Reference Measurements and Materials (IRMM) Commission of the European Communities-JRC
B-2440 Geel, Belgium

Standards for CSIA

Arndt Schimmelmann, Ph.D.
Indiana University
Department of Geological Sciences
Biogeochemical Laboratories
1001 East 10th Street
Bloomington, IN 47405–1405, USA
E-mail: aschimme@indiana.edu
http://mypage.iu.edu/~aschimme/hc.html

ISOFLEX Europe
Grote Beer 569
3067 TR Rotterdam,
The Netherlands
Tel: +31 10-744-2266
Fax: +31 10-785-4044
http://www.isoflex.com/index.html

Euriso-Top
Parc des Algorithmes, Bâtiment Homère
Route de l'Orme
F-91194 Saint-Aubin Cedex
Tel: +33 1 69 41 95 96
Fax: +33 1 69 41 93 52
E-mail: eurisotop@eurisotop.com
http://www.eurisotop.com/

Chiron AS
Stiklestadveien 1
7041 Trondheim
Norway
Tel: +47 73-874490
Fax: +47 73-874499
E-mail: chiron@chiron.no
http://www.chiron.no

REFERENCES

1. *Guide to the Expression of Uncertainty in Measurement;* ISO, Geneva, 1993; ISBN: 92-67-10188-9.
2. *Analytical Methods Committee technical brief No.13 Terminology – the key to understanding analytical science. Part 1: Accuracy, precision and uncertainty*, Royal Society of Chemistry, London, 2003.
3. *Uncertainty of Measurement – Part 3: Guide to the expression of uncertainty in measurement (GUM:1995);* ISO, Geneva, 2008.
4. M. Gröning, in *Handbook of Stable Isotope Analytical Techniques*, ed. P. A. De Groot, Elsevier, Amsterdam, 2004, vol. 1, pp. 874–906.
5. P. D. P. Taylor, P. De Bièvre and S. Valkiers, in *Handbook of Stable Isotope Analytical Techniques*, ed. P. A. De Groot, Elsevier, Amsterdam, 2004, vol. 1, pp. 907–927.
6. T. B. Coplen, in *Elemental and Isotope Ratio Mass Spectrometry*, ed. D. Beauchemin and D. Matthews, Elsevier, Amsterdam, 2010, pp. 802–810.

7. H. Kipphardt, in *Handbook of Stable Isotope Analytical Techniques*, ed. P. A. De Groot, Elsevier, Amsterdam, 2004, vol. 1, pp. 928–943.
8. *Isotope Aided Studies of Atmospheric Carbon Dioxide and Other Greenhouse Gases, Phase II, IAEA-TECDOC-1269*, International Atomic Energy Agency, Vienna, 2002.
9. W. A. Brand, L. Huang, H. Mukai, A. Chivulescu, J. M. Richter and M. Rothe, *Rapid Commun. Mass Spectrom.*, 2009, **23**, 915–926.
10. W. Brand, *Personal Communication*.
11. T. B. Coplen, *Pure Appl. Chem.*, 1994, **66**, 273–276.
12. T. B. Coplen, *Geochim. Cosmochim. Acta*, 1996, **60**, 3359.
13. T. B. Coplen, W. A. Brand, M. Gehre, M. Gröning, H. A. J. Meijer, B. Toman and R. M. Verkouteren, *Rapid Commun. Mass Spectrom.*, 2006, **20**, 3165–3166.
14. T. B. Coplen, W. A. Brand, M. Gehre, M. Gröning, H. A. J. Meijer, B. Toman and R. M. Verkouteren, *Anal. Chem.*, 2006, **78**, 2439–2441.
15. E. Barkan and B. Luz, *Rapid Commun. Mass Spectrom.*, 2005, **19**, 3737–3742.
16. J. K. Böhlke and T. B. Coplen, in *Reference and Intercomparison Materials for Stable Isotopes of Light Elements, IAEA-TECDOC-825*, International Atomic Energy Agency, Vienna, 1995.
17. Z. Sharp, *Principles of Stable Isotope Geochemistry*, 1st edn., Pearson/Prentice Hall, Upper Saddle River, NJ; London, 2007.
18. S. Epstein and T. Mayeda, *Geochim. Cosmochim. Acta*, 1953, **4**, 213–224.
19. R. Gonfiantini, W. Stichler and K. Rozanski, in *Reference and Intercomparison Materials for Stable Isotopes of Light Elements, International Atomic Energy Agency, IAEA-TECDOC-825*, International Atomic Energy Agency, Vienna, 1995.
20. H. Craig, *Science*, 1961, **133**, 1833–1834.
21. F. L. Mohler, *Science*, 1955, **122**, 334–335.
22. F. L. Mohler, *Isotopic Abundance Ratios Reported for Reference Samples Stocked by the National Bureau of Standards. National Bureau of Standards (USA), Technical Note No*, 1960, **51**, 8.
23. International Atomic Energy Agency and International Union of Geodesy and Geophysics, *Isotopes in Hydrology: proceedings of the symposium*, International Atomic Energy Agency, Vienna, 1967.
24. T. B. Coplen and R. N. Clayton, *Geochim. Cosmochim. Acta*, 1973, **37**, 2347–2349.
25. R. Gonfiantini, *Nature*, 1978, **271**, 534–536.

26. G. Hut, *Consultants' Group Meeting on Stable Isotope Reference Samples for Geochemical and Hydrological Investigations, Sept 16–18, 1985, Report to the Director General*, International Atomic Energy Agency, Vienna, 1987.
27. *IAEA, Isotope Hydrology Laboratory, Info Sheet-VSMOW2-SLAP2 Information Sheet on the new International Measurement Standards VSMOW2 and SLAP2*, 2007.
28. Y. Lin, R. N. Clayton and M. Gröning, *Rapid Commun. Mass Spectrom.*, 2010, **24**, 773–776.
29. *Commission on Isotopic Abundances and Atomic Weights (CIAAW); Reference materials;* http://www.ciaaw.org/.
30. R. M. K. Verkouteren, D.B., *Value Assignment and Uncertainty Estimation of Selected Light Stable Isotope Reference Materials: RMs 8543-8545, RMs 8562-8564, and RM 8566, NIST Special Publication 260-149*, National Institute of Standards and Technology, Gaithersburg, Maryland, 2004.
31. W. A. Brand, T. B. Coplen, A. T. Aerts-Bijma, J. K. Böhlke, M. Gehre, H. Geilmann, M. Gröning, H. G. Jansen, H. A. J. Meijer, S. J. Mroczkowski, H. P. Qi, K. Soergel, H. Stuart-Williams, S. M. Weise and R. A. Werner, *Rapid Commun. Mass Spectrom.*, 2009, **23**, 999–1019.
32. R. Gonfiantini, *Advisory Group Meeting on Stable Isotope Reference Samples for Geochemical and Hydrological Investigations, IAEA, Vienna, 19–21 September 1983, Report to the Director General*, International Atomic Energy Agency, Vienna, 1984, 77.
33. T. B. Coplen and H. Qi, *Rapid Commun. Mass Spectrom.*, 2010, **24**, 2269–2276.
34. T. B. Coplen, C. Kendall and J. Hopple, *Nature*, 1983, **302**, 236–238.
35. T. B. Coplen and H. Qi, *Forensic Sci. Int.*, 2012, **214**, 135–141.
36. T. B. Coplen, J. A. Hopple, J. K. Böhlke, H. S. Peiser, S. E. Rieder, H. R. Krouse, K. J. R. Rosman, T. Ding, J. Vocke, R.D., K. M. Révész, A. Lamberty, P. Taylor and P. De Bièvre, *U.S. Geological Survey Water-Resources Investigations Report 01–4222*, 2001, 98.
37. J. K. Böhlke, S. J. Mroczkowski and T. B. Coplen, *Rapid Commun. Mass Spectrom.*, 2003, **17**, 1835–1846.
38. E. A. Wachter and J. M. Hayes, *Chem. Geol.*, 1985, **52**, 365–374.
39. International Atomic Energy Agency, *IAEA-TECDOC-825: Reference and Intercomparison Materials for Stable Isotopes of Light Elements. Proceedings of a consultant meeting in Vienna*, Vienna, 1993.

40. C. R. McKinney, J. M. McCrea, S. Epstein, H. A. Allen and H. C. Urey, *Rev. Sci. Instrum.*, 1950, **21**, 724–730.
41. H. C. Urey, H. A. Lowenstam, S. Epstein and C. R. McKinney, *Geol. Soc. Am. Bull.*, 1951, **62**, 399–416.
42. H. Craig, *Geochim. Cosmochim. Acta*, 1953, **3**, 53–92.
43. S. Epstein, R. Buchsbaum, H. Lowenstam and H. C. Urey, *Geol. Soc. Am. Bull.*, 1951, **62**, 417–426.
44. S. Epstein, R. Buchsbaum, H. A. Lowenstam and H. C. Urey, *Geol. Soc. Am. Bull.*, 1953, **64**, 1315–1325.
45. S. Epstein and H. A. Lowenstam, *J. Geol.*, 1953, **61**, 424–438.
46. H. Aspaturian, California Institute of Technology Archives, Pasadena, 1991.
47. R. A. Werner and W. A. Brand, *Rapid Commun. Mass Spectrom.*, 2001, **15**, 501–519.
48. H. Craig, *Geochim. Cosmochim. Acta*, 1957, **12**, 133–149.
49. A. O. Nier, *Phys. Rev.*, 1950, **77**, 789–793.
50. R. Gonfiantini, *J. Hydrol.*, 1984, **72**, 205.
51. L. Stalker, A. J. Bryce and A. S. Andrew, *Org. Geochem.*, 2005, **36**, 827–834.
52. Q. L. Zhang and W. J. Li, *Chin. Sci. Bull.*, 1990, **35**, 291.
53. M. Berglund and M. E. Wieser, *Pure Appl. Chem.*, 2011, **83**, 397–410.
54. *EUROPEAN COMMISSION Joint Research Centre IRMM Geel, Belgium, Certified Reference Materials 2012.*
55. C. Guillou, G. Remaud and M. Lees, *EUR 20064-The Certification of the Carbon-13 and Deuterium Isotopic Ratio Content and the Alcoholic Grade of Ethanol from Wine, BCR-656 (96% vol.). The Carbon-13 Isotopic Content of Sugar, BCR-657. The Oxygen-18 Isotopic Ratio Content of Water from Wine, BCR-658 (7% vol.) BCR-659 (12% vol.). The Carbon-13 and Deuterium Isotopic Ratio Content and the Alcoholic Grade of Ethanol from Wine BCR-660 (12 vol.%)*, European Commission, Luxembourg, 2001.
56. H. Craig and H. C. Urey, *Isotopic and Cosmic Chemistry*, North-Holland Pub. Co., Amsterdam, 1964.
57. A. Schimmelmann, *Personal Communication*.
58. H. P. Qi, T. B. Coplen, H. Geilmann, W. A. Brand and J. K. Böhlke, *Rapid Commun. Mass Spectrom.*, 2003, **17**, 2483–2487.
59. G. Junk and H. J. Svec, *Geochim. Cosmochim. Acta*, 1958, **14**, 234–243.
60. A. Mariotti, *Nature*, 1983, **303**, 685–687.
61. C. Kendall and E. Grim, *Anal. Chem.*, 1990, **62**, 526–529.

62. J. K. Böhlke and T. B. Coplen, *Interlaboratory Comparison of Reference Materials for Nitrogen-Isotope-Ratio Measurements*, in *Reference and Intercomparison Materials for Stable Isotopes of Light Elements*, International Atomic Energy Agency, *IAEA-TECDOC-825*, International Atomic Energy Agency, Vienna, 1995, 51–66.
63. G. Junk and H. J. Svec, *Geochim. Cosmochim. Acta*, 1958, **14**, 234–243.
64. P. De Bièvre, S. Valkiers, H. S. Peiser, P. D. P. Taylor and P. Hansen, *Metrologia*, 1996, **33**, 447–455.
65. H. R. Krouse and T. B. Coplen, *Pure Appl. Chem.*, 1997, **69**, 293–295.
66. T. B. Coplen and H. R. Krouse, *Nature*, 1998, **392**, 32.
67. J. MacNamara and H. G. Thode, *Phys. Rev.*, 1950, **78**, 307–308.
68. M. L. Jensen, *Biogeochemistry of Sulfur Isotopes*, u.p., [S.l.], 1962.
69. G. Beaudoin, B. E. Taylor, D. Rumble and M. Thiemens, *Geochim. Cosmochim. Acta*, 1994, **58**, 4253–4255.
70. A. Amrani, A. L. Sessions and J. F. Adkins, *Anal. Chem.*, 2009, **81**, 9027–9034.
71. B. W. Robinson, *Sulphur Isotope Standards*, in *Reference and Intercomparison Materials for Stable Isotopes of Light Elements, International Atomic Energy Agency, IAEA-TECDOC-825*, International Atomic Energy Agency, Vienna, 1995, 39–45.
72. H. P. Qi and T. B. Coplen, *Chem. Geol.*, 2003, **199**, 183–187.
73. Q. L. Zhang and T. Ding, *Chin. Sci. Bull.*, 1989, **34**, 1086–1089.
74. Y. K. Xiao, Z. Yinming, W. Qingzhong, W. Haizhen, L. Weiguo and C. J. Eastoe, *Chem. Geol.*, 2002, **182**, 655–661.
75. T. B. Coplen, J. K. Böhlke, P. De Bièvre, T. Ding, N. E. Holden, J. A. Hopple, H. R. Krouse, A. Lamberty, H. S. Peiser, K. Révész, S. E. Rieder, K. J. R. Rosman, E. Roth, P. D. P. Taylor, R. D. Vocke Jr and Y. K. Xiao, *Pure Appl. Chem.* 2002, **74**, 1987–2017.
76. H. G. M. Eggenkamp and M. L. Coleman, *Chem. Geol.*, 2000, **167**, 393–402.
77. D. Paul, G. Skrzypek and I. Forizs, *Rapid Commun. Mass Spectrom.*, 2007, **21**, 3006–3014.
78. T. B. Coplen, *Rapid Commun. Mass Spectrom.*, 2011, **25**, 2538–2560.
79. P. M. Grootes, W. G. Mook and J. C. Vogel, *Zeitschrift Für Physik*, 1969, **221**, 257–273.
80. T. B. Coplen, *Personal Communication*.
81. R. M. Verkouteren and J. N. Lee, *Fresenius J. Anal. Chem.*, 2001, **370**, 803–810.

82. T. B. Coplen, *Chem. Geol.*, 1988, **72**, 293–297.
83. W. A. Brand and T. B. Coplen, *Fresenius J. Anal. Chem.*, 2001, **370**, 358–362.
84. D. A. Merritt, W. A. Brand and J. M. Hayes, *Org. Geochem.*, 1994, **21**, 573–583.
85. W. A. Brand, *J. Mass Spectrom.*, 1996, **31**, 225–235.
86. W. Meier-Augenstein, *Anal. Chim. Acta*, 2002, **465**, 63–79.
87. W. Meier-Augenstein, *Rapid Commun. Mass Spectrom.*, 1997, **11**, 1775–1780.
88. J. T. Brenna, T. N. Corso, H. J. Tobias and R. J. Caimi, *Mass Spectrom. Rev.*, 1997, **16**, 227–258.
89. A. L. Sessions, *J. Sep. Sci.*, 2006, **29**, 1946–1961.
90. A. Schimmelmann, A. Albertino, P. E. Sauer, H. P. Qi, R. Molinie and F. Mesnard, *Rapid Commun. Mass Spectrom.*, 2009, **23**, 3513–3521.
91. B. E. Kornexl, R. A. Werner and M. Gehre, *Rapid Commun. Mass Spectrom.*, 1999, **13**, 1248–1251.
92. F. Serra, A. Janeiro, G. Calderone, J. M. Moreno-Rojas, C. Rhodes, L. A. Gonthier, F. Martin, M. Lees, A. Mosandl, S. Sewenig, U. Hener, B. Henriques, L. Ramalho, F. Reniero, A. J. Teixeira and C. Guillou, *J. Mass Spectrom.*, 2007, **42**, 361–369.
93. G. Calderone, F. Serra, M. Lees, A. Mosandl, F. Reniero, C. Guillou and J. M. Moreno-Rojas, *Rapid Commun. Mass Spectrom.*, 2009, **23**, 963–970.
94. D. Juchelka, T. Beck, U. Hener, F. Dettmar and A. Mosandl, *HRC-J. High Resolut. Chromatogr.*, 1998, **21**, 145–151.
95. D. Juchelka, A. Hilkert and M. Krummen, *Thermo Scientific Application Note*, **30108**.
96. R. J. Caimi, L. A. Houghton and J. T. Brenna, *Anal. Chem.*, 1994, **66**, 2989–2991.
97. D. Hunkeler, R. U. Meckenstock, B. Sherwood Lollar, T. C. Schmidt and J. T. Wilson, *USEPA Guideline EPA 600/R-08/148*, 2008.
98. W. A. Brand, *Advances in Mass Spectrometry*, 1998, **14**, 661–686.
99. B. Sherwood Lollar, S. K. Hirschorn, M. M. G. Chartrand and G. Lacrampe-Couloume, *Anal. Chem.*, 2007, **79**, 3469–3475.
100. D. A. Merritt and J. M. Hayes, *Anal. Chem.*, 1994, **66**, 2336–2347.
101. J. Schmitt, B. Glaser and W. Zech, *Rapid Commun. Mass Spectrom.*, 2003, **17**, 970–977.
102. H. S. Dempster, B. Sherwood Lollar and S. Feenstra, *Environ. Sci. Technol.*, 1997, **31**, 3193–3197.

103. G. F. Slater, H. S. Dempster, B. Sherwood Lollar and J. Ahad, *Environmental Science and Technology*, 1999, **33**, 190–194.
104. W. Brand, in *Handbook of Stable Isotope Analytical Techniques*, ed. P. A. De Groot, Elsevier, Amsterdam, 2004, vol. 1, pp. 835–856.
105. K. Habfast, *Modern Isotope Ratio Mass Spectrometry*, 1997, 11–82.
106. A. T. Cawley and U. Flenker, *J. Mass Spectrom.*, 2008, **43**, 854–864.
107. M. Blessing, M. A. Jochmann and T. C. Schmidt, *Anal. Bioanal. Chem.*, 2008, **390**, 591–603.
108. T. B. Coplen, *Laboratory Information Management System (LIMS) for Light Stable Isotopes: U.S. Geological Survey Open-File Report 00–345*, 2000, 121.

CHAPTER 6
Applications of Compound-specific Stable Isotope Analysis

6.1 SCOPE

Applications of stable isotope analysis have become numerous over recent years. Most of that work though is still utilizing bulk analysis methods, either the classical dual-inlet mode if the utmost precision is required, or the more recent continuous-flow mode in combination with EAs that allows much higher sample throughput and smaller sample sizes. Since in SIA at natural abundance we always rely on the comparison of isotope compositions of (many) samples this advantage can hardly be overvalued.

Bulk analysis methods are not within the scope of this book, thus we also restrict the exemplary applications discussed below to CSIA that utilizes hyphenation with a chromatographic method and on-line conversion to measurement gases. Although in principle CSIA can also be performed by off-line isolation of target analytes from the matrix of a sample and separation from interfering compounds followed by bulk isotope analysis of the pure compound, this approach, with a few exceptions in food science, is not discussed in the frame of this book. However, this does not imply by any means that we disregard the exciting applications and developments in the area of BSIA. With regard to the often extensive sample preparation prior to any isotope analysis both approaches may require similar steps.

Many CSIA applications rely on *KIEs* that differ among biochemical pathways and result in compounds of varying isotope composition. CSIA is then looking at isotopic differences in these compounds often expressed as a $\Delta\delta$-value. With regard to carbon isotope analysis, this is most obvious in the case of carbon dioxide fixation pathways by C_3- and C_4-plants (see Section 6.2.2), which is the basis for many adulteration studies. Furthermore, depending on external parameters such as humidity, temperature and sunlight availability, the extent of isotope fractionation even for the same pathway may differ, thus sometimes allowing further constraints on product provenance. An example of these secondary effects on isotopic composition from the investigation of a large number of oil samples is shown in Figure 6.1.

In contrast to plants, animals fully rely on external carbon sources from plant or animal matter, thus their carbon composition is governed by their diet. For this reason, one of the tenets in food chemistry has been coined: "you are what you eat (isotopically)".

Nitrogen is well known to reflect trophic levels in food webs and shows a distinct enrichment in marine foodstuffs. However, nitrogen isotope analysis is still a domain of BSIA with only a small, but steadily rising, number of CSIA applications on individual nitrogen-containing organic compounds. Typical ranges of terrestrial δ^{13}C- and δ^{15}N-values within ecosystems are depicted in Figure 6.2.

In the case of hydrogen and oxygen, incorporation in organic compounds typically reflects the local water isotope signature. The Global Network of Isotopes in Precipitation (GNIP, see link list at the end of the chapter) has provided global, spatially resolved data on the isotopic

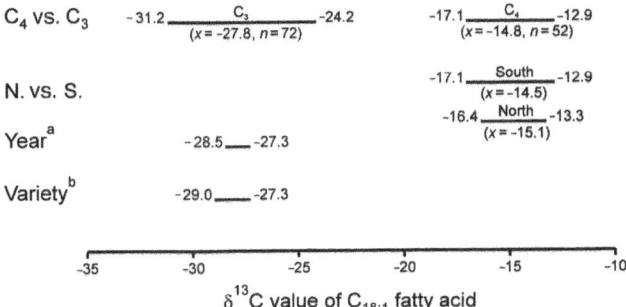

Figure 6.1 Variations in δ^{13}C-values of oleic acid due to plant type, origin, and year of harvest. a) Year-to-year variability is shown for variety Idol (winter oilseed rape) grown in five consecutive years. b) Variability between varieties for 1992/1993.
(Graphic reprinted with permission from Woodbury et al.[1])

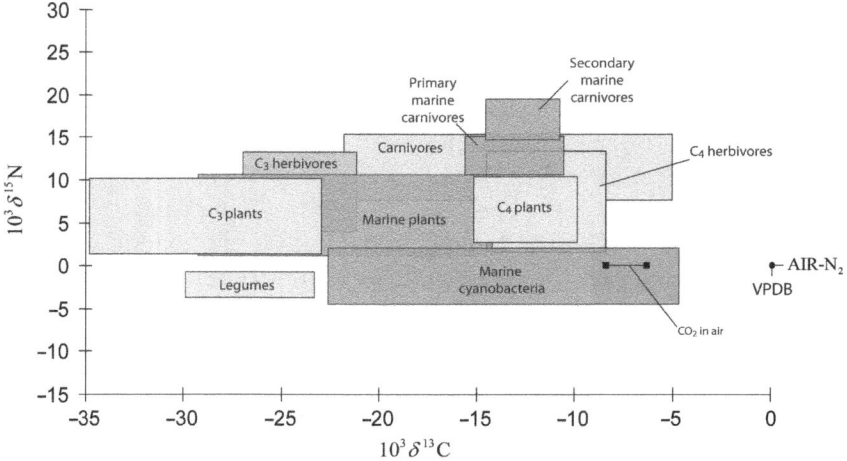

Figure 6.2 Carbon and nitrogen stable isotope ranges in ecosystems.

composition of rain water for decades, which can be linked via the meteoric water line(s) to climate, in particular temperature. As a rule of thumb, δ^2H and $\delta^{18}O$ values decrease with increasing distance to coast, latitude and altitude. Thus, H and O isotope data can provide further evidence for product provenance. In addition to the isotope signature of the source water, fractionation during processes within the plant (transpiration, biosynthesis) influences the final isotope composition in organic compounds produced by the plant. Furthermore, for oxygen one needs to keep in mind that atmospheric sources enriched in ^{18}O compared with source water are also relevant for plant build-up of organic compounds, thus hydrogen is considered to be a better indicator of provenance.[2] On the other hand, for hydrogen a careful validation is required if it exchanges with surrounding water, thus losing the unique isotope signature of the compound. This is especially pronounced for oxygen-bound hydrogen in compounds such as acids, alcohols, phenols and sugars.

Figure 6.3 gives a more detailed overview of isotope signatures of H, O and C in reference materials (SLAP, V-SMOW, V-PDB, see also Chapter 5), large pools of the corresponding elements (oceans, atmosphere), and corresponding isotope ratios in glucose and ethanol. For all elements, incorporation in organic molecules is accompanied by a clear shift in isotope composition. Differences may be conserved though in subsequent biochemical/biosynthetic steps (e.g., fermentation of glucose to ethanol). Table 6.1 summarizes parameters that influence isotopic composition for the four elements discussed above.

Applications of Compound-specific Stable Isotope Analysis

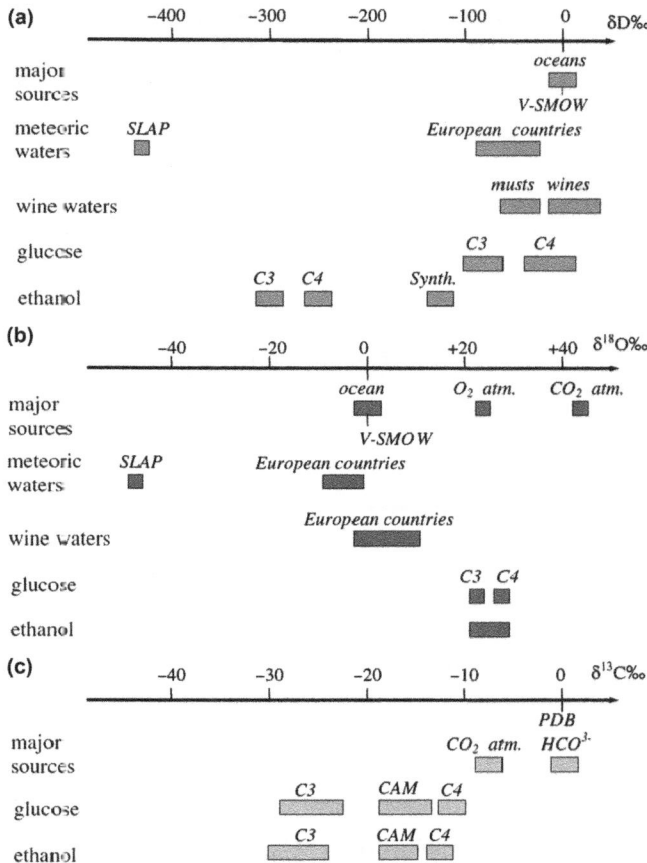

Figure 6.3 Isotope signatures of H, O and C in large reservoirs of the elements, reference materials, and organic compounds relevant in food science. (Graphic reprinted with permission from Calderone and Guillou.[3])

Whereas in these examples organic compounds differ in their isotopic composition as a result of processes prior to their (bio)synthesis, sometimes the process itself rather than the mere products are of interest and can be monitored if data on the *KIEs* involved are available from controlled laboratory studies. Examples are abundant in biochemistry although these have been studied primarily by using isotopically labelled compounds. Natural abundance studies are increasingly used in environmental research, where based on *KIEs* unequivocal evidence for transformation processes may be provided and, in some instances, even the extent of degradation can be estimated.

We have selected six application areas of particular interest for society or related to our own work, i.e., food authenticity, forensics,

Table 6.1 Overview of parameters influencing isotopic composition in plants for the bioelements measured by CSIA.

Element	Primary pool(s)	Primary influencing parameter on isotope composition	Secondary influencing parameter on isotope composition
C	Atmospheric CO_2, geogenic bicarbonate, soil organic matter	Photosynthetic pathway	Climate (humidity, temperature, sunlight), cultivation practice, plant variety, ripening stage, secondary metabolism
H	Ocean water	Distance from coast, latitude, altitude	Temperature
O	Ocean water, atmospheric oxygen	Distance from coast, latitude, altitude	Temperature
N	Atmospheric N_2, soil	Agricultural practice, environmental and climatic parameters	

archaeology, doping control, environmental science, and extraterrestrial materials. Of course, this is not a comprehensive coverage of all areas where CSIA makes important contributions and, depending on their background, some readers may miss applications in geochemistry, food web studies, nutrition, ecology, microbiology, or soil science. For some of these areas though, dedicated and excellent textbooks exist which we refer to.[4–9] However, all of these books cover CSIA only to a small extent.

6.2 AUTHENTICITY OF FOOD AND RELATED COMMODITIES

6.2.1 Scope

The term food authenticity refers to whether the food purchased by a consumer matches its description. Food control must be able to verify these descriptions since fraud in the food sector is abundant and occurs in many ways. This might include (i) dilution with water, (ii) undeclared adulteration with other, cheaper materials, and (iii) non-compliance with the declared geographical, animal or plant origin, all of which are typically done to maximize profit by giving false descriptions. In the following, we will discuss examples of tracing the geographical origin and, in particular, detecting food adulteration based on compound-specific isotope information. With regard to tracing origin and detecting water dilutions, there is abundant literature utilizing bulk isotope measurements. However, this is not within the scope of this book, thus the reader is referred to recent overview articles.[10–13] On the other hand, we also have incorporated studies on essential oils even if these are not solely used for dietary purposes since this is an important and well-covered area in authenticity studies. For the classification of adulteration studies there are various possibilities. One could primarily distinguish (i) the isotopes measured, (ii) the analytical techniques in use, (iii) the target analytes addressed, or (iv) the foodstuff being investigated. We chose to primarily categorize studies with regard to foodstuff although there is a clear relation with the target analytes, i.e., in each foodstuff category one or a few analyte classes are targeted. These are listed in tables summarizing relevant studies of the past 20 years in Sections 6.2.3 to 6.2.8. Note, however, that such a list can never be truly comprehensive and, rather, should provide a selective overview.

First of all, though, a more detailed description of isotope fractionation during primary carbon fixation in photosynthesis is necessary as background, since many of the following applications utilizing carbon isotope signatures are ultimately based on this fundamental process. Actually, this goes way beyond applications in food authenticity and thus is also a basis for applications described in Sections 6.3 to 6.6.

6.2.2 Isotope Fractionation during Carbon Fixation in Photosynthesis

In plants, carbon is primarily made available by photosynthesis. During photosynthesis, carbon dioxide diffuses through the stomata of plant leaves (see Figure 6.4), becomes dissolved in the sub-stomatal cavity and is then fixed by an enzyme in adjacent chloroplasts.

Figure 6.5 depicts the processes involved in photosynthetic carbon fixation by the three domains of plants, i.e. C_3, C_4 and CAM plants that are further explained in the following.

In C_3 plants, CO_2 is fixed in a primary step to ribulose 1,5-bisphosphate (RuBP, containing five carbon atoms) catalysed by the enzyme ribulose 1,5-bisphosphate carboxylase oxygenase, most commonly known as RuBisCO. This step is dominating overall isotope fractionation during photosynthesis and depletes the $^{13}C/^{12}C$ ratio by $\sim -20‰$, i.e., from the mean isotopic composition of atmospheric CO_2 ($-8‰$) to an average $-28‰$ in C_3 plants (see also Figure 6.7). The carboxylated ribulose 1,5-bisphosphate instantaneously breaks up into two molecules of glycerate 3-phosphate. This molecule contains three carbon atoms and is responsible for the name C_3 plants. The photosynthetic cycle of C_3 plants has first been described by Melvin Calvin and co-workers and is therefore sometimes also called the Calvin cycle.

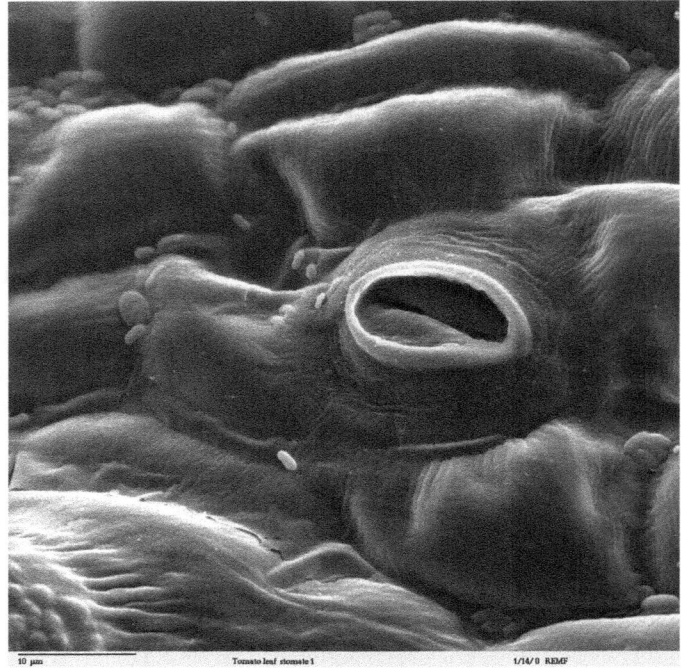

Figure 6.4 Scanning electron microscope picture of a stoma in a tomato leaf. (Picture taken at the Dartmouth Electron Microscope Facility and made available in the public domain by Louisa Howard.)

C_4 plants are believed to have evolved later than C_3 plants in order to better cope with higher temperatures and less humid atmosphere. The reason is that RuBisCO at higher temperatures tends to favour oxygenation of RuBP rather than carboxylation, leading to loss of substrate and energy in a photorespiration process. Thus, C_4 plants have developed a more complex carbon dioxide fixation step that incorporates another primary step. In this primary step, carbon dioxide is fixed to phosphoenolpyruvic acid (PEP, containing 3 carbon atoms) catalysed by the enzyme PEP carboxylate to form oxaloacetate, which contains 4 carbon atoms and thus is responsible for the name C_4 plants. This primary step involves a much smaller isotope fractionation (-2 to $-6‰$)[14] than fixation by RuBisCO. PEP carboxylate is not sensitive to oxygen presence and thus also works efficiently at low CO_2 concentrations. The acid formed is then transferred to the bundle sheath cells where CO_2 is released again and only there fixed to RuBP by RuBisCO in a subsequent Calvin cycle. This more complex carbon dioxide fixation pathway was discovered by M. D. Hatch and C. R. Slack and is

Applications of Compound-specific Stable Isotope Analysis 237

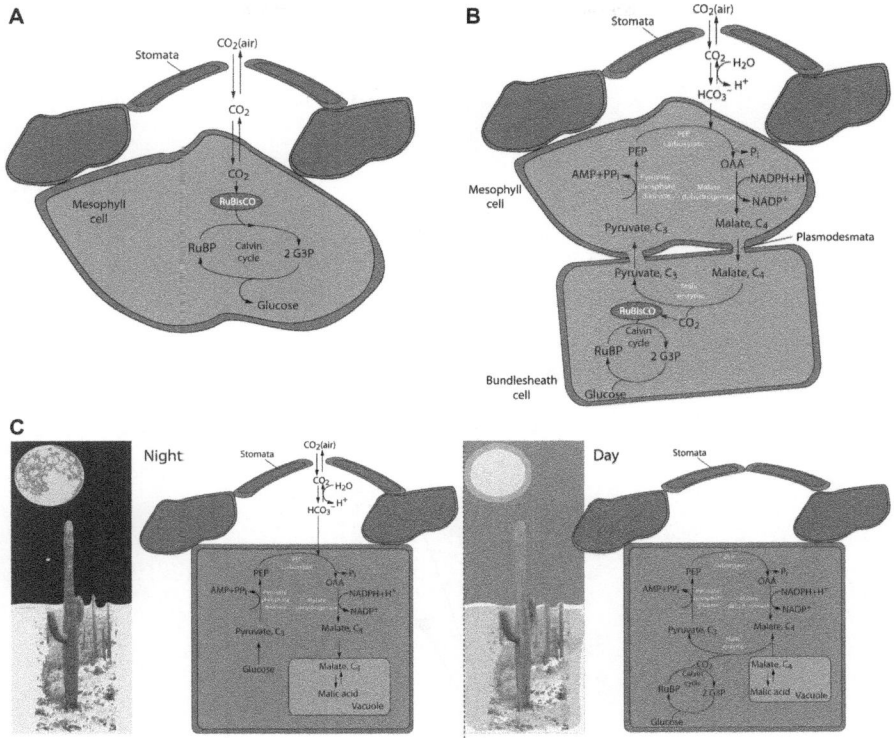

Figure 6.5 Sketch of carbon fixation in plants during photosynthesis. A: C_3 plants, B: C_4 plants, C: CAM plants (night and day cycle shown).

therefore sometimes also called the Hatch–Slack cycle. The disadvantage of the C_4 cycle is that driving it requires more energy, thus at modest temperatures and humidities, C_3 plants naturally prevail. Around 95% of all plant species utilize the C_3-fixation pathway.

A rather small number of plant species uses a third pathway, the so-called Crassulacean acid metabolism to fix carbon dioxide (CAM, named after the acid metabolism by *Crassulaceae*). As in C_4 plants in the primary step of CO_2 fixation, oxaloacetate is formed but in CAM plants this is further converted into malate by another enzyme. Malate is then transported into vacuoles and stored as malic acid. In CAM plants primary CO_2 fixation only operates during the night since during the day the stomata are closed, which efficiently prevents water loss by evapotranspiration. During the day the stored malate is shuttled back to cell chloroplasts and enzymatically split into pyruvate and CO_2. The latter then goes into the Calvin cycle described above. Thus instead of spatial separation of primary carbon fixation and the Calvin cycle as in C_4

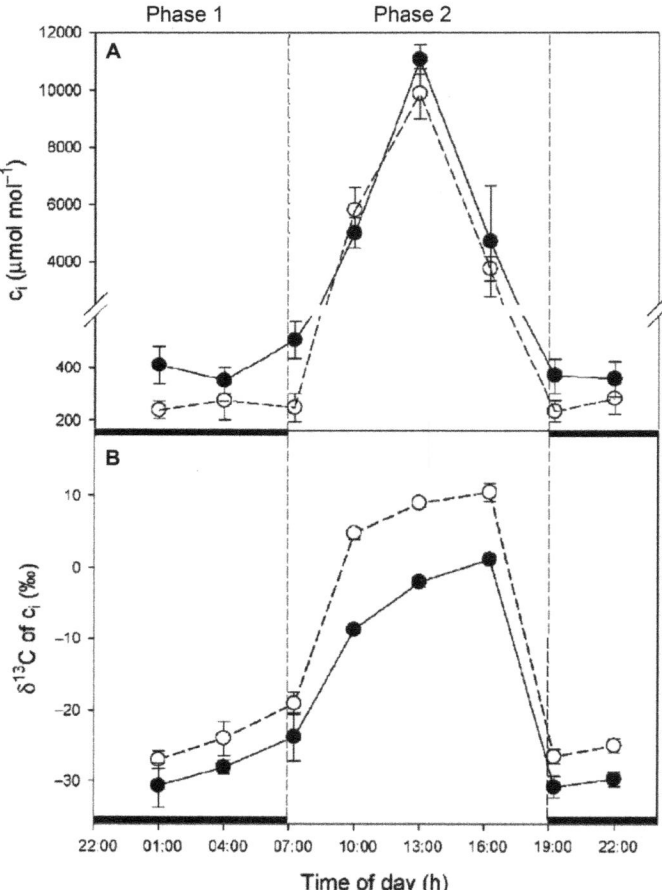

Figure 6.6 Diurnal variation of intercellular concentration (A) and isotopic composition (B) of CO_2 in cladodes of the cactus pear *Opuntia ficus indica*, an important CAM plant crop in Mexico. Phase 1: Night time, stomata open. Phase 2: Day time, stomata closed. Open and closed symbols are for plants grown with different external, atmospheric CO_2 levels and isotope compositions but show the same trend.
Modified from Nogués et al.[14]

plants, these processes are separated in time in CAM plants. The corresponding initial isotope fractionation may cover a broad range, overlapping with both C_3 and C_4 plants (see Figure 6.2). The diurnal cycle is well reflected in the intercellular carbon dioxide concentration and isotope composition within the plant as shown in Figure 6.6. When stomata are closed during the day (Phase 2), CO_2 concentrations rise until noon due to malate conversion, and then decrease again due to fixation in the Calvin cycle. With a small time delay, residual CO_2

becomes enriched in ^{13}C due to the isotope fractionation in CO_2 fixation.

CAM plants are even better adapted for extremely dry environments than C_4 plants due to their restricted water loss. On the other hand, carbon fixation is small, limiting plant growth since CO_2 fixation is only done during the night.

Table 6.2 gives an overview of plant species important for human use with regard to their photosynthetic pathways.

The different *KIEs* involved in the primary CO_2 fixation steps of the three photosynthetic pathways are reflected in the biomass isotope composition, with C_3 plant material being the most depleted in ^{13}C. As mentioned previously, both environmental parameters and intra-species differences influence fractionation to some extent. This is partially due to effects on stomata opening and thus CO_2 uptake. For example, stomata open wider under colder and/or more humid conditions. Thus, even for a given plant species no single isotope fractionation factor can be given, and a rather broad carbon isotope range is observed. This is shown for C_3 and C_4 plants in Figure 6.7.

Plant carbon isotope composition of the human diet is reflected in the body tissue, including tooth enamel, collagen, etc. Thus, it is possible to distinguish humans (or animals) consuming primarily C_3 plants from those relying mostly on C_4 plants. Since in the US maize as a C_4 plant is a much more common part of the human diet than in Europe, this is sometimes generalised to the ability to distinguish 'average' Europeans and Americans by their isotope signatures. However, this distinction clearly depends on the dietary behaviour rather than the location or

Table 6.2 Classification of plants with regard to carbon dioxide fixation pathway.

C_3 plants	C_4 plants	CAM plants
Cereals: wheat, barley, oat, rye, rice	Sugar cane	Pineapple
Sugar beet	Maize/corn	Vanilla
Potato	Millet	Agave
Grape	Sorghum	Cacti/Succulents
Citrus fruit	Sedge	Orchid
Cotton	Tropical grasses	
Trees (including maple, olives)		
Groundnut/peanut, coconut		
All *Cruciferae* including rape		
Soybean		
Sunflower		
Cocoa		
Sesame		
Most vegetables		

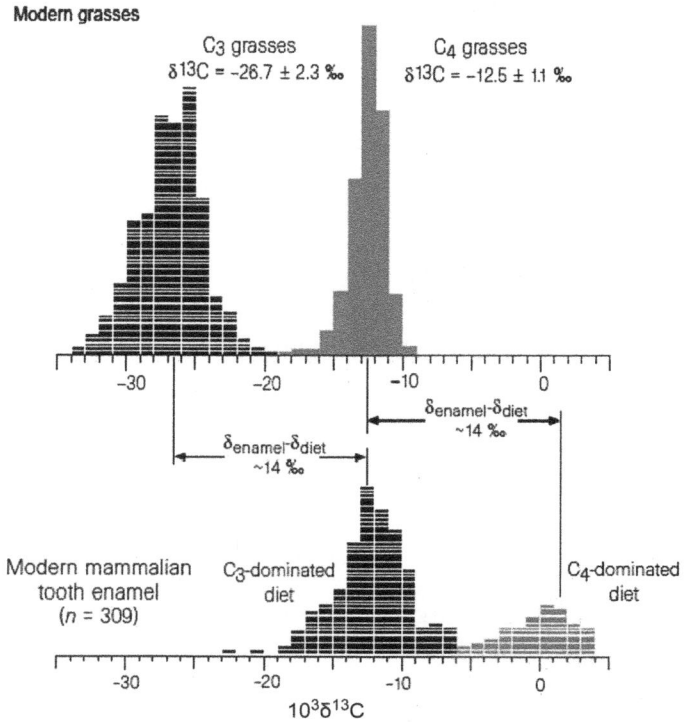

Figure 6.7 Range of $\delta^{13}C$-values observed for C_3 and C_4 grasses and in tooth enamel of mammalians living primarily on a C_3- or C_4-based diet. (Modified from Cerling et al.[15])

origin of an individual. During mammalian metabolism and tooth enamel building the average isotope composition is shifted by ~14‰ regardless of the source value, thus preserving the initial isotope spacing (see Figure 6.7).

6.2.3 Aromas, Flavours and Essential Oils

The first group of foodstuffs discussed are aromas, flavours and the related group of essential oil components. In this area, many technical developments have been demonstrated, and recommended CSIA strategies as well as isotope data presentation possibilities can be discussed best. Since flavours and essential oils typically contain volatile species, application of GC-based IRMS methods is often suited for their investigation. In the case of flavours often a range of structurally related compounds are measured. One important class of flavours are structurally related terpenes and terpenoids. In order to extract the flavour components from fruit and to purify the extracts, extensive sample preparation is necessary. The isotopic composition of flavours even for the same plant type is influenced by differences in the primary carbon fixation to a small extent (see above for an explanation of this variability). In order to eliminate these fluctuations the group of Mosandl has suggested the use of an appropriate internal isotopic standard (i-IST) thus creating an isotopic fingerprint for compounds of secondary biogenetic pathways. Measured δ-values for the target analytes are then recalculated as differences with regard to the i-IST ($\Delta\delta$ values), an approach frequently used also in doping analysis and archaeometry. Criteria for selection of a suitable i-IST are as follows:[16,17]

 (i) The substance selected as i-IST should be a characteristic genuine compound of less importance in terms of sensorial relevance.
 (ii) The compound must be available in sufficient amounts and free of isotopic discrimination during sample clean-up.
 (iii) The selected compound should be biogenetically related to the compounds under investigation.
 (iv) Chemical purity and inertness during storage and/or technical processes are mandatory.

By measuring a number of authentic samples an authenticity range can be derived for each measured isotope in one or several components of a product of interest. A recent example of such an authenticity range is shown in Figure 6.8. It can be clearly seen that for both commercial samples the isotopic composition of some of the terpenes is outside the range of authentic mandarin oil and thus has been adulterated, in particular γ-terpinene for Co4 and α-farnesene for Co5.

Although no normalization to an i-IST has been used, an even more sophisticated data plot combining isotope and concentration information for several vanilla components is shown in Figure 6.9.

Although such profiles can be useful since it is immediately visible in which variable a measured sample differs from the authenticity range, it becomes difficult to decipher small differences. Furthermore, more than three parameters are impossible to incorporate. Therefore, rather than plotting data directly, they are often processed further in a chemometric analysis such as principal component analysis (see later for examples).

If more samples are compared to define authenticity ranges, crossplots of just two components are also frequently used as shown in Figure 6.10 for two major components in lavender oils.

If possible, it is recommended that more than one isotope should be utilized to further constrain sources of flavour compounds. This

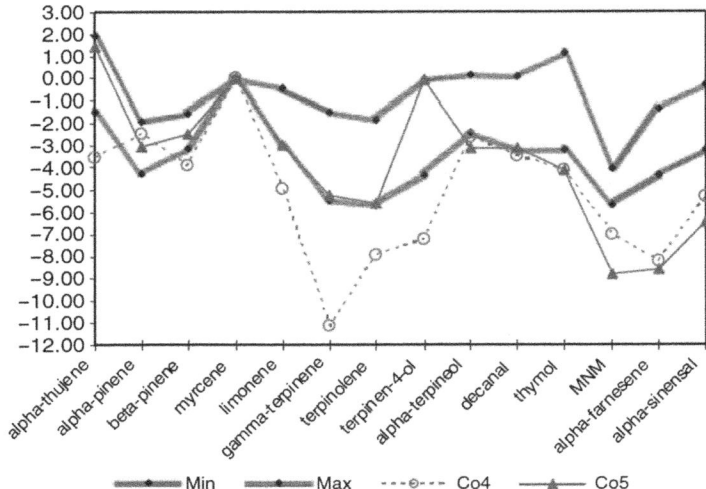

Figure 6.8 Authenticity range of carbon isotopic composition for mandarin oil using myrcene as i-IST. Y-axis thus represents $\Delta\delta$-values with respect to myrcene. Co4 and Co5 are two commercial samples.
(Graphic reprinted with permission from Schipilliti et al.[18])

is essential in samples with only one major compound that can be subjected to CSIA and thus might be even more common in other application areas. Data presentation in these cases is exemplified in Figure 6.11 for authenticity testing of cinnamon by combined carbon and hydrogen isotope analysis of cinnamaldehyde. The most precious Ceylon variety of cinnamon can clearly be distinguished from all other samples. The least valuable wood cinnamon is also separated from the other cinnamaldehyde sources. Commercial cinnamon oils are in the same authenticity range as wood cinnamon, indicating that this source has been used for further processing.

Table 6.3 gives an overview of CSIA studies in aroma and flavour analysis. In many cases, substantial sample preparation of fruit parts is required to separate flavour components from more abundant matrix components. This frequently involves solvent extraction, liquid-liquid extraction, steam distillation and sometimes further clean-up steps. In most cases, prior to injection of a final extract, the solvent has to be partially removed by distillation. All of these steps could, in principle, be associated with an isotope fractionation. However, experience and previous validation measures have shown that during carefully controlled operational procedures in foodstuff analysis, isotope fractionation is marginal. Nevertheless, the most time-consuming step in CSIA of flavours typically is not the GC-IRMS run but sample preparation. Due

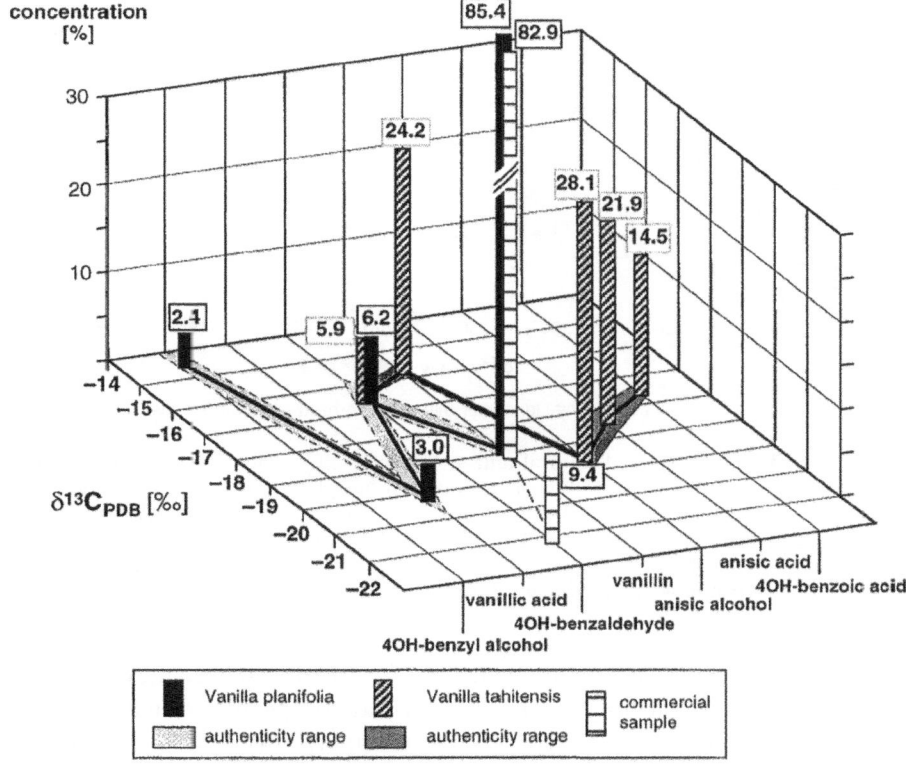

Figure 6.9 3D plot of isotope and concentration data for vanilla components as an example of a complex authenticity profile.
(Graphic reprinted with permission from Kaunzinger et al.[19])

to the often strongly varying concentrations of flavour components of interest and remaining matrix components, careful setting of backflush times is required to avoid excess sample constituents entering the combustion or pyrolysis reactor. Sometimes, reanalysis of samples at different dilutions or with different injection settings (volume, split ratio) is necessary. However, one needs to be careful with the latter option since the use of different split ratios has been shown to potentially influence the measured δ-values (see Chapter 4). As is obvious from Table 6.3, in most studies only carbon isotope analysis or a combined analysis of carbon and hydrogen has been utilized. So far, only in very few cases have nitrogen or oxygen isotope analyses been used for flavour authenticity control.

In many cases, CSIA can help to distinguish natural components produced by the plant of interest from synthetic compounds. In these cases, though, often enantioselective analysis of chiral compounds may

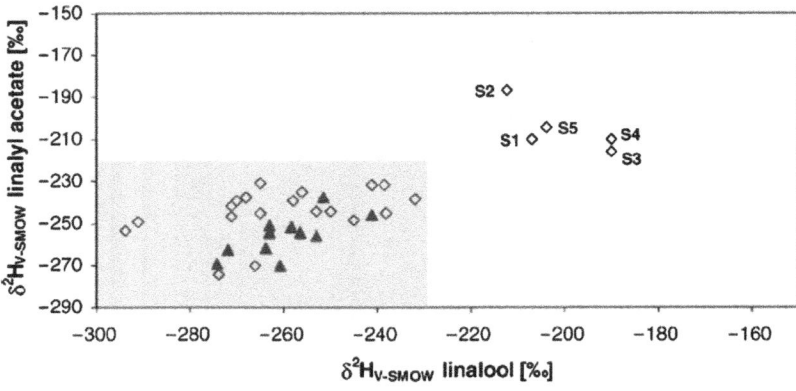

Figure 6.10 Distinction of natural and synthetic lavender oils by a crossplot of δ^2H isotope composition of linalool and linalyl acetate. The grey shaded area highlights the authenticity range, i.e., the range found in natural lavender oils. Commercial samples S1 to S5 are clearly outside this range and thus mis-labelled.
(Graphic adapted from Bilke and Mosandl.[20])

be of equal importance since synthetically produced flavours are typically obtained as racemates whereas natural compounds consist of only one enantiomer or show a specific enantiomeric ratio. Use of enantioselective analysis is less suitable though if the flavour compound can be produced equally natural biotechnologically. For *Prunus* fruit it has been demonstrated that even in such cases it is possible to distinguish natural 'ex plant' aroma components (γ-lactones, δ-lactones) from equally 'natural' ones obtained biotechnologically from microorganisms, or synthetic ones, by their isotopic compositions.[28]

To combine the advantages of enantioselective and isotope analysis, as early as 1990 the group of Mosandl introduced the coupling of enantioselective GC separation with IRMS. With the example of γ-decalactone they have shown that the isotope ratio of individual enantiomers differs and thus a further distinction of racemic γ-lactone and biotechnologically produced pure R-enantiomer is possible.[32] A further example is the separation and isotope analysis of ethyl 2-methylbutanoate declared to be of natural origin that is depicted in Figure 6.12.

The differences in isotope composition of enantiomers can be utilized to maximize information content in adulteration studies, as has been demonstrated for linalool from coriander[17] as depicted in Figure 6.13. However, in this example no clear conclusions were possible with the limited number of samples investigated.

Applications of Compound-specific Stable Isotope Analysis

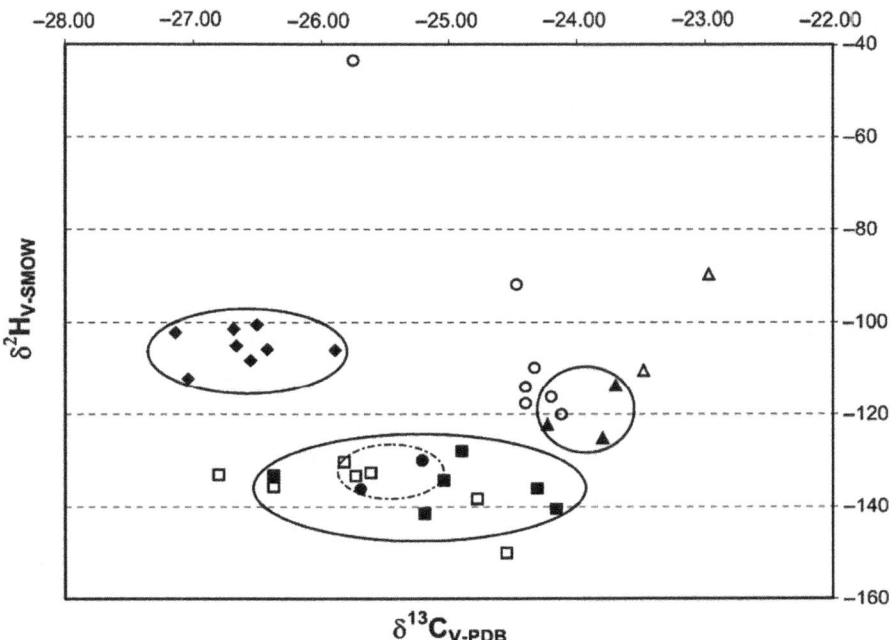

Figure 6.11 Dual isotope plot of carbon and hydrogen in cinnamaldehyde from various cinnamon varieties (Ceylon, cassia and cassia vera) and commercial sources. Key: Ceylon (diamonds), cassia (closed squares), cassia vera (closed circles), wood cinnamon (closed triangles), cinnamon powder (open squares), commercial cinnamon oils (open circles), commercial cinnamaldehydes (open triangles).
(Graphic reprinted with permission from Sewenig et al.[21])

Mosandl and co-workers have also introduced the use of multi-dimensional gas chromatography (MD-GC) coupled on-line with IRMS detection (see also Chapter 7). The system used consisted of two independent ovens and a heart-cut transfer from the first to the second oven. Thus, no comprehensive 2D analysis was possible but an improved separation of a small number of target analytes, including separation of enantiomers in the second dimension, was achieved. It was shown with standard mixtures of several components relevant in the analysis of essential oils that there is no intrinsic isotope fractionation involved in MD-GC-IRMS. However, due to the partial chromatographic isotopologue separation, heart-cut times need to be carefully controlled in order to guarantee total transfer of a peak to the second dimension. As expected, dramatic shifts up to 24‰ towards more enriched compounds are observed with premature cuts and, conversely, shifts up to −60‰ were seen with delayed cuts.[34]

Table 6.3 Compilation of CSIA studies of aroma and flavour components (in alphabetical order).

Source (oil)	Purpose	Target analyte	Isotopes used	Sample preparation and analytical technique (if not GC-IRMS)	Reference
Bergamot	Authenticity control, Detection of adulteration	linalyl acetate, linalool, α-pinene, β-pinene, limonene, sabinene, myrcene, γ-terpinene, neryl acetate	C	Solvent extraction of fruit peels, Split injection	22
Cactus pear fruit	Fruit characterization	1-Hexanol, E-2-Hexenol, E-2-Nonenol, E,Z-2,6-Nonadienol	C	Splitless injection of pentane-dichloromethane extract	23
Cinnamon	Differentiation of cinnamon varieties	Cinnamaldehyde	C, H	SD, no further information, Splitless injection	21
Coriander	Distinction of natural and synthetic flavour compounds	limonene, γ-terpinene, p-cymene, linalool, geraniol, myrcene, geranyl acetate, β-pinene, camphene, terpinolene, sabinene	C	SD, LLE with pentane-diethyl ether, flash chromatography, Split injection	17
Lavender	Distinction of natural and synthetic flavour compounds	linalool and linalyl acetate	H	SD of oils, diethyl ether extraction of lavender, Splitless injection	20
Lavender	Authenticity assessment	linalool and linalyl acetate	C, H, O	As before, Splitless injection	24
Lemon	Authenticity control	β-pinene, limonene, γ-terpinene, nerol, geraniol, neral, geranial, neryl acetate, geranyl acetate	C	Solvent Extraction, Preparative high resolution segment chromatography, Split injection	16
Mandarin	Detection of adulteration, e.g., by sweet orange oil	α-thujene, β-pinene/sabinene, myrcene, limonene, octanal, γ-terpinene, terpinolene, linalool, methyl N-methylanthranilate, α-sinensal	C	Solvent extraction of fruit peels, steam distillation, LLE, preparative layer chromatography for clean-up, Split injection	25
Mandarin	Distinction of natural and synthetic flavour compounds	methyl N-methylanthranilate, methyl anthranilate	C, N	Ditto	26
Mandarin	Detection of adulteration, e.g., by sweet orange oil	α-thujene, α-pinene, β-pinene, myrcene, limonene, γ-terpinene, terpinolene, terpinen-4-ol, α-terpineol, decanal, thymol, methyl N-methylanthranilate, α-farnesene, α-sinensal	C	n-Hexane dilution of oils, Split injection	18

Source	Application	Compounds	Isotopes	Method	Ref
Numerous	Distinction of natural and synthetic flavour compounds	Methyl cinnamate	C, H	SDE, Splitless injection of pentane-diethyl ether extract	27
Prunus fruit (peach, apricot, nectarine)	Distinction of natural and synthetic flavour compounds	γ-decalactone, δ-decalactone	C, H	SDE, Splitless injection of pentane-diethyl ether extract	28
Pineapple	Pineapple fruit characterization	Methyl 2-methylbutanoate, ethyl 2-methylbutanoate, methyl hexanoate, ethyl hexanoate, 2,5-dimethyl-4-methoxy-3[2H]-Furanone	C, H	LLE with pentane dichloro methane, Split injection	29
Raspberry	Distinction of natural and synthetic flavour compounds	(E)-α-ionone, (E)-β-ionone	C, H	SDE, Splitless injection solvent not indicated	30
Vanilla	Distinction of natural and synthetic vanillin and of the varieties V. planifola and V. tahitensis	Vanillin, 4-hydroxy benzylalcohol, 4-hydroxy benzaldehyde vanillic acid, vanillyl alcohol, anisic alcohol, anisic acid, 4-hydroxy benzoic acid	C	Vanilla extraction by ethanol/water, filtration, LLE with pentane-diethyl ether, Split injection	19
Vanilla	Distinction of natural and synthetic vanillin	Vanillin	C, H	Vanillin isolation as before, Split injection for total vanillin; Headspace injection for methoxy group after cleavage and transformation to methyl iodide	31

LLE: Liquid-liquid extraction, SD: steam distillation, SDE: Simultaneous distillation extraction; note that number of entries in the table for a specific source or oil does not reflect its economic importance.

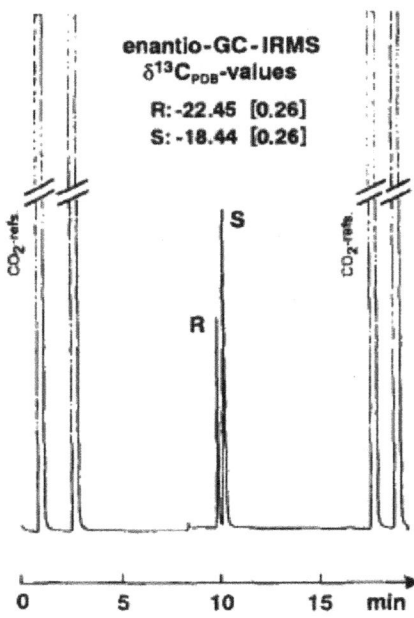

Figure 6.12 Enantiomeric separation of ethyl 2-methylbutanoate demonstrating significantly different isotopic compositions of the R- and S-enantiomers. (Graphic reprinted with permission from Karl et al.[33])

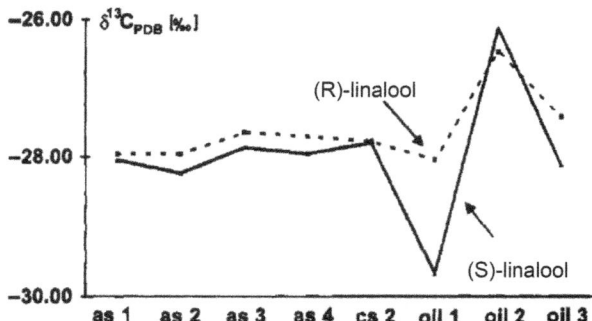

Figure 6.13 Carbon isotope composition of (R)- and (S)-linalool in coriander. Key: as: authentic samples, cs: commercial spice, oil: commercial coriander oils. (Graphic adapted from Frank et al.[17])

An interesting example of position-specific isotope analysis has recently been presented for vanillin authenticity investigations. Based on initial work by Keppler and co-workers on isotope analysis of plant methoxyl groups[35] this group has realized the potential of the developed method for vanilla profiling. Vanillin also contains a methoxy group that can be selectively cleaved by hydriodic acid to yield gaseous methyl iodide in a nucleophilic substitution reaction. Subsequently, carbon or hydrogen isotope analysis is done on the methyl iodide which can easily be separated from the matrix by headspace analysis. A comprehensive validation has shown that despite the harsh conditions, no substantial isotope fractionation occurs, thus the methyl group in methyl iodide is isotopically identical to the vanillin methoxy group.[31] Dual isotope plots of $\delta^2 H$ vs. $\delta^{13}C$ for both bulk vanillin and the methoxy group are shown in Figure 6.14. It is obvious from the figure that (i) for both hydrogen and carbon different isotopic ranges are observed for bulk vanillin and its methoxy group, clearly indicating a non-statistical intramolecular isotope distribution (see also Schmidt),[36] and (ii) the methoxy isotope composition is as well suited for a differentiation of sources as that of bulk vanillin. The major advantage of the position-specific method is not directly visible here: in fraudulent synthetic vanillin production, incorporation of ^{13}C-enriched precursors have been used more and more to adjust the bulk isotope composition of resulting vanillin, which is most easily done on the carbonyl and methoxyl functional groups. Thus, a more sophisticated method for authenticity control is required.

6.2.4 Alcoholic Drinks

For the isotopic analysis of alcoholic drinks including beer, wine and spirits, many continuous-flow methods for bulk samples using EA-IRMS have been described in the past. Recently, Jochmann *et al.* have shown that LC-IRMS can also be applied in flow injection mode, i.e. utilizing direct injection without column separation for the rapid classification of spirits. Another advantage of the method is the very small consumption of sample. As little as 10 µL can be directly injected and enable a reliable $\delta^{13}C$ determination.[37]

Carbon dioxide from alcoholic beverages and also bottled water samples can be directly measured with a commercial GC-IRMS system without modifications, keeping the combustion and reduction oven temperature at 100 °C since no further reaction is necessary. A polar column is incorporated to keep ethanol (which could interfere at m/z 46 measurements) from entering the interface. This method allows the very rapid screening of samples and has been shown to be applicable to a number of real samples, for example, enabling a differentiation of beers prepared from C_3 and C_4 plant fermentation or distinguishing naturally sparkling and artificially carbonated water.[38]

Most CSIA applications in authentication of alcoholic drinks are based on ethanol, with a few aiming at glycerol and/or minor fusel

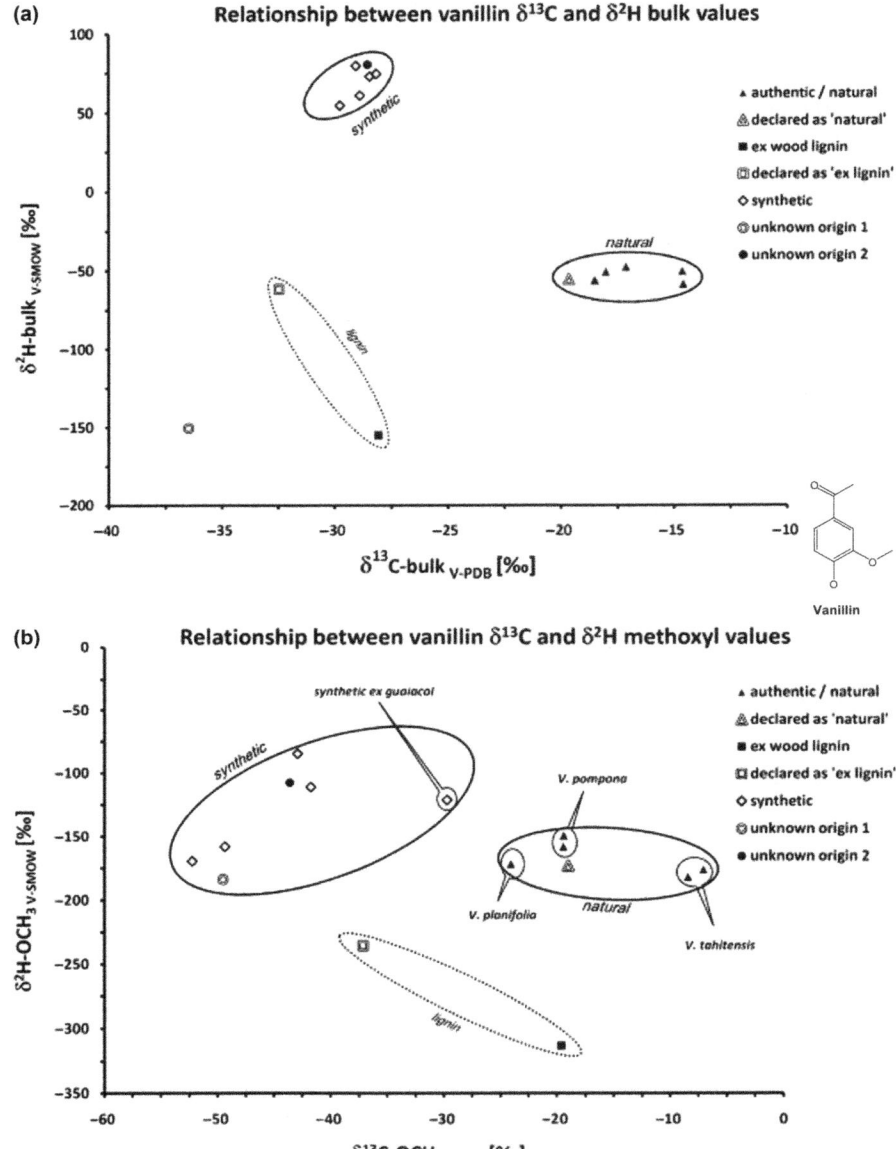

Figure 6.14 Dual isotope plots of carbon and hydrogen in vanillin from various authentic vanilla varieties (*V. pompona*, *V. planifolia*, *V. tahitensis*) and commercial sources. a) bulk vanillin; b) vanillin methoxy group. Vanillin structure with highlighted methoxy group added to original figure. (Graphic adapted from Greule et al.[31])

alcohols. In addition to carbon isotope analysis, oxygen isotope analysis on ethanol and glycerol has proved useful for creating multiple isotope authenticity ranges (see below). In contrast, hydrogen isotope analysis of glycerol was found to be useless due to exchange with water.[39] An overview of relevant studies in the area is given in Table 6.4.

Wine produced within the European Community is probably the best-investigated commodity in terms of isotopic characterization. Since 1990, wines originating from the different member states are annually analysed by official laboratories and results fed into a European wine database. The isotopic characterization includes $\delta^{18}O$ measurements of wine water, SNIF-NMR for $\delta^{2}H$ and EA-IRMS for $\delta^{13}C$ of wine ethanol after distillation.[44] For ethanol measurements in wine samples also direct splitless injections into GC and LC are amenable. Both methods and the official EA-based method after ethanol distillation did agree well within the analytical error. No bias of any method was observed. Both methods have also been validated as part of an inter-laboratory test.[42] Although the authors did not directly recommend one of the methods, the GC method was approximately four times faster, and did not involve use of further reagents as for the post-LC conversion of analytes to CO_2. Later on, the method was extended to wine glycerol with the same agreement among various methods.[43] Both $\delta^{13}C$ and $\delta^{18}O$ in glycerol can be useful markers for detecting adulteration in wine by synthetic or C_4-derived glycerol. It is, however, not suited to identify glycerol from C_3 plant sources other than grapes. Neither is it possible to distinguish grape varieties or vintages as shown in Figure 6.15.[39]

Besides ethanol and glycerol, minor so-called fusel alcohols are also present in alcoholic beverages. These do not necessarily derive from sugar fermentation but might, instead, be formed by other biosynthetic pathways. 2- and 3-methylbutanol, for example, are transformation products of the amino acids isoleucine and leucine, repectively. Their isotopic composition can be used to distinguish C_3 and C_4 plant-derived spirits, which has been applied for authenticity control in apple products such as cider and calvados.[47] Interestingly, 3-methylbutanol in all fermented products was more depleted in ^{13}C than 2-methylbutanol by up to 8‰, which reflects typical differences in their precursor amino acid isotope composition in plant tissues.[48]

Despite the different origin, recently a correlation in the isotopic composition of 2- and 3-methylbutanol (not chromatographically separated) with ethanol has been shown. Other fusel alcohols that had been co-extracted with ethanol from wine samples by liquid-liquid extraction were less correlated, which was partly attributed to unresolved poor repeatability of isotope measurements of some compounds.[44] Although

Table 6.4 Compilation of CSIA studies of alcoholic drinks.

Alcoholic beverage	Purpose	Target analytes	Isotopes used	Sample preparation and analytical technique (if not GC-IRMS)	Reference
Wine	Origin control	Ethanol	C, O	Headspace, split injection GC-IRMS,	40, 41
Wine	Method comparison	Ethanol	C	LC-IRMS, direct split injection GC-IRMS	42
Wine	Adulteration by synthetic or C_4-derived glycerol	Glycerol	C, O, H useless due to H exchange in the alcohols	Precipitation, subsequent dissolution in ethanol, split injection	39
Wine	Method comparison	Ethanol, glycerol	C	LC-IRMS, direct split injection GC-IRMS	43
Wine	Method comparison	Ethanol, 2-methylpropanol, 2- and 3-methylbutanol, butan-2,3-diol, 2-phenyl-1-ethanol, ethanoic acid, glycerol	C	LLE with cyclohexane-*tert*-butanol, split injection	44
Tequila	Differentiation of authentic and commercial tequilas and, within these, groups of white, rested and aged tequila	Ethanol	C, O	HS-SPME, GC-IRMS	45
Scotch whisky	Differentiation of Scotch and maize-derived whisky	Acetaldehyde, ethyl acetate, *n*-propanol, iso-butanol, amyl alcohols	C	Direct injection of sample in split mode	46
Apple derived products	Characterization	2-methylbutanol, 3-methylbutanol	C	LLE with pentane-diethyl ether, split injection	47

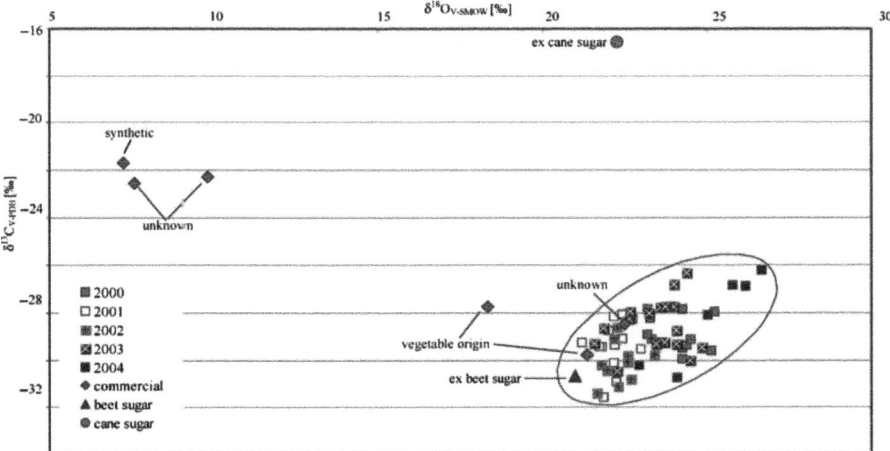

Figure 6.15 Dual isotope plots of carbon and oxygen in glycerol from wines of various vintages, commercial glycerol, and glycerol from fermented beet sugar and cane sugar.
(Graphic reprinted with permission from Jung et al.[39])

the method is more laborious than the direct sample injection described above, the authors claim advantages, in particular, for sweet wines where the co-injected sugars may cause problems.

For alcoholic products other than wine, there are so far hardly any applications reported in the international literature. The few exceptions include ethanol analysis in tequila[45] and minor component analysis in whisky.[46] The former is also one of the very few examples in food studies, where microextraction techniques (here, solid-phase microextraction, SPME) have been used in conjunction with CSIA although for ethanol analysis in spirits this might not be required in terms of sensitivity.

As an alternative to the presentation of multicomponent isotope data shown in Section 6.2.3, radar or star plots can also be used, as shown in Figure 6.16.

6.2.5 Honey

Honey fraud may involve (i) the deliberate addition of an adulterating sugar or syrup or (ii) non-compliance with an origin name; this can involve substitution or mixing of honeys from several locations. Bulk isotope analysis is well established in honey control. Isolated honey protein is used as an internal standard which improves the detection of C_4 sugar addition. Recently though, it has been shown that an LC-based method utilizing isotope analysis of individual sugars in honey samples

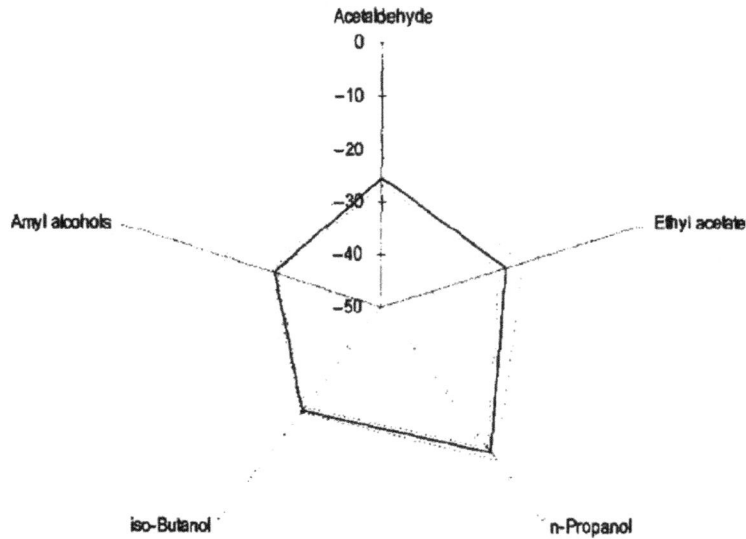

Figure 6.16 Radar plot of multicomponent isotope data of minor components in whisky. (Graphic reprinted with permission from Parker et al.[46])

is superior to bulk methods in detecting adulteration by beet sugar, cane sugar or cane syrup.[49] In fact, the described method has the potential to identify adulteration by beet sugar, revealing isotopic differences between the three major sugars fructose, glucose and sucrose in honey samples (see Figure 6.17). This result was confirmed by a follow-up study incorporating nearly 500 honey samples and thus providing a large database for comparison.[50] In this comprehensive approach δ^{13}C-values of bulk honey, isolated protein, fructose and glucose as well as of disaccharide and trisaccharide fractions are reported. Isotope analysis of carbohydrates, for example, in honey samples is, so far, the only area where commercial laboratories routinely utilize LC-IRMS.

6.2.6 Juice

In juice authentication, common analytes are sugars and ascorbic acid. However, there have been only very few reports on CSIA applications in this area. As discussed briefly for honey above, carbon isotope analysis allows distinguishing sugars derived from C_3 and C_4 plants and thus is also able to detect adulteration of juices with cane- or corn-derived sugars. However, it is typically insufficient to reveal fraudulent addition of C_3-derived sugars such as from beet. LC-IRMS similar to the method described above for honey has not yet been used in juice authentication.

Applications of Compound-specific Stable Isotope Analysis 255

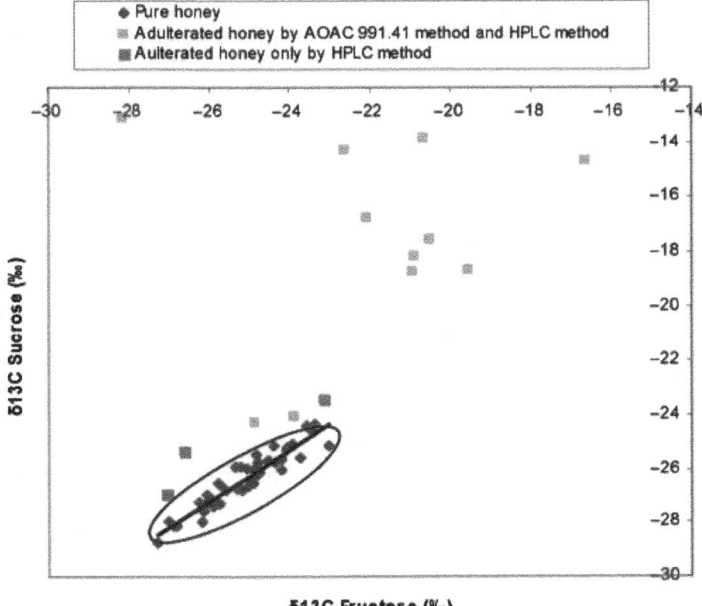

Figure 6.17 Carbon isotopic composition of sucrose vs. fructose clearly showing detection of adulteration even in cases where the official bulk method fails.
(Graphic reprinted with permission from Cabanero et al.[49])

Due to the typically very different climatic conditions in which citrus fruit grows, in comparison with sugar beet, hydrogen isotope analysis might offer an alternative in such cases. This requires getting rid of the exchangeable hydrogen bound to oxygen prior to hydrogen isotope analysis, which can be done, for example, by nitration. The nitrated sugars from orange juice and commercial beet sugars, indeed, consistently showed a huge isotopic difference of at least 50‰ in H but no discernable isotopic difference in C.[51] Similar results have been obtained for sugar beet syrup additions to apple juice. Here, a more rigorous conversion of fructose to hexamethylentetramine after comprehensive extraction and isolation of fructose from juice has been validated and used. Despite the harsh conditions, no significant isotope fractionation or isotopic exchange with water was observed.[52]

Another important topic in fruit juice analysis is detection of adulteration by addition of synthetic L-ascorbic acid (vitamin C). Distinction of natural and synthetic L-ascorbic acid was possible down to 20% synthetic compound in a mixture since commercial L-ascorbic acid had an average δ^{13}C value of -11.3‰ whereas L-ascorbic acid in authentic orange juice was around − 20.7‰. Here, the principle of internal isotope standardisation discussed above for flavours was invoked, utilizing the (bulk) sugar isotope composition for standardisation.[53] This application utilised off-line fractionation followed by EA-IRMS measurement.

6.2.7 Oils

In addition to essential oil components, covered in Section 6.2.3, oil analysis, mostly of the fatty acid isotope composition via GC-IRMS of the corresponding FAMEs, is also quite common. FAMEs analysis has been reviewed previously by Meier-Augenstein[54] (see also Section 6.4 since fatty acids in preserved oil residues are also important target compounds in archaeological uses of CSIA). Table 6.5 summarizes adulteration and provenance studies of modern vegetable oils. Adulteration can either be done by addition of (cheaper) oils from other plants or by a lower grade quality oil of the same plant. The latter issue has been thoroughly addressed for olive oil, for example, by Angerosa et al.[55] With very few exceptions, CSIA investigations have been limited to carbon so far.

Certain compound class fractions show significant deviations from whole olive oil measurements. Glycerol is more depleted, whereas sterols are more enriched in ^{13}C compared with the bulk oil.[56] Furthermore, the ripeness status of olives does not have a substantial influence on isotopic composition. The aliphatic alcohol fraction in pomace oil is more depleted in ^{13}C than in refined or virgin olive oil, which allows differentiation if the difference between the isotope composition of the whole oil, as measured by EA-IRMS (mostly representative of the triacylglycerol fraction), and the isotope composition of the aliphatic alcohol fraction is calculated. The calculated difference always results in positive values for virgin and refined oils and negative values for pomace oils. A fraction as small as 3% of added dewaxed pomace oil was successfully detected.[55] Similarly, measuring FAMEs of maize and rapeseed or groundnut oil, it was possible to detect adulteration of maize oil by only 5% of the C_3 derived oil. A controlled adulteration study has shown a good agreement of predicted values based on an isotopic mass balance and experimental data.[61] Even various C_3 plant oils could be distinguished by the isotopic composition of specific fatty acids.[62] For example, linolenic acid in poppy seed oil was consistently less depleted in ^{13}C than that in rapeseed oil. Flaxseed oil was intermediate to these with a small overlap of isotopic ranges, thus no conclusive distinction based only on this one parameter was possible.

This is a rather general conclusion: in many cases, detection of adulteration requires the combination of several parameters. This can be

Table 6.5 Compilation of CSIA studies of oils.

Oil type	Purpose	Target analytes	Isotopes used	Sample preparation and analytical technique (if not GC-IRMS)	Reference
Olive oil	Adulteration	Aliphatic alcohols, glycerol, sterols	C	EA-IRMS after fractionation	56
Olive oil	Detection of adulteration by pomace oil	Aliphatic alcohol fraction	C	EA-IRMS after fractionation	55
Olive oil	Geographical origin	Aliphatic alcohol fraction	C	EA-IRMS	57
Olive oil and others	Adulteration	Fatty alcohols, glycerol after reductive ester cleavage	C, O	Splitless injection of ethanolic and diethyl ether phase	58
Olive oil	Adulteration	Fatty acids as FAMEs, in particular $C_{16:0}$ (palmitic acid) and $C_{18:1}$ (oleic acid)	C	Alkaline hydrolysis, methylation, LLE of FAMEs with hexane, splitless injection	59
Olive oil, pumpkin seed oil	Adulteration	Fatty acids as FAMEs, in particular $C_{16:0}$ (palmitic acid) and $C_{18:1}$ (oleic acid)	C	As before	60
Maize oil	Adulteration by C_3 oil (rapeseed or groundnut)	Fatty acids as FAMEs, in particular $C_{16:0}$ (palmitic acid), $C_{18:1}$ (oleic acid) and $C_{18:2}$ (linoleic acid)	C	Saponification, methylation, LLE of FAMEs with diethyl ether, solvent exchange to n-hexane, injection not specified	61
Rape, flax, and poppy seed oil	Differentiation of oil types	Fatty acids as FAMEs, in particular $C_{16:0}$ (palmitic acid), and $C_{18:3}$ (linolenic acid)	C (plus bulk C,H,O)	Saponification, methylation, LLE of FAMEs with hexane, split injection	62
Camelina sativa oil	Geographical origin, adulteration	Fatty acids as FAMEs	C	Not indicated	63
Camellia seed oil, perilla seed oil and flax seed oil	Detection of adulteration by cheaper oils (soy, maize)	Fatty acids as FAMEs	C	One-pot saponification/methylation with sodium methoxide in petroleum ether, splitless injection	64

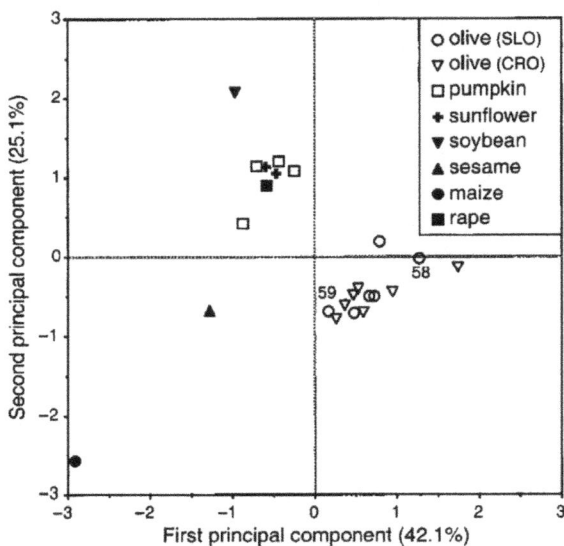

Figure 6.18 Results of principal component analysis of oil samples based on fatty acid composition, bulk isotope analysis and individual isotope composition of palmitic and oleic acid. Results are shown as score scatter plots of the first two principal components.
(Graphic reprinted with permission from Spangenberg et al.[60])

done most stringently with chemometric approaches such as principal component analysis that helps in clustering samples of different origin or processing. An example is shown in Figure 6.18 that demonstrates the classification of oils based on the first two principal components.

6.2.8 Natural Stimulants: Cocoa, Coffee, Tea

In the area of natural stimulant analysis by CSIA there have been surprisingly few applications published so far. Weinert et al. have investigated major volatile compounds in black teas (Ceylon, Assam, Darjeeling), i.e., linalool, *trans*-2-hexenal, *cis*-3-hexenol, *cis/trans*-linalool oxides (furanoid, fur.) and geraniol, out of the ~500 reported volatiles in tea. The samples were investigated in order to evaluate if CSIA could help in authenticity testing after previous tea testers had found some samples suspect of adulteration by methyl salicylate. Indeed, this could be confirmed for many of the suspected samples when measured carbon isotope data for methyl salicylate were plotted vs. an i-IST to eliminate differences from variability in the primary carbon dioxide fixation step.[65]

Spangenberg et al. determined carbon isotope ratios for fatty acids (as FAMEs) in cocoa butter and cocoa butter equivalents (CBEs).

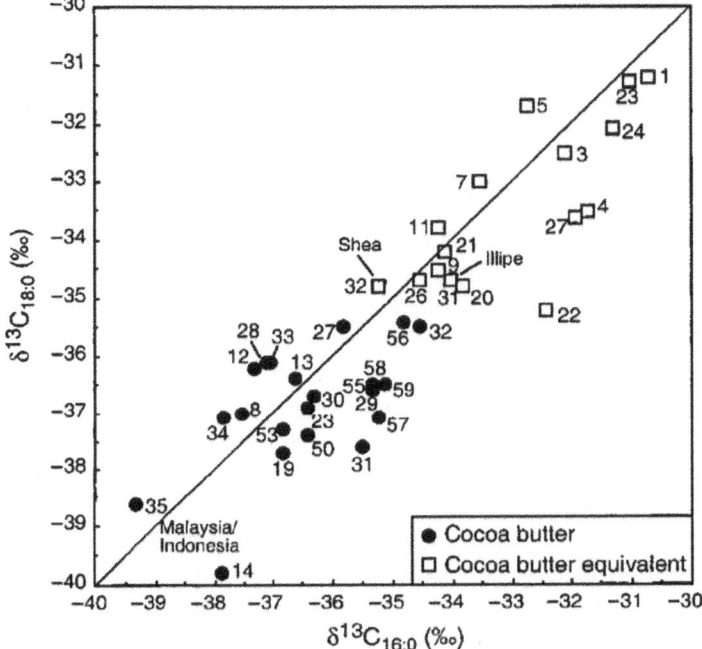

Figure 6.19 Differentiation of authentic cocoa butter and CBEs by a crossplot of isotopic composition of stearic vs. palmitic acid (measured as FAMEs). (Graphic reprinted with permission from Spangenberg and Dionisi.[66])

By profiling the fatty acid isotopic composition of stearic and palmitic acid in a crossplot they were able to differentiate quite well original cocoa butter and equivalents (see Figure 6.19). The fatty acid crossplot approach is further explained in Section 6.4. Furthermore, as in related work of the group discussed in Section 6.2.7, principal component analysis taking into account concentration and isotopic data of four fatty acids helped to quantify the addition of CBEs to real cocoa butter.[66]

Recently, Zhang *et al.* investigated sources of caffeine in tea, coffee and energy drinks by LC-IRMS applying high temperature HPLC (see also Chapter 3). It was shown that natural caffeine can be clearly distinguished from synthetically produced caffeine by its carbon isotope ratio. Several mislabelled products have been identified.[67]

6.3 FORENSICS

The following section on forensics focuses on the use of SIA in criminal investigations, including arson, origin of illicit drugs, and explosives.

In these areas SIA may help to establish a link between traces or residues of a material found or seized and a production process or source area, by providing isotopic profiles of compounds involved.[68] In turn, this may help to elucidate trafficking of such materials relevant for prosecution. Food related studies, doping analysis, and environmental forensics are discussed in Sections 6.2, 6.5 and 6.6. IRMS use in forensics (both BSIA and CSIA) has been reviewed recently.[69,70] In a recent article on the differentiation of black powder samples by EA-IRMS for carbon and nitrogen, Gentile *et al.* nicely illustrated how stable isotope data can be integrated into an overall forensic investigation framework.[71] It needs to be pointed out, though, that in a forensic investigation very strict requirements have to be met with regard to compound identification. Thus, for data to be used in a court of law, it is necessary to assure compound identity beyond mere peak retention time matching. This can only be achieved by combined GC-MS/IRMS instruments with partial split to the MS. At best, this is an MS with MS^n capabilities such as an ion trap to allow for a further identification point by fragmentation of selected mother ions.[70] However, to our knowledge such instruments are hardly in use and reports published in the open literature are very scarce,[72] none of which was in the area of forensic investigations. As in the other sections of Chapter 6, we limit our further discussion to the use of CSIA although, so far, reports on bulk isotope analysis using elemental analysers are more frequent. The main application area of CSIA in forensics discussed in open literature is in tracing the origin of illicit drugs such as heroin and 3,4-methylenedioxy *N*-methylamphetamine (MDMA), better known as 'Ecstasy' (see molecular structures in Figure 6.20 and Figure 6.21).

Rather early in the development of GC-IRMS a first application has shown the potential use of carbon isotope composition in heroin for tracing its country of origin. At least it was possible to distinguish three

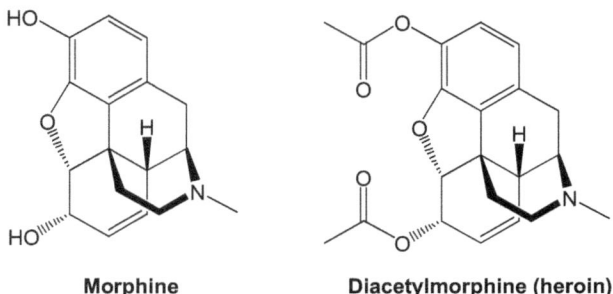

Figure 6.20 Molecular structures of morphine and diacetylmorphine (heroin).

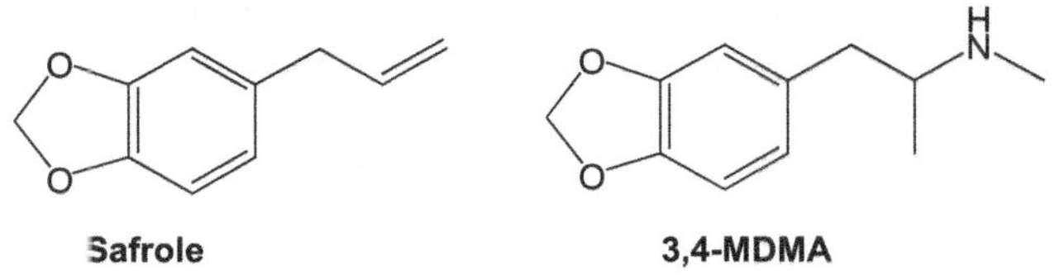

Figure 6.21 Molecular structures of safrole and 3,4-methylenedioxy *N*-methylamphetamine (MDMA, Ecstasy).

source areas.[73] However, later works have partially questioned this finding and suggested utilizing nitrogen isotope composition for that purpose or, even better, combining both data sets in a dual isotope plot. Although not based on GC-IRMS but bulk isotope analysis, the same conclusion was also drawn for the origin of coca leaves.[74]

Heroin is manufactured from natural morphine by acetylation. Thus, the stable isotope composition in C, O and H is affected by the morphine source and the acetic anhydride used in the process, which may explain the controversial initial findings on the capability of carbon stable isotope analysis in provenance studies. Therefore, Besacier *et al.* suggested measuring the stable isotope composition of heroin after deacetylation to morphine and have shown that the original carbon isotope composition is preserved. Furthermore, measurement of both morphine and heroin can even yield information on the source of acetic anhydride used in morphine processing.[75] Later on, Casale *et al.*, mostly of the US Drug Enforcement Administration, confirmed these findings in a comprehensive study of isotope effects in coca and morphine processing.[76] After a huge heroin seizure on a cargo ship in Australia, the group has been able to rule out known provenances of morphine with the help of isotope data in morphine and heroin and comparison to isotope data of samples of known origin from South America, south-west Asia, south-east Asia, and Mexico.[77]

In drug manufacturing, excipients or 'cutting agents' are used, either as diluent or adulterant. These are mixed with heroin, which typically makes up only a small fraction of the final product. Common cutting agents are caffeine and acetaminophen. As found in other application areas, the information content with regard to source differentiation can be increased if multiple components are analysed. Thus, a simultaneous isotope analysis of cutting agents along with morphine and heroin could prove useful. For both acetaminophen and caffeine this has been shown previously.[78,79] Furthermore, Idoine *et al.* have used carbon and hydrogen isotope data of both heroin and caffeine to maximize the potential to differentiate among various heroin seizures.[79]

MDMA is mostly produced from safrole, a natural product in sassafras plants by various synthetic pathways. Due to its use in MDMA production, safrole trade is heavily regulated in most countries. The final isotope composition in MDMA is influenced by the natural precursor, the reagents used in synthesis and potential *KIEs* during processing.[68,80]

As for heroin/morphine, multi-isotope investigations are most useful for differentiating sources of MDMA.[81–83] This is shown as dual isotope plots in Figure 6.22 for carbon, hydrogen and nitrogen isotope data.

Based on isotope measurements of a large set of seized tablets, Palhol *et al.* pointed out that nitrogen isotope data in particular are useful in distinguishing MDMA tablets from various seizures. Carbon isotope data were in a rather narrow range and identical to that of potential precursors (safrole, isosafrole). With pyrolysis-IRMS they also investigated the potential use of $\delta^{18}O$ analysis but found little variation. Since oxygen in the safrole ring system is not involved in further chemical reactions, there should also be no further fractionation in the manufacturing process.[83] Schneiders *et al.* have recently investigated CSIA including $\delta^{13}C$, $\delta^{2}H$ and $\delta^{18}O$ measurements, for differentiation of sources of 1-phenyl-2-propanone (P2P), a common precursor in the clandestine production of the stimulants amphetamine and methamphetamine. For seven seizures from illicit production it was indeed possible by a dual isotope plot of $\delta^{2}H$ vs. $\delta^{13}C$ to group these into four different clusters. However, the investigation of 27 legally manufactured P2P samples from 9 batches produced over five years has shown that intra-batch variations can be substantial, thus hampering source–product relationships based on isotope data.[84]

Recently, CSIA has also been used to provide proof of administration of synthetic γ-hydroxybutyric acid (GHB). GHB is frequently a target analyte in forensic cases investigating drug-facilitated sexual assault. However, it is also produced endogenously and inter-individual variability of GHB levels is large. Thus, concentration data for GHB alone are insufficient to verify claims of administration of the drug. Similar to synthetic steroids, exogenous GHB is depleted in ^{13}C and thus the distinction seems possible by the $\delta^{13}C$ ranges measured. Saudan *et al.* reported a method for GC-IRMS of GHB after intramolecular esterification to the corresponding γ-butyrolactone (GBL) to avoid introduction of extraneous carbon by a derivatizing reagent. They reported $\delta^{13}C$ ranges of -32.1 to $-42.1‰$ for subjects exposed to GHB, whereas in the control group a range from -23.5 to $-27.0‰$ was found.[85] Interestingly, GBL is also a precursor for GHB, which is transformed in the human body to GHB and thus has become popular recently as a

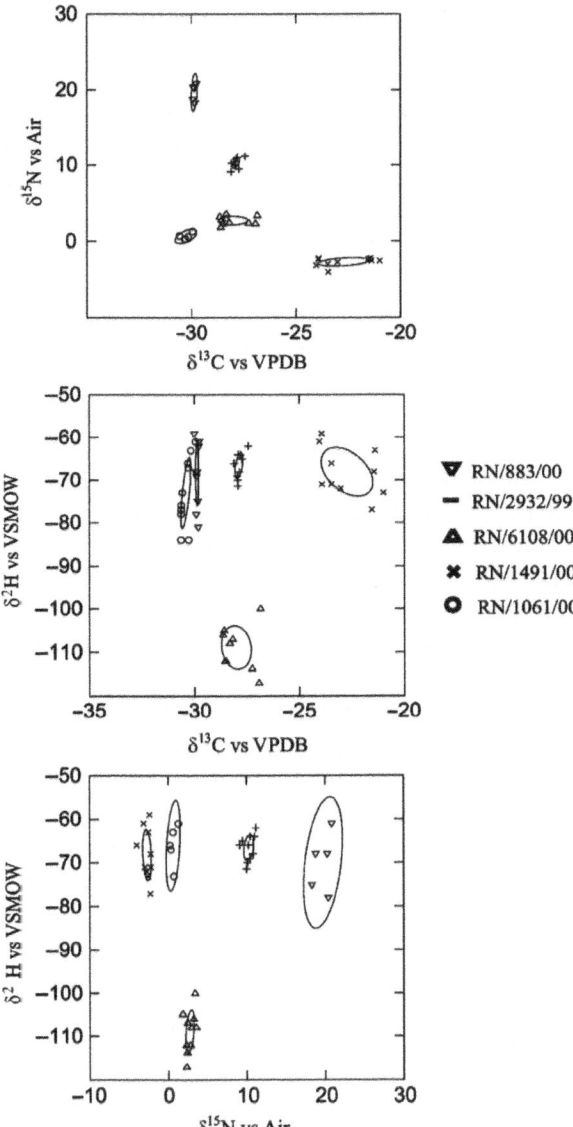

Figure 6.22 Dual isotope plots of five batches of MDMA-containing tablets. $\delta^{13}C$ and δ^2H data are from GC-IRMS measurements, whereas $\delta^{15}N$ data have been obtained from EA-IRMS bulk measurements assuming that the extracted tablets contain no impurities substantially contributing to the overall $\delta^{15}N$ reported. Note that batch RN/1491/00 contained the N-demethylated 3,4-methylenedioxyamphetamine (MDA) rather than MDMA. This is also an active ingredient and similarly regulated. (Graphic reprinted with permission from Carter et al.[81])

drug itself. Recently, the first isotopic discrimination between seized GBL samples based on GC-IRMS has been reported.[86]

As mentioned previously, applications of CSIA in other forensic investigations have, so far, rarely been openly published. Whether this is due to the delicate nature of some of the data generated, to the lack of time in corresponding forensic laboratories, or indeed due to the still emerging use of CSIA in this area, we do not know. We can confirm from own contacts though that there is considerable interest in CSIA use in state offices of criminal investigation. In our own preliminary investigations we have demonstrated the usefulness of adding isotope data to a database on suitable parameters to distinguish gasoline samples. In case of remains at an arson site, very useful isotope markers for correlating trace residuals with suspected sources are naphthalene, and 2- and 1-methylnaphthalene since their carbon isotope composition is hardly affected even after prolonged periods of evaporation.[87]

We would like to conclude this section with one of the few examples of a published criminal investigation case that bridges well to the next section on archaeology. In this case study, fatty acids from residual fats (measured as FAMEs) were analysed, which are also primary targets in archaeological stable isotope applications. Here, samples from a disinterred grave were investigated using several analytical techniques including a preliminary investigation of the usefulness of CSIA for providing a further line of evidence for clandestine burial of a human being. To that end the isotopic composition of fatty acids in soil samples containing adipocere, a partially decomposed human adipose fat, were compared with those from reference animal and human fats as shown in Figure 6.23. Later on, adipose fat from the victim became available as well and the fatty acid isotopic composition could be compared with the previously measured samples. In contrast to Figure 6.25, isotope data on palmitic and stearic acid are presented as $\varDelta\delta$-values. This has been done to eliminate potential shifts in both fatty acid isotopic compositions due to differences in dominating diet (C_3 or C_4).[88]

6.4 ARCHAEOLOGY

As suggested above, there is a strong connection between archaeological and forensic applications since major research questions are often quite similar, for example when trying to locate the origin and/or more recent residence of a person whose remains are to be investigated. The major difference here is on the time scale: in forensics more recently deceased persons are usually of interest in criminal investigations, whereas in archaeology remains typically date back at

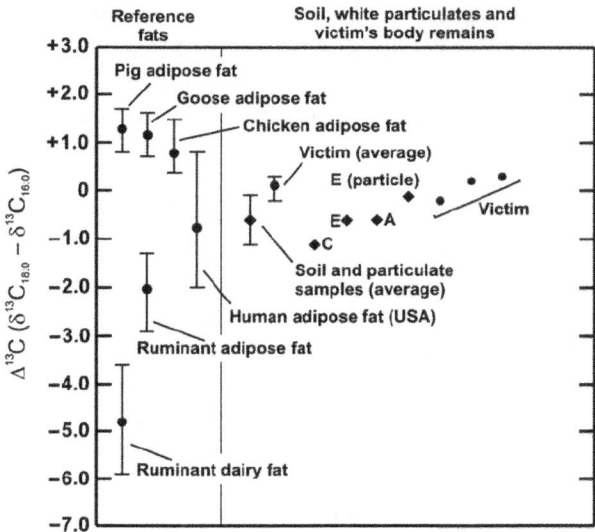

Figure 6.23 Comparison of differences in isotopic composition in stearic and palmitic acid for human and animal reference fats (left-hand side) and soil samples of the suspected temporary grave site (data points A, C, E), separated white particulates associated with sample E and adipose fat from body remains of the victim found later (all right-hand side). Averages of $\Delta\delta$-values on the right-hand side agree rather well and are in the (rather broad) range of human reference fat whereas fats originating from domesticated animals are clearly different.
(Graphic reprinted with permission from Bull et al.[88])

least a few hundred years. As in forensics it needs to be emphasized that stable isotope data on their own provide an important but often insufficient line of evidence. Further evidence from other measurements or observations is necessary to validate or falsify hypotheses on ancient human activities.

Numerous applications have shown the potential of bulk stable isotope analysis of bone collagen, tooth enamel, bioapatite or preserved hair samples in archaeology (see, for example, the thorough review of Lee-Thorp on calcified tissues[89]). However, the use of CSIA has only more recently been explored further. Recently, Evershed published an excellent review on the importance of organic residue analysis in modern archaeology and coined the expression 'Archaeological Biomarker Concept' in which molecular and stable isotope mass spectrometry play equally important roles (Evershed 2008). In there, he also gives an example of the usefulness of methods in archaeology for more recent forensic studies.

A major use of organic residue analysis is in paleodiet studies, i.e., studies devoted to food sources in ancient times. Furthermore, information on the use of different organic substances for illumination and on tracking trade pathways can be obtained, all of which have already been demonstrated. In many such cases modern materials are investigated as reference along with residues obtained from artefacts. This is done to constrain ranges of isotope compositions to be expected under the assumption that the chosen target analytes have been sufficiently preserved in order to keep their isotopic composition constant over extended periods of time. In these cases, a correction due to higher ^{13}C contents in atmospheric carbon dioxide in preindustrial times is necessary. To that end, ice core data of CO_2 can be used.[90] Sometimes, simulated ageing studies may also help to explain shifts in molecular and isotope composition of biomarkers observed in real artefact investigations.

The most frequently used substance class in such studies are fatty acids, which are dominant lipid components derived from hydrolysis of triglycerides (also called triacylglycerols or TAG) and preserved in artefacts such as potsherds. Thus, for a long time their extraction and subsequent concentration analysis has been used in archaeology (see Reber and Evershed).[91] Consequently, CSIA of fatty acids also raised interest soon after commercial systems became available in the late 1980s. Meier-Augenstein has provided an excellent review on analytical aspects of fatty acid isotope analysis[54] and the reader is referred to this article for further information. In general, fatty acids are measured as their methyl esters (FAMEs), mostly after derivatization with a commercially available BF_3-methanol reagent (see Chapter 4). For FAMEs, isotopically referenced compounds are available from Arndt Schimmelmann (see Chapter 5), which can help substantially in maintaining accuracy of measurement results when cross-checked regularly. A detailed account of necessary corrections due to derivatization (not limited to FAMEs) is given by Docherty *et al.*[92] In order to avoid derivatization, Sephton *et al.* have suggested an alternative approach for fatty acid isotope analysis based on off-line catalytic hydropyrolysis to the corresponding *n*-alkane.[93] However, so far this approach has neither been demonstrated for complex samples nor widely adopted.

Meier-Augenstein gave also some application examples from various areas including early work in archaeology. Dudd and Evershed, in the late 20th century, showed by a thorough comparison with modern animal fat equivalents that a distinction of livestock fats stemming from pigs or ruminants such as cattle, sheep and goats is possible in organic residues from pottery sherds and other artefacts.[94,95] The chemistry of animal fats in archaeology has been discussed in detail by Evershed and Dudd.[96]

Unsaturated fatty acids are at least partially oxidized over archaeologically relevant time scales. Concurrently, short-chain fatty acids are more water-soluble and volatile and thus typically washed out or evaporated.[97] Therefore, longer-chain saturated fatty acids (starting from lauric acid, $C_{12:0}$), dominate in residues and are most frequent target analytes in CSIA. One may ask why some fatty acids remain intact after archaeological time frames at all, in particular since microbial degradation of these compounds is rather fast. Figure 6.24 sketches the main reason: absorbed lipids reside in nano-scale pores in the pottery clay that are too small for microorganisms to enter. This retards degradation of the less water-soluble fatty acids for a sufficiently long period.

Over recent years the approach to distinguishing animal fats in artefacts has been further explored. Particularly useful is the isotope analysis of n-hexadecanoic acid (palmitic acid, $C_{16:0}$) and n-octadecanoic acid (stearic acid, $C_{18:0}$). In a crossplot of $\delta^{13}C$ data for these two compounds one can nicely distinguish porcine adipose fats, ruminant adipose fats and ruminant milk fats as demonstrated in Figure 6.25. Remains of the latter in vessels can be used to provide evidence for ancient dairying and thus allows us to establish the time when mankind started to raise ruminants for use of their milk in the fifth and fourth millennia BCE in mid and northern Europe.[97–99] Recent analysis of organic residues from the Anatolian region have shown, though, that in the Near East as early as in the seventh millennium BCE cattle were already domesticated for dairying.[100] The same classification as discussed above for distinguishing animal fats may also help to identify fats or oils that have been used in illumination lamps (e.g., Copley *et al.*).[101]

The approach of plotting isotope data of $C_{16:0}$ vs. $C_{18:0}$ has been extended recently to horse milk. Although $\delta^{13}C$ analysis of the two fatty acid markers already allowed a distinction between ruminant and horse derived lipids, only with the help of additional $\delta^{2}H$ analysis was it possible to distinguish horse milk and adipose fat. The former was a strong line of evidence to support the hypothesis of horse domestication as early as 3500 BCE in the region of today's northern/central Kazakhstan.[103] A final example of the palmitic/stearic acid carbon isotope composition crossplot is the characterization of 'bog butters', white fat masses found buried in bogs in Scotland and elsewhere. From historical records it is clear that they have been deposited intentionally, presumably for preservation under the unique conditions of peat bogs.

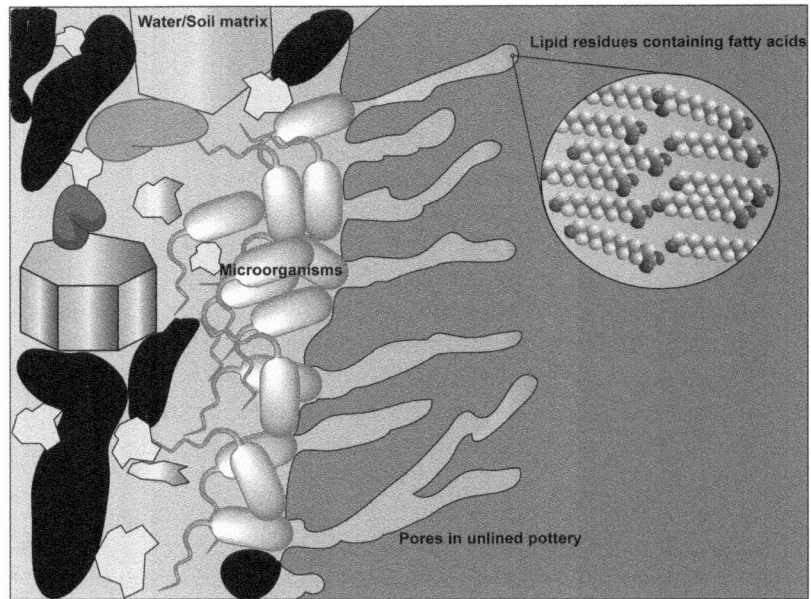

Figure 6.24 Sketch of a cross-section through a porous pottery clay surface showing absorbed lipid residues in the pores and larger matrix components not able to enter.

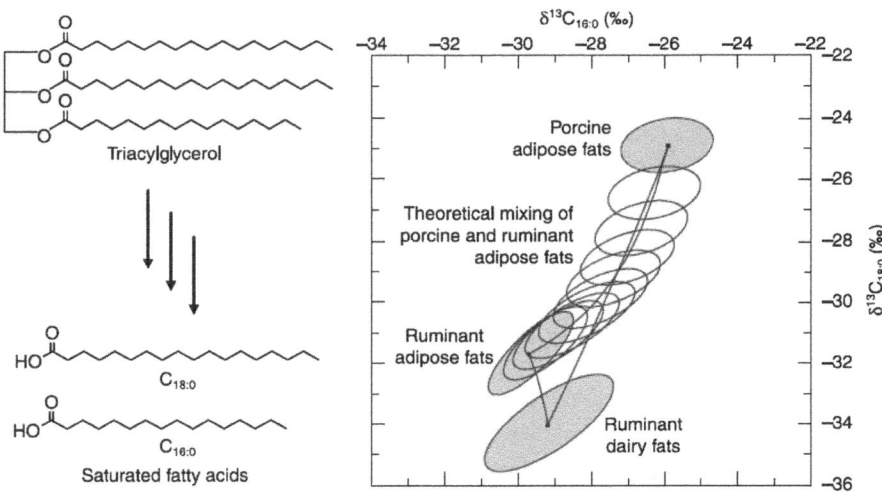

Figure 6.25 Generation of palmitic and stearic acid via hydrolysis of triglyceride (triacylglyerol) and a crossplot of their $\delta^{13}C$-values highlighting the differentiation of various lipid precursors. Ellipses comprise 1σ confidence ranges. (Graphic reprinted with permission from Evershed.[102])

Using CSIA it was, for the first time, unequivocally shown that at most sites, dairy fats have been buried, but at a few sites, the waxy mass originated from ruminant adipose fats.[104] Gregg and Slater recently discussed some limitations of this approach when applying the discussed classification to potsherd samples from the Middle East.[105] Finally, it needs to be emphasized that additional lipid biomarker profiling, typically by LC-MS, may help to corroborate conclusions by stable isotope evidence for certain paleodiets.[102,106]

The second major area of CSIA applications in archaeology is amino acid isotope analysis. Carbon and nitrogen isotope composition of the abundant body protein collagen, which is preserved in bones over archaeological time scales, has frequently been utilized in archaeology to obtain information on protein sources in paleodiets from skeletal remains. It was recognized early on, though, that isotope composition of individual amino acids in bone collagen could extend the information content from bulk isotope values for the total collagen alone while, at the same time, contaminations blurring the isotope information are removed.[48] For example, bulk collagen isotope data cannot be used to distinguish between C_4 and marine dietary proteins.[107] Collagen is made up primarily of the three amino acids glycine, proline and hydroxyproline but contains other amino acids as well, thus separation of many amino acids after acidic hydrolysis of collagen is required. Therefore, several procedures to determine $\delta^{13}C$-values for individual amino acids from bone collagen have been developed. Since amino acids are not amenable to GC analysis, derivatization of amino acids prior to GC-IRMS has been introduced with trifluoroacetate/isopropanol (TFA/IP) as the most common two-step derivatization reagent.[108] Neither the hydrolysis step nor subsequent derivatization induce a major isotope fractionation, as a thorough comparison of bulk collagen isotope composition with that of individual amino acids by means of an isotope mass balance has shown (see below for limitations though).[109] In this study, it was shown by a comparison of modern and ancient bone collagens that the isotope composition is preserved over archaeological times although the general validity of this conclusion is not yet clear since there is conflicting evidence from a bone collagen decomposition study.[110] Preparation of TFA/IP derivatives is still widely used today although over time the catalyst in the combustion oven inevitably will be poisoned due to formation of extremely stable metal fluorides.[111] Recently, derivatization of amino acids to the *n*-acetyl methylester was reported to be superior since the number of extraneous carbon atoms introduced in the analytes is smaller.[112] Successful applications of these and related derivatization methods have been reported for $\delta^{13}C$ and

δ^{15}N analysis of individual amino acids allowing discrimination of marine and terrestrial food sources (e.g., Naito et al.).[113]

However, any derivatization introduces extraneous carbon, thus requiring a correction by an isotope mass balance. Furthermore, *KIEs* in the derivatization reaction might not be negligible. As discussed thoroughly by Docherty et al. with an error propagation analysis, the resulting measurement error in amino acid δ^{13}C-values in TFA/IP derivatives can be as high as 1.4‰.[92] At least partly, the analytical error could also be due to the above-mentioned poisoning of the oxidation catalyst by fluorine.

Due to the problems inherently accompanying derivatization, in particular of small molecules such as amino acids, there has been tremendous interest in utilizing LC for separation of underivatized amino acids. A first approach utilised off-line separation with cation-exchange columns prior to bulk analysis of the isolated amino acids.[114] Only in 2004, though, a commercial interface for LC-IRMS became available. Rather quickly, this interface has been adopted in amino acid investigations in an archaeological context. The major limitations of the LC-IRMS interface have been thoroughly discussed (see Chapter 3). However, amino acid separations are possible using pH gradients, thus the restriction of purely aqueous mobile phases in LC-IRMS is not that much of a problem here. McCullagh et al. were the first to successfully demonstrate this with hydrolysates from archaeological bone collagen samples (see Figure 6.26) using a mixed-mode reversed-phase and ion-exchange stationary phase.[115]

Analysis time is very long and can even exceed three hours for a single run,[116] thus restricting the number of samples to be analysed to one per day if appropriate replicates are run. Furthermore, retention times need to be carefully checked since literature examples sometimes show severe shifts.[116]

Marine and terrestrial carbon sources lead to $\Delta\delta^{13}$C-values in individual amino acids ranging from 3 to 13‰.[116] $\Delta\delta^{13}$C-values for the pair glycine and phenylalanine ($\Delta\delta^{13}$C$_{Gly-Phe}$) have been suggested as an indicator distinguishing terrestrial (average $\Delta\delta^{13}$C$_{Gly-Phe}$ = 5‰) and marine (average $\Delta\delta^{13}$C$_{Gly-Phe}$ = 12‰) protein sources in food.[107] A recent study on Korean bone remains has shown similar shifts between terrestrial and marine food sources but different absolute values (see Figure 6.27). Since $\Delta\delta^{13}$C$_{Gly-Phe}$-values alone were not as clear as in the previous study on South African bone remains, adding threonine δ^{13}C-values as a further isotope marker was suggested. By presenting these two parameters in a crossplot, a clear distinction of marine, terrestrial and omnivorous food sources was possible (Figure 6.27).

Applications of Compound-specific Stable Isotope Analysis 271

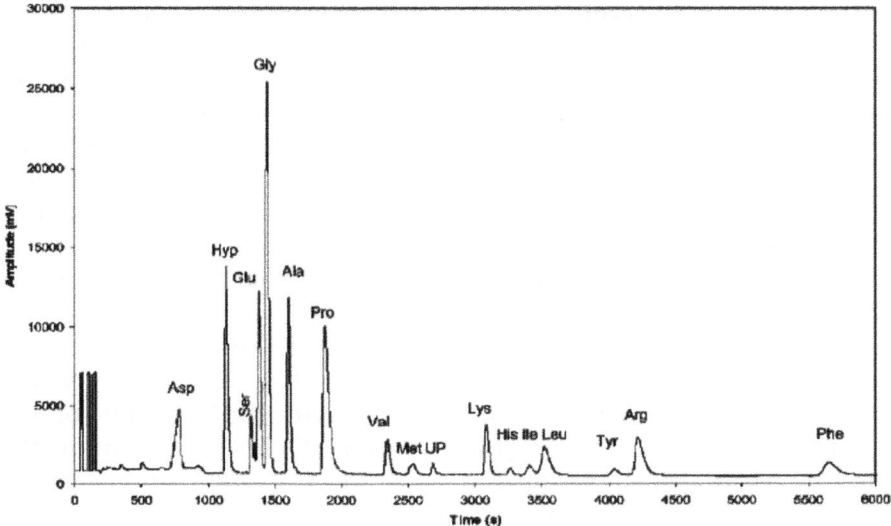

Figure 6.26 LC-IRMS chromatogram of amino acids from an archaeological bone collagen sample using a pH gradient for separation.
(Graphic reprinted with permission from McCullagh et al.[115])

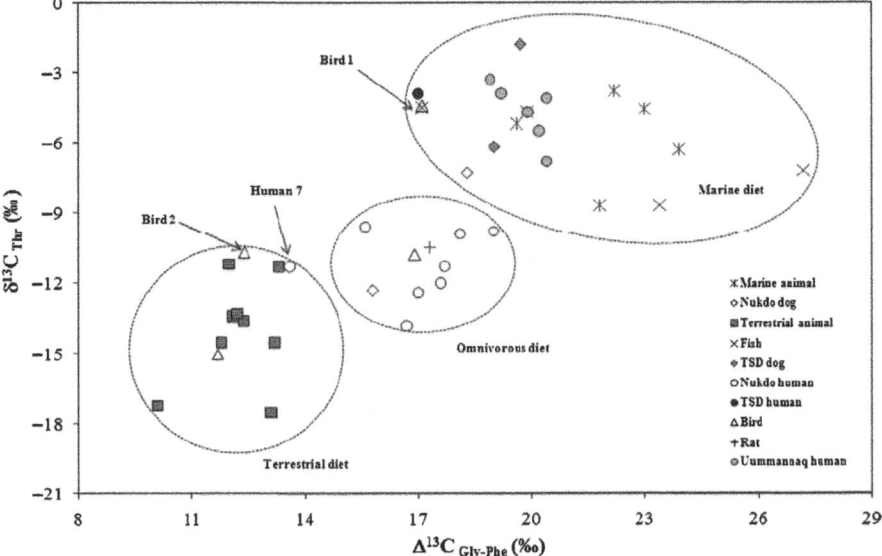

Figure 6.27 Isotope crossplot showing the isotopic composition of the amino acid threonine (Thr) vs. difference in isotopic composition of the amino acids glycine and phenylalanine.
(Graphic reprinted with permission from Choy et al.[116])

Raghavan *et al.* recently have shown that the isotopic composition of amino acids in hair could provide a useful additional or alternative proxy since patterns in bone collagen and in hair proteins are similar but body turnover times are very different. Hair is easier to abstract from archaeological remains without damage and comparison with modern humans is much easier.[117] With regard to SIA of individual amino acids we should finally note that there is considerable and increasing interest in other areas such as environmental science.[118]

Another useful biomarker in skeletal remains is the dietary steroid precursor cholesterol (see Figure 6.28).

Cholesterol levels are sufficiently high in bones and, particularly, in teeth, and cholesterol is preserved sufficiently long to allow for CSIA after considerable sample preparation and trimethylsilylation of the compound.[119] Due to the higher turnover of cholesterol during the lifetime of individuals compared with bone collagen or bioapatite, cholesterol δ^{13}C-values reflect the more recent diet of the being before death. Thus, it might also prove a useful biomarker in forensic studies, although its use in that area has not yet been described in literature. Furthermore, cholesterol δ^{13}C-values reflect dietary lipid and carbohydrate fraction, thus providing complementary information to bulk isotope analysis of collagen which, rather, reflects dietary proteins. Stott and Evershed have shown from the analysis of a large number of femur bone pieces at a coastal excavation site in the UK that marine foodstuffs were an important part of the diet during the time of burials starting in the 9th century. This is indicated by enrichment in ^{13}C compared with cholesterol from a C_3-plant-based diet. These authors could also confirm relatively homogeneous δ^{13}C-values of cholesterol over an entire femur

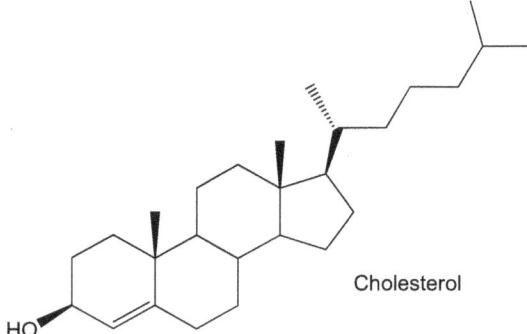

Figure 6.28 Molecular structure of cholesterol. Note the structural similarity with the anabolic steroids depicted in Figure 6.29.

6.5 DOPING CONTROL

and, even, little difference among different bones from complete skeletons. However, inter-individual differences were pronounced.[119]

Active sports and participating as a spectator are among the most popular recreational activities but nowadays are also a huge business. Thus, there is a large incentive for individuals to excel by administering synthetic non-own-body (exogenous) substances; so-called doping. Although doping was already known in ancient times its widespread and systematic use in many sports started only after the Second World War and is not limited to human sports but is also prevalent in horseracing. Nowadays, the list of forbidden substances established by the World Anti-Doping Agency (WADA) has become excessively long and comprises more than 100 compounds altogether (status 2010). The current list includes 59 anabolic agents such as the steroid testosterone (see Figure 6.29) which has been banned since 1974, with the Olympic Games in Montreal 1976 being the first major sports event with thorough control of anabolic agents. Misuse of these is still the reason for most adverse analytical findings. For example, in the 2004 Olympic Games in Athens out of 23 reported positive findings, 16 were attributed to anabolic steroid misuse.[120] Thus, a reliable method for proving steroid misuse has long been sought,[121] which is complicated by the fact that some of these steroids are also produced naturally in the body (such as testosterone) and thus it is not possible to distinguish natural and artificial steroids even by state-of-the-art analytical methods such as

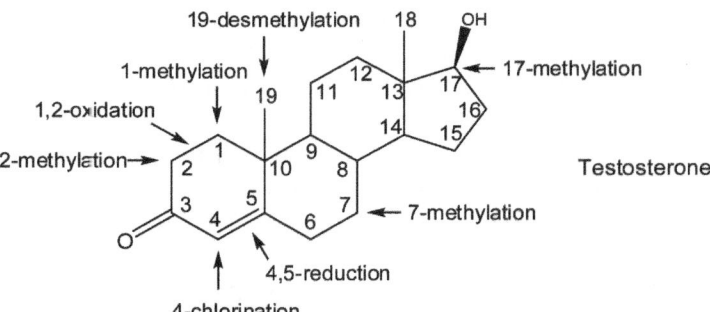

Figure 6.29 General structure of Anabolic Androgenic Steroids (carbon numbers refer to standard nomenclature). The depicted structure without modifications represents testosterone. Epitestosterone has exactly the same structure but the stereochemistry of the OH- group at C-17 is reversed. (Adapted from and redrawn after http://www.dopinginfo.de/rubriken/00_home/00_tes.html.)

hyphenated chromatography with mass spectrometric detection. For many years, the only method available for proving testosterone misuse has been the comparison of testosterone and epitestosterone levels in urine samples. Epitestosterone is a structurally identical steroid with different stereochemistry only in the OH- group at C-17 that is not a metabolite of testosterone, see Figure 6.29. If the ratio of these levels exceeded 6:1 (in 2004 lowered to 4:1), administration of synthetic testosterone was suspected and the doping control laboratory would report the case to the sports authorities for further investigation.

However, this approach has many disadvantages: (i) inter-individual variability is high, thus hampering the use of a single threshold value and ultimately leading to frequent false positive or false negative results,[122] (ii) it only works for a few hours after administration, and (iii) co-administration of synthetic epitestosterone could be used to conceal doping. For the latter reason, administration of epitestosterone is also forbidden but very difficult to prove just based on concentration levels in urine.[123] It would be possible to partially overcome the variability issue by subject-based threshold values but these require long-term monitoring and are not yet applied officially.[122] Figure 6.30 depicts the general pathway of testosterone production in mammals (endogenous) and the main precursors of synthetic testosterone (exogenous) used in doping.

The ability to distinguish natural and synthetic steroids by SIA results from the different precursors. Synthetic steroids are produced from C_3-plant precursors, in particular soy, that give rise to compounds rather depleted in ^{13}C. An overview of isotopic compositions of 26 synthetic steroids available commercially has resulted in a mean value of $-30.1‰ \pm 2.8‰$.[124] As long as the endogenous steroids show a different isotope composition, i.e., are more enriched in ^{13}C, one can distinguish both sources. This is generally fulfilled since typical ranges for endogenous steroids are $-19‰$ to $-24‰$. The isotope signature of the endogenous steroid, in the case of doping, will then be a mixture of the natural and synthetic signature as described by an isotope mass balance.

Individual isotopic composition of steroids can be influenced primarily by diet and maybe other factors. Gender and age of athletes, though, have been shown not to affect isotopic composition of steroids significantly.[125] Due to the inter-individual differences, it is obligatory to measure both suitable endogenous reference steroids (ERC, e.g., precursors generally neither involved in doping nor formed by transformation of an administered steroid) and metabolites of testosterone in the same method since the isotope composition of ERC will not be affected by administering synthetic steroids and thus can be used as perfect

Applications of Compound-specific Stable Isotope Analysis 275

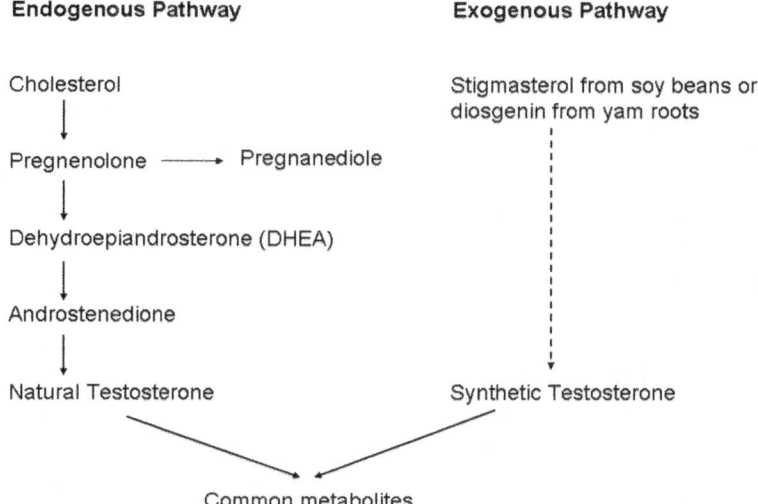

Figure 6.30 Synthetic pathway of testosterone.

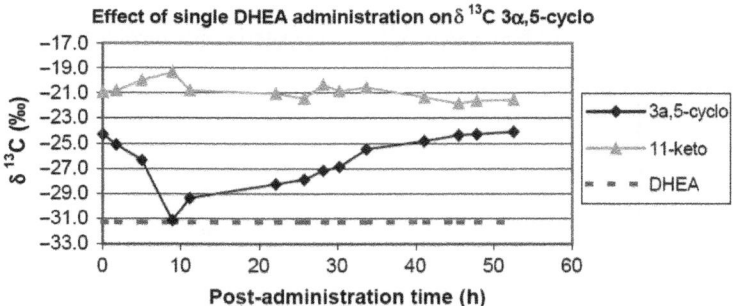

Figure 6.31 Kinetic profile of isotope values in an ERC (here: 11-keto) and a metabolite (here: 3a,5-cyclo) of dehydroepiandrosterone (DHEA). (Graphic reprinted with permission from Cawley et al.[126])

internal standards. In contrast, metabolites of testosterone will carry the shifted isotope composition. The full clearance of steroids is not very fast, thus the shifted isotope composition in steroids and their metabolites can be observed up to days after administration depending on the specific compounds investigated – another major improvement over the testosterone to epitestosterone (T/E) ratio method which just incorporates testosterone itself. As an example, Figure 6.31 shows the development of δ^{13}C-values over time after single administration of exogenous dehydroepiandrosterone with a δ^{13}C-value of −31.0‰.

SIA can also be used to distinguish endogenous and synthetic epitestosterone,[127] which is important since epitestosterone can be used to disguise testosterone misuse (see above).

Today, doping control analyses for steroid misuse typically comprise the following compounds: androsterone, etiocholanolone, 5β-androstane-3α,17β-diol (5β-adiol), 5α-androstane-3α,17β-diol (5α-adiol), DHEA, testosterone, epitestosterone, and as potential ERC 5β-pregnane-3α,20α-diol (5β-pdiol), 16(5α)-androstenol, and 11-ketoetiocholanolone (11-keto). Following the WADA recommendations, doping by an endogenous steroid is shown when a difference $\Delta\delta^{13}$C of 3‰ or more between the ^{13}C/^{12}C ratio of testosterone or a testosterone metabolite relative to an ERC has been found and/or the δ^{13}C-value of the target metabolites is $< -28‰$. Figure 6.32 demonstrates that in a study with controlled administration of testosterone, the $\Delta\delta^{13}$C between the two testosterone metabolites 5α-adiol and 5β-adiol on the one hand, and the ERC 5β-pdiol on the other, always exceeded this limiting value.

Due to the principal advantages of the steroid misuse test based on SIA, WADA endorsed its use starting from the 1998 Olympic

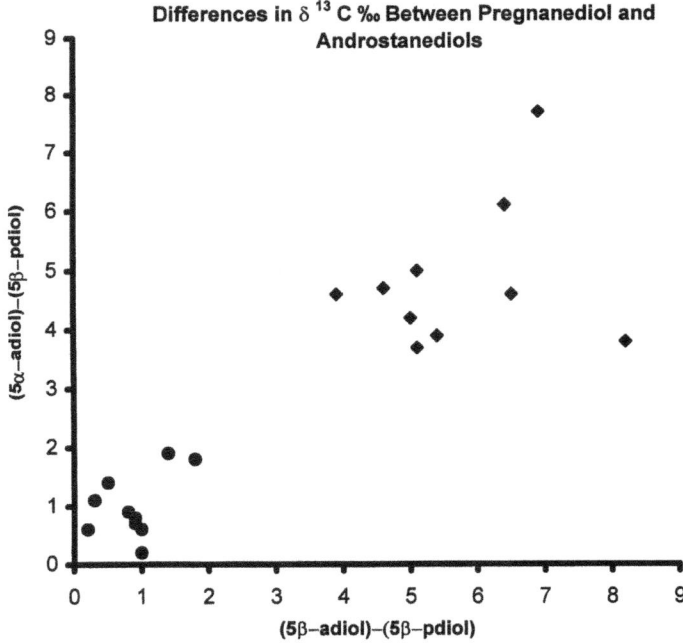

Figure 6.32 Differences in δ^{13}C in the two testosterone metabolites 5α-adiol and 5β-adiol and the ERC 5β-pdiol.
(Graphic reprinted with permission from Aguilera *et al.*[128])

Games in Nagano. As of 2010, 35 anti-doping laboratories are accredited by WADA, many of which utilize IRMS for detecting steroid misuse.

Nevertheless, due to the huge number of samples to be processed, often in a short period of time during major sports events, one of the first indicators of testosterone misuse is still the T/E level of 4 with IRMS being used mostly for confirmatory analysis. Further indicator values triggering a subsequent analysis by IRMS are set for the glucuronide concentration of testosterone, epitestosterone, androsterone, etiocholanolone and DHEA.[127] The reason for IRMS analysis being more laborious compared with concentration analysis is not in the separation and detection steps of analysis. Rather, it is the need for dedicated sample preparation that removes interfering co-eluting matrix components as much as possible from the samples that is responsible. Since the IRMS in contrast to organic MS detectors is inherently a non-selective detector of any carbon-containing species eluted at a given time co-elution will otherwise falsify obtained results for the target steroids. A more detailed discussion of sample preparation procedures is given in Chapter 4 and exemplified for steroid analysis below.

Finally, without any judgement of the individual's guilt, it is noteworthy to read about some potential pitfalls in carrying out IRMS analysis from the Floyd Landis case (Tour de France 2006) heavily debated in public.[129]

It is interesting to note that doping is not restricted to sports but also occurs frequently in raising livestock, as a growth promoter in order to support faster weight gain. Within the European Union the use of steroids for this purpose was banned in 1988. Thus, testing for the forbidden use of steroids in this area by urine analysis of cattle or pigs is also an important issue related to food control and was the subject of the recent European research project ISOSTER. Since both the steroid metabolism and the urine matrix are not identical to human beings, methods established in doping analysis cannot be immediately transferred for such applications and need to be adapted and validated thoroughly. Thus, currently no threshold values for $\Delta\delta^{13}C$ unequivocally distinguishing natural and synthetic steroids is possible, as has been discussed for cholesterol in bovine urine recently.[130]

6.5.1 Analytical Procedure for Isotope Analysis of Steroids

A very useful and comprehensive overview of IRMS use in steroid doping analysis, including a recommended measurement protocol, has recently been provided by Cawley and Flenker.[131]

The sole matrix of relevance in doping control analysis for steroids is urine. Urine is a rather complex matrix containing high amounts of salts, urea, creatinine, proteins, etc. Thus, a complex sample preparation and clean-up procedure is required before subsequent analysis by GC-IRMS.[132] A typical sample preparation scheme demonstrating the numerous necessary steps is shown in Figure 6.33. The steroids are excreted as glucuronides that are extracted from the raw urine on SPE cartridges in a first step (note that in more recent work this first step is sometimes omitted.[127] In SPE, mostly standard C_{18} materials are used. The eluent is evaporated to dryness, reconstituted and the conjugates enzymatically cleaved. In a second SPE step the free steroids are extracted, often fractionated by suitable elution conditions and derivatized with an acetic anhydride-pyridine mixture. Sometimes, the SPE step prior to derivatization is replaced by HPLC fractionation. The solution containing the derivatized steroids is evaporated again to dryness, reconstituted in acetonitrile-water and subjected to a third SPE or HPLC purification step. After washing, elution and a third evaporation step to dryness the derivatized steroids are redissolved (mostly in

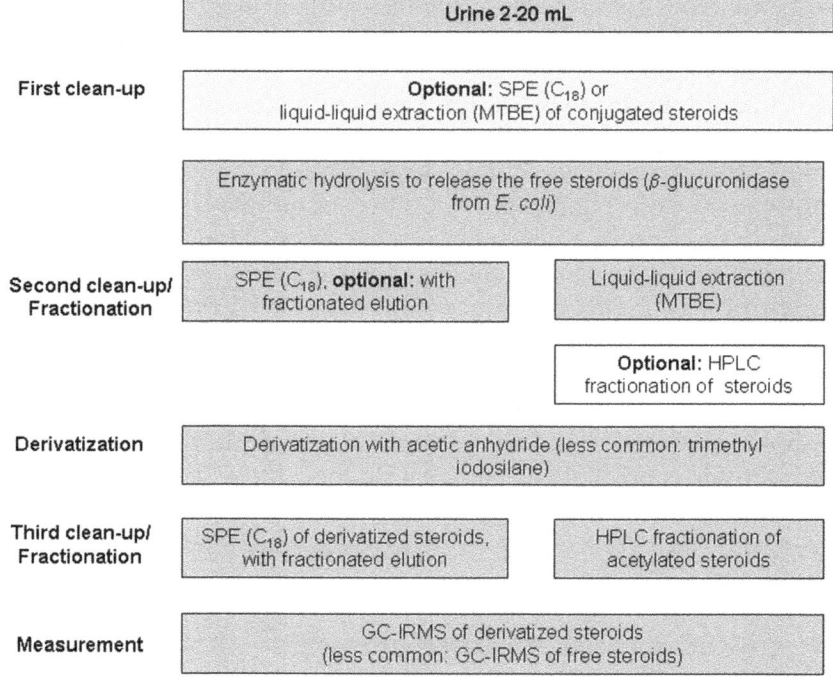

Figure 6.33 Typical steroid analysis procedure for GC-IRMS analysis.

cyclohexane) and injected in splitless mode. A full isotopic method validation has shown no isotopic discrimination in this complex scheme despite the various phase transfer and potential volatilization steps.[128] Using a slightly modified method, a sample throughput of 10 urine specimens plus two quality control samples is possible per day.[132] Other methods used in sample preparation include liquid-liquid extraction, RP-HPLC, and immunoaffinity chromatography.

As mentioned above, steroids are usually derivatized to a form more amenable for GC analysis during sample preparation although the mono-hydroxylated species could also be measured underivatized. Derivatization is generally done by a modification of hydroxyl moieties via acetylation or trimethylsilylation. The external carbon introduced into the derivative before combustion should be corrected for by an isotope mass balance. The isotope shift between the measured derivative and the originating steroid may be as large as 5‰.[132] Furthermore, acetylation might involve a *KIE* since a carbon-containing bond is formed in the rate-limiting step of derivatization. However, ERC and exogenous steroids are influenced nearly to the same extent, thus it is not clear if in practice a correction for the derivatization reagent is always necessary. Of course, a comparison of isotope values from different sources requires the same data treatment.

As always is the case in SIA, there is a need for dedicated reference materials and standards in steroid isotope analysis. Recently, Zhang *et al.* proposed a protocol for the generation of steroid isotope standards helpful in method validation within a laboratory but also a step towards a harmonization of stable isotope data among different laboratories and analysts.[133]

6.5.2 Further Developments

Recently, Sacks *et al.*[134] have shown the possibility of separating 10 underivatized steroids tested for in doping control with a fast GC approach in ~10 min. They utilized a 20 m, 0.15 mm, 0.60 μm VF-5ms column for the separation and achieved mostly comparable $\delta^{13}C$-values to a conventional method with the additional advantage of an improved separation of DHEA and androsterone. In follow-up work requiring fast detector response and corresponding data processing this group has shown the potential of comprehensive GC × GC for improved separations of steroids.[135] Although promising, this method is not yet suitable for routine doping control analysis since data processing is neither fully automated nor commercially available (see Chapter 7 for a more thorough discussion of GC × GC-IRMS).

Concurrently, Flenker *et al.* also showed advantages of GC-IRMS analysis of underivatized steroids that was made possible by a modification of the GC-interface to the combustion oven in order to reduce the number of valves and connectors.[136] They were also the first to evaluate the use of large-volume injection with solvent removal in the injector, which provides two advantages: (i) higher sensitivity and (ii) reduction of the number of unions and valves in the system that could adversely affect peak shape and leak tightness since no backflush valve is required anymore.

An alternative option for analysis of non-derivatized steroids is the hyphenation of HPLC with IRMS. Although first attempts exist, no reports are yet published in literature, thus hinting at remaining problems in the method. Hydrogen isotope analysis of steroids also has only recently been explored and its potential use in doping tests is still unclear since exogenous steroids show a large range of $\delta^2 H$ values mainly influenced by the local water isotope composition.[137]

6.6 ENVIRONMENTAL SCIENCE

6.6.1 Scope

In the past 15 years, applications of CSIA in environmental science have gained momentum and are nowadays widespread but also initiating further developments in instrumentation and data interpretation. Emphasis in this chapter will be on sources and fate of environmental contaminants in aqueous systems, although we will not limit ourselves exclusively to aqueous compartments. The vast majority of all studies described in this chapter have been done on carbon isotope analysis. Thus, whenever not explicitly stated otherwise, 'isotope analysis' refers to carbon SIA. It has been demonstrated in other application sections and will be shown in Section 6.6.2 though that dual (or two-dimensional) isotope investigations are an exceptionally powerful tool for source discrimination. Very recently, dual isotope plots have also been used more frequently in transformation studies as will be discussed in Section 6.6.4.

6.6.2 Source Values of Organic Compounds

A prerequisite for distinguishing sources of anthropogenic organic compounds based on isotope data is an overview of potential source values. Unfortunately, in many environmental fate studies, no isotope data for reference compounds are given, which otherwise could be added to an isotope database of commercial chemicals of relevance as contaminants. Although availability of such a database as open source would be extremely helpful, such an endeavour is beyond the scope of this book. We need to restrict ourselves here to selected studies that have specifically investigated isotope compositions of individual organic compounds in industrial products. These are compiled in Table 6.6 with a clear majority of studies on volatile and semi-volatile halogenated compounds.

Differences in isotope signatures in commercial products may stem from the isotope composition of precursors and/or isotope fractionation during synthesis processes. It has been shown for some chlorinated solvents that this can indeed be used to distinguish different producers,[138,139,142] in particular if dual isotope information is used. An example of differentiation of various manufacturers of chlorinated solvents by a dual isotope plot is shown in Figure 6.34.

As shown in Figure 6.34, a very low $\delta^{13}C$ has been observed for chloroform (but only from one manufacturer), and a low $\delta^{37}Cl$ value for 1,1,1-TCA (again, only from one manufacturer).[142] Chlorinated methanes are always rather depleted in ^{13}C, due to their production from largely ^{13}C depleted methane.[161] Individual congeners in commercial PCB mixtures may show differences as high as 8‰ in carbon isotope composition depending on the product. Some products show a trend of more depleted ^{13}C with increasing number of chloro substituents. However, this trend was less pronounced for PCB mixtures with low chlorine content.[146]

Hydrogen isotope data for pure compounds of relevance in environmental isotope studies are still rather scarce. For chlorinated hydrocarbons this might be due to inherent problems in their hydrogen isotope analysis due to the formation of HCl (see Chapter 3). Commercial TCE was found to be strongly enriched in ^2H which has been ascribed to the manufacturing process which involves many steps of chlorination and dehydrochlorination potentially leading to isotope fractionation.[143]

Beneteau et al.[139] investigated the same chlorinated solvents from the same manufacturers as van Warmerdam et al. four years previously[138] but from new batches. Differences between these batches were as high as 4.4‰ for carbon and 2.5‰ for chlorine. Such large differences could easily be misinterpreted as evidence of different producers. However, with this one exception it remains unclear to date if and by how much the isotope profile of a product changes over time. Such information is of great interest though not only in environmental science but also in the area of product authenticity.

Table 6.6 Overview of industrial product studies aiming at providing isotope ranges for environmentally relevant contaminants. Most of the values have been obtained by off-line procedures or EA-IRMS.

Compounds	Isotope (s)	Isotope range in ‰	No. of manufacturers	Reference
Halogenated Compounds				
TCE, PCE, 1,1,1-TCA	C, Cl	C: −33.8 to −23.2 Cl: −2.9 to +4.1	2	138
TCE, PCE, 1,1,1-TCA	C, Cl	C: −35.8 to −27.6 Cl: −2.5 to +3.8	2	139
DCE, TCE, PCE, 1,1,1-TCA, CM, DCM, CF, CT	C, Cl	C: −58.4 to −24.1 Cl: −2.9 to +1.6	1	140
TCE, PCE, 1,1,1-TCA, DCM, CF	C	C: −51.7 to −24.1	Up to 3	141
TCE, PCE, 1,1,1-TCA, DCM, CF	C, Cl	C: −51.7 to −24.1[a] Cl: −2.7 to +3.4	4	142
TCE, 1,1,1-TCA,	C, H, Cl	C: −31.6 to −25.8 H: 459 to 682 (TCE), −23.1 to +22.2 (TCA) Cl: −3.2 to +3.9	5	143
DCE, TCE, PCE	Cl	−3.2 to +5.9	6	144
1,2-dichloroethane	H	−92 to −65	3 (6 batches)	145
PCBs	C	−28.6 to −18.9	14 commercial mixtures	146
PCBs	C	−33.1 to −14.5	4 commercial Aroclor mixtures	147
PCBs	C	−34.4 to −22.0	18 technical mixtures	148
PCBs	Cl	−3.5 to −1.3	3 commercial mixtures	149
Chlorinated SVOC (pesticides, PCBs)	C, Cl	C: −31.8 to −23.6 Cl: −5.1 to +1.2	7	150
31 chlorinated SVOC (pesticides including DDT and metabolites, bromophenols)	C	C: −38.2 to −22.9 H: −325 to −1.0	9	151
DDT	Cl	−5.4 to −3.5	3	152
17 polybrominated diphenylether (BDEs) congeners	C	−34.4 to −26.7	1	153
Gasoline Components				
BTEX	C	−28.6 to −26.0	3	154
BTEX	C	−29.4 to −23.9	Up to 6	155
Benzene	C, H	C: −27.9 to −24.9 H: −94.4 to −27.9	3 (4 batches)	156
MTBE	C, H	C: −30.7 to −28.8 H: −56.8 to −36.9	3	157
MTBE	C, H	C: −33.0 to −27.4		

Applications of Compound-specific Stable Isotope Analysis 283

Table 6.6 (Continued).

Compounds	Isotope(s)	Isotope range in ‰	No. of manufacturers	Reference
			From gasoline samples and source zones of MTBE plumes	Cited in 158
MTBE	C, H	H: −125 to −80.0 C: −28.8 to −27.6	4	159
19 gasoline components	C	H: −103 to −60.8 −34.0 to −16.4	28 gasoline samples worldwide	159
Others				
Trinitrotoluene (TNT)	C, N	C: −26.3 to −22.1[b] N: −5 to +10[b]	5 sources (mostly military)	160

[a]Partially including C data from ref. 141.
[b]Data estimated from Figure 2 in ref. 160.
Abbreviations: BTEX: group of benzene, toluene, ethylbenzene and the three xylene isomers; VOC: volatile organic compounds; SVOC: semi-volatile organic compounds; MTBE: methyl tert-butyl ether; TCE: trichloroethene; PCE: tetrachloroethene (perchloroethylene); 1,1,1-TCA: 1,1,1-trichloroethane; DCE: dichloroethene; CM: chloromethane; DCM: dichloromethane; CF: trichloromethane (chloroform); CT: tetrachloromethane (carbon tetrachloride); PCBs: polychlorinated biphenyls; DDT: dichlorodiphenyltrichloroethane.

6.6.3 Source Apportionment

With few exceptions sources of compounds are distinguished based on their isotopic composition as a result of photosynthesis, secondary biogenetic processes or industrial synthesis. Source apportionment studies require that isotope compositions remain stable over time or change only insignificantly. Time scales for this requirement may reach from days in doping control to at least millions of years in paleoenvironmental studies. Such source apportionment studies are also important in environmental science, since they may provide evidence of the origin of contaminants and will be discussed in the following with three exemplary compound classes, i.e., PAHs, *n*-alkanes and chlorinated hydrocarbons.

Polycyclic Aromatic Hydrocarbons. PAHs are ubiquitous environmental contaminants. Both their widespread occurrence from anthropogenic and natural sources and their high toxicological relevance make them one of the most investigated compound classes in environmental chemistry. Dominant sources of PAHs in the environment are petrogenic, i.e.,

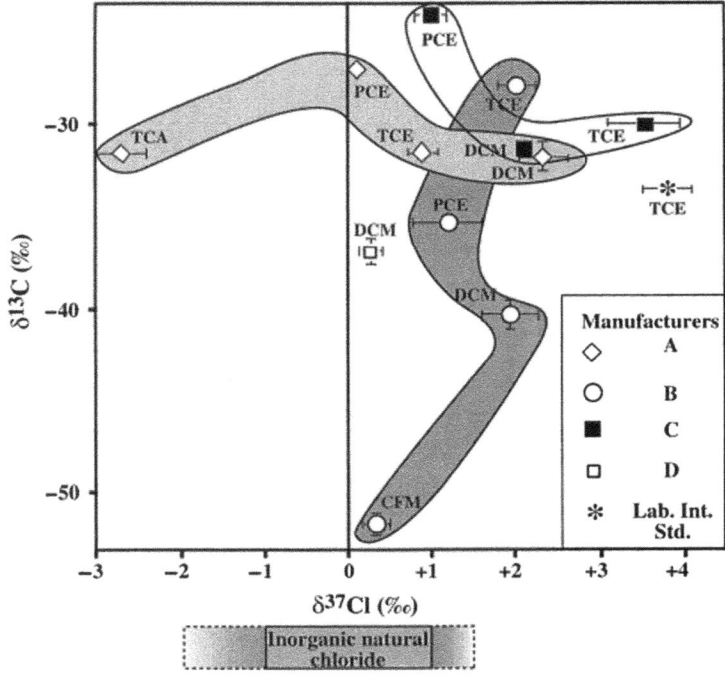

Figure 6.34 Differentiation of manufacturers by carbon and chlorine isotope analysis of pure products. CFM: chloroform, other abbreviations as in Table 6.6. (Graphic reprinted with permission from Jendrzejewski et al.[142])

derived from slow maturation of organic matter over long time scales and suitable temperature and pressure conditions, and pyrogenic, i.e., derived from incomplete combustion of recent and fossil organic matter.[162]

Table 6.7 gives a comprehensive overview of PAH isotope signatures (both total range and selected individual compounds) reported to date.

Pyrogenic PAHs are typically less depleted in ^{13}C than their petrogenic counterparts which can be used to differentiate the sources, mostly in combination with other indicators (molecular ratios of alkylated homologues to parent PAHs, other molecular indices, ratios of low molecular weight to high molecular weight compounds) and chemometric data analysis.[162,164,167,179,180] At Lake Erie it was possible to distinguish three areas of different contamination history. Furthermore, it could be shown that the main immission pathway for PAHs was fluvial input.[179] In further studies, various sources of PAHs in sediments along highly industrialized rivers and estuaries have been characterized.[162,167,180] To give a few examples, less negative δ^{13}C-values were found for three-ring PAHs originating from aluminium smelting in one area along the St. Lawrence river;[167] coal transport and use was found to have considerably contributed to PAH contamination in addition to former wood-treatment facilities;[162] and pyrogenic rather than petrogenic sources of PAHs were found to be dominant in harbour sediments.[164] Even changes in contributions of pyrogenic vs. petrogenic sources of PAHs over time in sediment records could be resolved.[180] In this case, $\delta^{13}C$ of pyrene was used as end member with a purely pyrogenic ($-24‰$) and purely petrogenic ($-29‰$) source assigned. By a simple mass balance assuming just these two sources, the contribution of both sources to total PAH load was calculated.

Mazeas and Budzinski have used isotopic information on n-alkanes (see below) and PAHs to investigate sources of crude oil from a tanker wreckage of major importance near the French Atlantic coast 1999. With strong support from these data they were able to show that tar balls stranded at several beaches along the Southern coast did not originate from the tanker oil spill (see Figure 6.35) but, rather, from other sources. Furthermore, they concluded that isotope composition of individual compounds is hardly affected over time and thus allows source discrimination even if the molecular distribution has already been altered considerably by weathering processes.[181]

Blessing used information on the isotope ranges expected for PAHs from petrogenic and pyrogenic sources as a further line of evidence in distinguishing major sources of PAHs at a contaminated field site of a former mineral oil processing facility. Possible sources for contamination at the site were heavy fuel oils, waste oil, creosotes, and petroleum fuel oils. The results show $\delta^{13}C$-values indicative for creosote as the most likely contamination source for the soil samples at the site (see Figure 6.36).[178]

Isotopically extremely light PAHs ($\delta^{13}C = -31$ to $-62‰$) in lagoon sediments near Ravenna led to the conclusion that emissions were dominated by a former plant that used biogenic methane ($\delta^{13}C = -69$ to $-73‰$) as feedstock rather than by operating plants using petrogenic feedstocks of much higher ^{13}C content.[177,182] However, the wide range of observed PAH isotopic composition and the enrichment in comparison to the source values could not be explained since both superimposition from a secondary source and biodegradation were believed to be insufficient to induce the observed shifts.[182]

As expected, PAHs emitted from traffic are pyrogenic and thus are less depleted in ^{13}C than petrogenic-derived ones. In gasoline exhaust, typical $\delta^{13}C$ ranges are $-23.5‰$ to $-18.6‰$, while in diesel exhaust typical values are in the range from $-24‰$ to $-23‰$. Furthermore, for both sources, a clear trend towards less ^{13}C depleted compounds is observed with an increase in molecular weight. In contrast, for

Table 6.7 PAH isotope signatures reported in primary and secondary sources and in environmental receptors.

Reference	Source/Receiving matrix	Total $\delta^{13}C$ range		Average $\delta^{13}C$-values and/or ranges of selected individual PAHs								
				Ace			Fln			Phe		
		Max	Min	Av.	Max.	Min.	Av.	Max.	Min.	Av.	Max.	Min.
	Primary Sources											
163	Crude oil	−26	−29							−27.9		
164	Crankcase oil	−26.2	−28.9				−28.4			−28.9		
165	Groundwater contaminated with creosote	−22.1	−25.3	−23.4	−22.5	−23.7	−23.9	−22.5	−25.3	−23.7	−23.3	−24.4
162		−23.8	−24.8									
166	Fireplace soot	−24	−26.4							−26		
164	Fireplace soot	−25	−26.7							−26.7		
167	Fireplace soot	−23	−27.5								−24.3	−26.2
168	Wood burnings	−27	−32									
169	Soot mixed coal/wood-burning fireplace	−28.9	−30.6							−29		
170	Tar low-temperature carbonization	−24	−25.8							−25.3		
170	Tar high-temperature carbonization	−25.1	−26.5							−25.2		
171	Coke oven and coal carbonization tar	−21	−24					−22.6	−23.7		−21.2	−23.9
171	Carburetted water gas process tar	−26.8	−30.3				−28.4			−28.6		
162	Coal	−23.1	−24.6									
172	Coal hydropyrolysis	−23.3	−26									
172	Coal fluidized-bed pyrolysis	−24.5	−29.4									
169	Soot coal-burning fireplace	−24.2	−26.1							−24.8		
170	Tar town gas processing	−27.5	−29.5							−27.5		
170	Process gas fluidized-bed combustion	−28.8	−31.6								−29.8	−30.8
162	Coal gasification	−26.7	−27									
183	Diesel particulates	−28.1	−30									
173	Diesel exhaust	−27	−22									
166	Car muffler soot	−22.6	−25.4							−25.4		
164	Car soot	−24.7	−27							−26.1		
173	Gasoline exhaust	−26	−13									
	Secondary Sources											
166	Sewage	−24.2	−25.3							−24.8		
166	Road sweeps	−23.8	−26.4							−24.6		
167	Road sweeps	−21.5	−24.9								−22.4	−23.3
358	Urban road dust	−24.8	−31.9	−24.8			−29.3			−27.4		
358	Rural dust	−26.1	−36.2	−26.6	−26.1	−27.0	−29.5	−28.9	−30.2	−26.3	−25.8	−27.3
168, 174	Urban aerosols	−23.7	−27									
173	Aerosols	−18	−28									
175	Burned biomass (C3)	−28	−28.4	−28.4			−28.2			−28.8		
175	Burned biomass (C4)	−15.8	−17.1	−16.6			−17.1			−16.5		
176	Field burned biomass (sugar cane, C4)	−22.9	−25.4							−24.5		
	Soil/Sediments											
163	Marine sediments	−25.2	−26.5							−25.6		
164, 166	Harbour sediments	−24.9	−26.4								−25.1	−25.6
179	Lake sediments	−22.1	−31.2		−24.4	−28.2		−24.1	−26.4		−24.4	−29.2
171	Lake sediments	−24.8	−29.2		−26.7	−27.3		−24.8	−27.8		−26.2	−28.1
167	River sediments	−21.9	−29.2								−21.9	−24.2
177	Sediment/Soil near methane utilization	−30.8	−62								−30.8	−59.5
170	Town gas soil	−27.8	−29.6							−27.8		
170	Soil samples near low-temperature carbonization unit	−24	−31.9							−25.7	−25.3	−26.1
171	Soil samples from former manufactured gas production site	−26.3	−32.2		−29	−29.8		−27.7	−29.9		−29.3	−29.7
178	Soil samples from a creosote contaminated site	−23.5	−25.0	−23.5	−22.2	−24.5	−24.3	−22.4	−26.5	−24.5	−23	−25

Ace: Acenapthene; Fln: Fluorene; Phe: Phenanthrene; Ant: Anthracene; Fth: Fluoranthene; Py: Pyrene; BbF/BkF: Benzo[*b*]fluoranthene/Benzo[*k*]fluoranthene (typically not chromatographically resolved); BaP: Benzo[*a*]pyrene.

Ant Av.	Max.	Min.	Fth Av.	Max.	Min.	Py Av.	Max.	Min.	BbF/BkF Av.	Max.	Min.	BaP Av.	Max.	Min.
−23.1	−24.2	−20.7	−26 −27 −24	−23.5	−25	−28.5 −28.7 −23.6	−23.2	−23.9	−27.8			−27.7		
−24 −25	−27	−27.7	−24.4 −25.2 −25.3	−25.7	−26.5	−24.2 −25.1 −25.3	−25	−26.3	−24.1 −26.4 −26.1	−24.3	−26.1	−24.8 −25.7 −25.7	−23	−25.3
			−30.6			−28.9			−29.4					
−24.5						−24			−24.6			−25.8		
−25.2						−25.6			−26.4			−26.5		
	−21.2	−23.7		−22.2	−23.9		−22.4	−23.8		−21	−22.5		−23.7	−23.8
−28.8			−29 −24.4			−28.7 −24.6			−23.1					
−28.6	−29.9	−31	−25.9			−26.1 −28.3	−30.2	−31.6	−24.2 −28.8	−29.7	−31.1	−28.9	−29.9	−31.6
			−26.8			−26.8			−26.7			−26.9		
−24.2 −25			−23.6 −25.2			−22.6 −24.7			−24.4 −26.1			−24.1 −25.7		
−24.2			−24.8 −24.7			−24.8 −23.8			−25.3 −26.4			−25.2		
−31.9 −35.9	−35.4	−36.9	−27.8 −27.4 −25	−23 −26.4 −24.2	−24 −30.6 −26.2	−27.9 −27.6 −25	−23 −26.7 −23.5	−24 −28.2 −26.7	−31.2 −25	−22.7 −30 −23.6	−23.3 −32.4 −26.2	−21.7	−23	
			−28.3 −16 −25.4			−28 −15.8 −22.9								
−24.9	−23.9 −26.2 −23.6 −30.8	−27.9 −27 −26.8 −61.7	−25.2	−25.4 −24.3 −25.7 −24 −33.4	−25.9 −28.7 −28.5 −27.5 −63.5	−26.5	−25.5 −23.9 −26.1 −23.6 −47.7	−26 −28.5 −28.4 −28.5 −62.3	−25.5	−25.8 −23.7 −26 −22.8	−26.4 −27.4 −27.7 −28.3	−26.4	−25.2 −23.7 −25.5 −23.2	−26.1 −28.8 −28.6 −29
−28.6 −25.8	−25.4	−26				−28.6 −25.4	−25	−25.6	−28.3 −27.6	−26.2	−28.6	−28.8 −29.8	−26.9	−30.5
	−28.4	−29.6		−30.5	−31.2		−30.3	−31		−29.7	−30.2		−30.3	−31.5
−24	−23.1	−26	−25	−24.5	−25.7	−24.9	−24.1	−26						

288　　Chapter 6

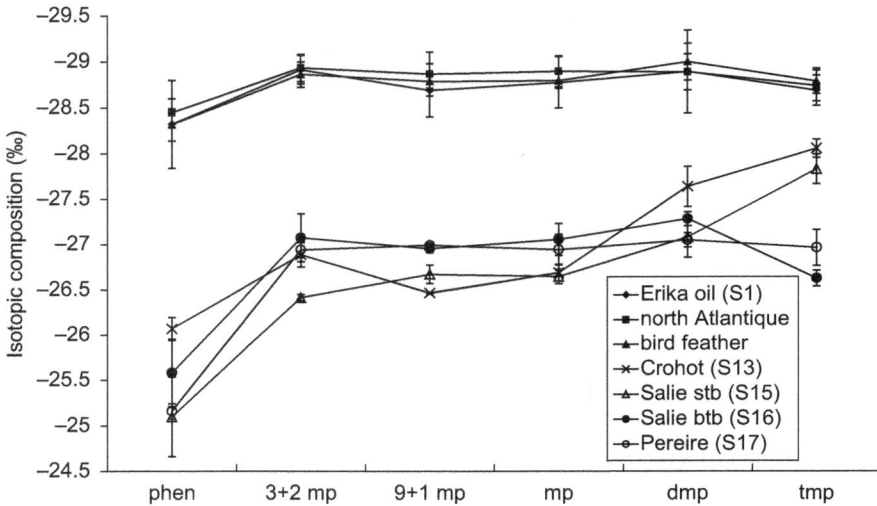

Figure 6.35　Isotopic composition in ‰ of phenanthrenes (mp, smp, tmp: methyl-, dimethyl-, and trimethylphenanthrene, respectively) in crude oil spilled by the tanker Erika in 1999, in oiled bird feathers and oil residues collected along the Atlantic coast of France.
(Graphic reprinted with permission from Mazeas and Budinski.[181])

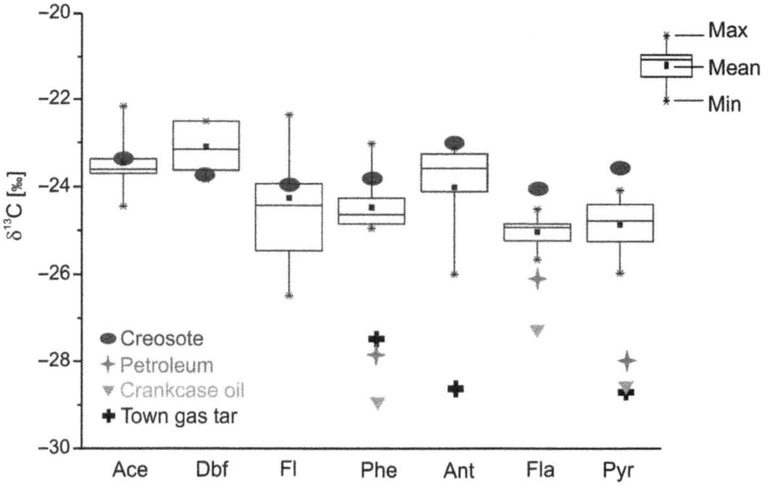

Figure 6.36　Box-whisker-diagram for individual PAHs of soil samples taken at a contaminated site compared to selected mean isotopic compositions of creosote, petroleum, crankcase oil and town gas process tar. Ace: Acenaphthen; Dbf: Dibenzofuran; Fl: Fluorene; Phe: Phenanthrene; Ant: Anthracene; Fla: Fluoranthene; Pyr: Pyrene.
(Graphic reprinted with permission from Blessing.[178])

atmospheric particles derived from C_3 wood burnings, a typical value of −28‰ is found and no trend with molecular weight is observed.[173,175] A completely reversed trend is found for particles derived from fluidized-bed coal pyrolysis. Here, higher molecular weight compounds are more depleted in ^{13}C, which is attributed to *KIEs* in the ring condensation reactions leading to the larger PAHs.[172] In contrast, in hydropyrolysis of coal, which is assumed to maintain major structural moieties of coal, this trend is not evident and no dependency on molecular size was observed, nor was this trend observed in fluidized-bed combustion of coal.[170]

Based on such considerations and further measurements of PAHs in atmospheric particles, Okuda *et al.* were able to exclude a major forest fire in Indonesia 1997 as the major contributor to atmospheric haze in neighbouring countries.[173] Figure 6.37 highlights the discrimination of major sources and particles collected from haze.

Zhang *et al.* have recently shown the application of individual PAH isotope analysis to distinguish two common sources of indoor pollution, namely cooking fumes and environmental tobacco smoke.[184] Carbon isotopic values of PAHs from both sources ranged from −21.8 to −29.3‰.

A further interesting topic in source discrimination using PAH isotope signatures is the differentiation of recent biogenic and pyrogenic origin of distinct PAHs, in particular perylene, in soils and sediments.

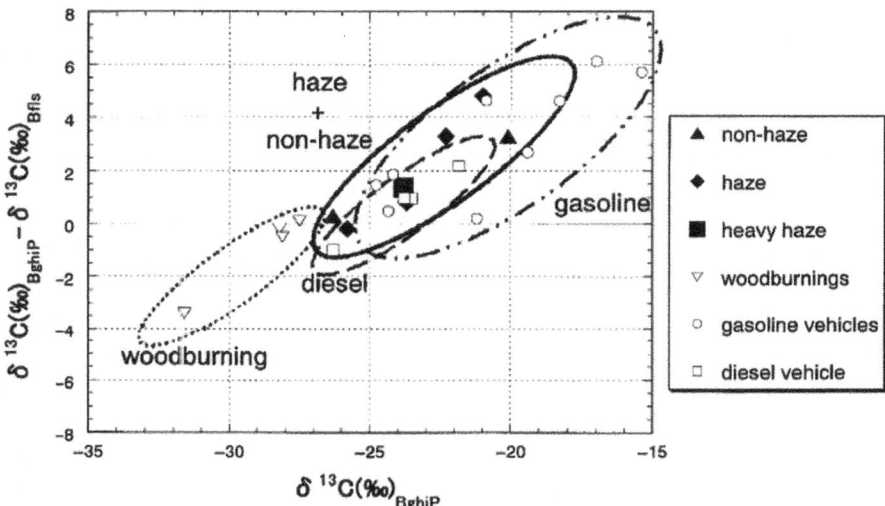

Figure 6.37 Discrimination of suspected sources of atmospheric haze in Malaysia based on isotopic composition of benzofluoranthenes (Bfls) and benzo(ghi)perylene.
(Graphic reprinted with permission from Okuda *et al.*[173])

Silliman et al. reported unusual concentration profiles of perylene in marine and lacustrine sediments, in particular in lower depths typically depleted in organic matter. They have corroborated their assumption of a biogenic (microbial) process responsible for perylene diagenesis by the isotope differences between perylene and average TOC.[185] In tropical soil and termite nest samples perylene was also found to be more depleted in ^{13}C than in temperate control soils and in studies of sediments known to be contaminated with pyrogenic PAHs. In combination with other supporting data this was attributed at least partially to a recent biological production of perylene from C_3 plants. In contrast, for naphthalene, although unusual concentration profiles were observed as well, no isotope discrimination was determined.[186] Recently, some preliminary indications of a biogenic production specifically of perylene in temperate sub-soils, based on concentration profiles of perylene and PAHs and the unusually low δ^{13}C-value of perylene, have been reported.[187]

n-Alkanes. In contrast to PAHs, coal-derived *n*-alkanes do not show significant alterations of isotopic composition with coal rank or conversion regime.[172] Furthermore, in contrast to the molecular distribution of individual species in petroleum products, which changes considerably over time and is thus used for classifications of degradation status, the isotope composition of individual compounds is hardly affected by weathering for carbon[188,189] and hydrogen.[190] Crude oils from various sources formed under different conditions can be distinguished by the carbon isotope profile of individual *n*-alkanes[189,191] although for smaller (C_4 to C_9) *n*-alkanes carbon isotope fractionation due to partial degradation can be substantial.[192] Thus, again it needs to be emphasized that additional information such as molecular distributions, biomarker concentrations and marker ratios are required in conjunction with isotope data. The potential role of CSIA of *n*-alkanes in such combined approaches has been discussed previously by various authors.[193,194] As pointed out by Mansuy et al., CSIA is particularly helpful for discrimination of sources if other information is not available, for example in the case of light fuel oils that lack the typically used biomarkers.[189]

An already classical environmental application example provided by Mazeas and Budzinski has been introduced in the section on PAHs above.[181] In addition to the isotopic profile of phenanthrenes, *n*-alkanes larger than C_{18} also agreed well among spilled crude oil, samples retrieved from the Northern Atlantic and examples of oiled bird feathers. However, tar ball samples from various beaches differed significantly from the isotope profiles in these samples, providing clear evidence of one or more additional crude oil sources (see Figure 6.38).

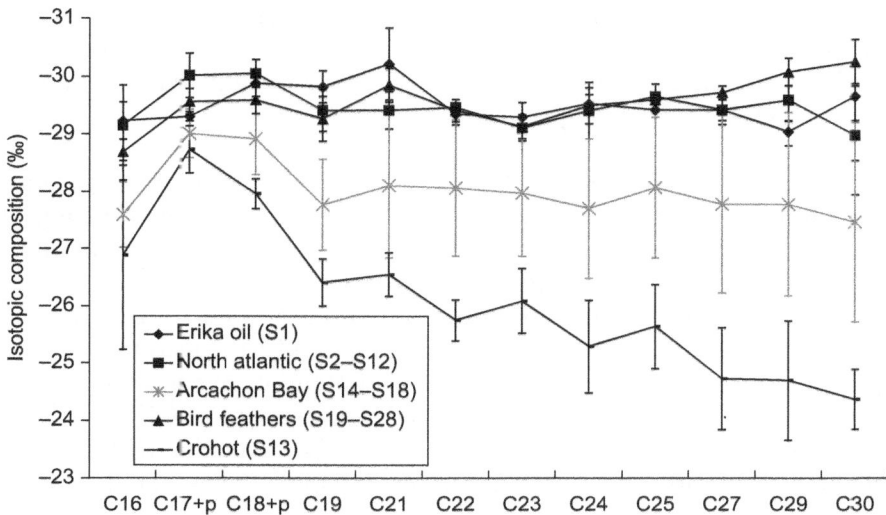

Figure 6.38 Carbon isotopic composition in‰ of n-alkanes in crude oil spilled by the tanker *Erika* in 1999, in oiled bird feathers and oil residues collected along the Atlantic coast of France.
(Graphic reprinted with permission from Mazeas and Budinski.[195])

The potential of such isotope correlations in *n*-alkanes of spilled oil and oiled bird feather samples had already been shown by Mansuy *et al.*[189]

n-Alkanes together with *n*-alkanoic acids have also been used as tracers in aerosols to distinguish contributions from C_3, C_4 and CAM plants. The aliphatic compounds show unique isotope profiles for each plant class due to the differences in photosynthetic pathways described in Section 6.2.2. For *n*-alkanes these are −39 to −31‰, −25 to −18‰, and −27 to −25‰ for C_3, C_4, and CAM plants, respectively. Isotope profiles of *n*-alkanes in the aerosols were consistent with predominantly C_3 plant origin and a minor contribution from CAM plants. In some samples, superimposition with petroleum-derived alkanes was found.[196]

Pond *et al.* suggested preferentially investigating hydrogen isotope profiles of the *n*-alkanes since its isotope range is much larger than for carbon. By covering a wide range of *n*-alkanes they have shown that both transformation and sources of crude oils may be evaluated by hydrogen isotope profiles of *n*-alkanes. For the lower molecular weight compounds (up to *n*-C_{18}) they showed substantial degradation accompanied by a large hydrogen isotope fractionation, whereas the larger *n*-alkanes showed constant hydrogen isotope values and thus may be useful source markers.[190] Meaningful data in this study have been generated under controlled laboratory conditions in order to study the weathering process. However, it remains to be shown if the same conclusive differences are seen for crude oils weathered in natural environments.

Chlorinated Hydrocarbons. Releases of chlorinated hydrocarbons from industrial activities such as metal degreasing and dry cleaning of clothing are among the most abundant anthropogenic contaminations on a local scale. Considering the intense work that has been carried out over the past 15 years on natural attenuation and isotope fractionation during transformation of such compounds it is astonishing that publications on source allocations of potential release sources at industrial sites are sparse. Hunkeler *et al.* investigated PCE releases at a dry cleaning site with presumed additional sources due to the width of the plume and various concentration hotspots. They could confirm the existence of at least three sources for PCE based on distinct differences in PCE carbon isotope signatures. Due to the aerobic conditions in the aquifer, no degradation of PCE occurred and the source signatures were preserved over more than 200 m flow distance.[197] The ability to distinguish sources of TCE in groundwater at a former manufacturing site despite partial degradation was demonstrated by Eberts *et al.*[198] Carbon isotope data of TCE and its primary degradation products could be utilized in conjunction with many other lines of evidence to correct a previously proposed site scenario that suggested a unique source of TCE and subsequent transport of pure phase compound as a dense non-aqueous phase liquid (DNAPL) into various sections of the affected aquifer. By means of an isotope mass balance of all products it was possible to infer PCE as the most likely source of pollution in one section of the aquifer and industrial TCE in another section. Recently, Blessing *et al.* studied an even more complex field site, with PCE contamination from previous multiple industrial use, water flow in several fractured bedrock aquifers with hydraulic connections via vertical faults, and differing geochemical conditions.[199] Despite degradation of PCE in anoxic parts of the aquifer (best represented by manganese dissolved concentration isolines), it was possible to delineate at least six different sources of PCE contamination, some of which were previously unsuspected of release. Similar to the previously discussed field site, conclusions on source delineation were only possible by the combination of historical, hydraulic, geochemical and concentration data with stable isotope data of PCE and its degradation products. Figure 6.39 depicts the final site scenario.

Applications of Compound-specific Stable Isotope Analysis 293

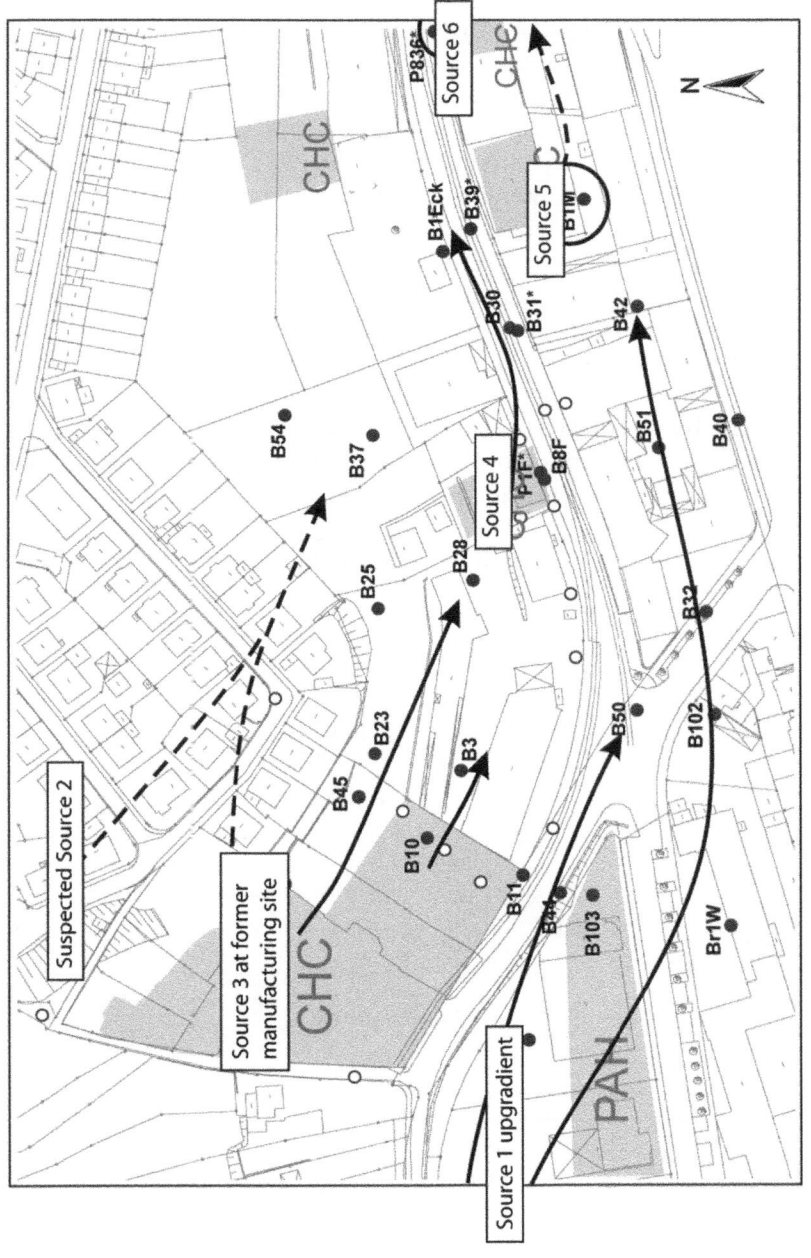

Figure 6.39 Delineation of at least six different sources of PCE at a multiple use industrial site with complex hydrogeology and geochemistry.
(Graphic modified from Blessing et al.[199])

Further Developments in Source Apportionment. Dual isotope approaches combining C and H, C and Cl, or C and N isotope analysis, which are becoming widely used in degradation studies, have hardly been applied in source apportionment studies so far and certainly will become more popular in this area as well. A first example for combined carbon and nitrogen isotope analysis has been given by Coffin *et al.* for distinguishing the source of explosive trinitrotoluene in groundwater at a military site although this was a preliminary data set as proof-of-concept.[160] Wang *et al.* were the first to combine carbon and hydrogen isotope analysis of *n*-alkanes and PAHs in a contaminated river to derive suggestions of different contaminant sources.[200] More recently, Mancini *et al.* were able to differentiate clearly two distinct sources of benzene at a contaminated site, also by combined carbon and hydrogen isotope analysis.[201] Mudge *et al.* have been able to classify fatty alcohol detergents in wastewater treatment plants and receiving compartments in a comprehensive evaluation of synthetic and natural surfactants as potential sources by a dual isotope plot of hydrogen vs. carbon shown in Figure 6.40.[202]

Finally, further statistical treatment of data certainly will gain importance since the multitude of data in field studies often cannot be

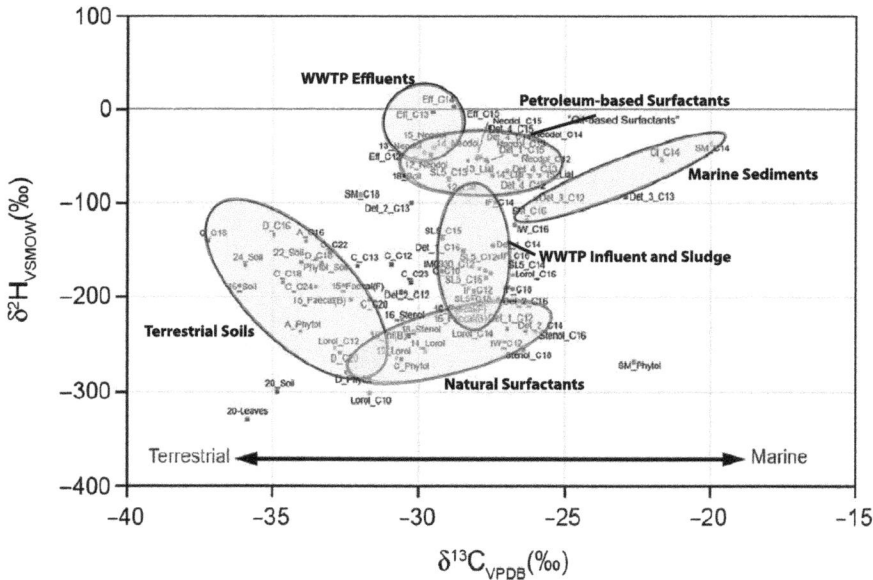

Figure 6.40 Dual isotope plot of δ^2H vs. δ^{13}C for C_{10} to C_{24} fatty alcohol (FA) detergents measured as trimethylsilyl ethers from various sources. (Graphic modified from Mudge *et al.*[202])

evaluated by intuition or experience alone anymore. Boyd *et al.* have detailed a step-by-step protocol utilizing a sample-wise principal component analysis (PCA, in this case of *n*-alkane carbon isotope data) followed by multivariate analysis of variance and a second PCA and/or hierarchical cluster analysis that helped to unravel the most likely source of a fuel oil spill in Elisabeth River, Virginia.[203] Such approaches require considerable additional expertise in chemometrics but will be invaluable in future field studies.

6.6.4 Use of Isotope Fractionation in Environmental Systems

In addition to source discrimination, there is a strong interest in monitoring isotope fractionation during environmental processes. Isotope fractionation can be used to identify or, under favourable conditions, even quantify processes such as diffusion and, in particular, abiotic and biotic transformation reactions in natural systems without the need for addition of reactive tracers. In contrast to source discrimination, a prerequisite is a measurable and constant isotope fractionation during the process(es) of interest (e.g., degradation of a compound by microorganisms).

Physical Processes. Vapour-liquid isotope fractionation for both carbon and hydrogen (and presumably also for the other elements considered in CSIA) is typically rather small[155,204] and only becomes relevant in the residual fraction at non-equilibrium if a large percentage (i.e., 95% or more) of the initial compound has been removed by evaporation. Both normal and inverse isotope effects have been reported for evaporation, making predictions difficult. Interestingly, and so far without a clear explanation, in vaporization experiments with TCE, for chlorine a normal and rather strong isotope effect was found but in the same experiments for carbon and hydrogen the observed isotope effect was inverse.[205] In the environment, isotope fractionation induced by evaporation may be relevant for VOC in specific scenarios such as shallow non-confined aquifers or during remediation by air sparging or soil bioventing[155,205,206] and may confound isotope evidence of degradation in such scenarios. However, as for most other physical processes, so far no field study has shown this effect clearly.

For dissolution of an organic compound from its pure phase in water, the few studies carried out so far, again, have indicated no discernable isotope effect in one-step equilibrium experiments.[204] However, mass-transfer limitations, for example, from a non-aqueous phase liquid to water can mask the inherent *KIE*,[207] as has been demonstrated with experimental data for microbial TCE reduction to *cis*-DCE.[208]

For sorption from aqueous phases many previous investigations have shown a non-discernable isotope effect for carbon[155,209,210] and hydrogen.[209] However, Kopinke et al. have been able to induce significant sorption-related isotope fractionation in multi-step sorption equilibrium or specifically designed column experiments at a mass loss exceeding 95%.[211] They found a correlation between the extent of isotope fractionation and sorption affinity, characterized by the partition constant between organic carbon and water (K_{OC}), although the number of compounds in this correlation was quite small. Numerical modelling of sorption-based isotope fractionation has quantitatively confirmed the effect for non-steady-state fronts (but not lateral fringes) of a contaminant plume.[212,213] However, to date, no clear evidence for sorption-related isotope fractionation at field scale can be shown. Thus, the general notion is still that sorption-related isotope fractionation may be of only limited relevance in field observations.

LaBolle et al. have discussed isotope fractionation during diffusion in the aqueous phase extensively and plea for a cautious consideration in field scale studies.[214] In general, high velocity flow in thin aquifers and heterogeneous flow systems tends to magnify fractionation. In a recent experimental study of transversal dispersion Rolle et al. have shown a rather strong fractionation of nondeuterated vs. perdeuterated ethylbenzene.[215] With a refined equation for transversal dispersion that takes into account calculated diffusion coefficients for both species in the pore diffusion and mechanical dispersion term, they were able to fit the experimental results well. However, the relative mass difference of the two species was large and although the authors point out, based on preliminary modelling, that fractionation might also be relevant for non-labelled species this has not yet been corroborated. In a source apportionment study, Hunkeler et al. observed an isotope shift in TCE of up to 2.4‰ near the fringe zone to an underlying aquitard.[197] They attributed this shift at least partially to diffusional isotope fractionation into the aquitard but were not able to exclude biodegradation in this fringe zone as the responsible process. Although it cannot be ruled out to superimpose fractionation or to confound source discrimination, so far, no clear evidence for diffusion-related fractionation in field scale studies has been shown. This is different for the unsaturated zone since gas-phase diffusion is known to induce a much clearer isotope effect as long as no steady-state is achieved (see also chapter 2). The relevance of diffusional transport that is accompanied by isotope fractionation has been shown in column experiments for methane[216] and typical fuel-related contaminants.[217] Results obtained in the latter study were confirmed in a field study with artificial placement of a contaminant source in the unsaturated zone above an unconfined aquifer.[218] In the column systems, Bouchard et al. found substantial isotope shifts due to diffusion along the column length during the first 12 hours after initiating the experiment. Once diffusion took place under quasi-steady-state conditions, a uniform isotope profile over the column length was established. However, over longer periods of time the isotope composition of the source and along the column shifted to less negative $\delta^{13}C$-values due to preferential loss of the compound without the heavy isotope. Observed isotope shifts during the experiment were inversely correlated with molecular size. Largest effects were found for n-pentane (-4.8‰) and barely recognizable ones for isooctane (-1.4‰). This can be explained by the decreasing molecular weight ratio between the isotopologues with increasing molecular size that is proportional to the diffusion coefficient ratio: for n-pentane the resulting diffusion coefficient ratio (expressed as diffusion coefficient of the heavy over diffusion coefficient of the light isotopologue) is -2.18‰, for isooctane -0.99‰. Recently, Kuder et al. also have emphasized the need to consider isotope effects for carbon and hydrogen during volatilization and subsequent gas-phase diffusion in the vadose zone. These were studied with MTBE as probe compound in column systems.[219] Although isotope enrichment was smaller than in biodegradation studies it could obscure data interpretation at field sites. The authors showed the suitability of dual isotope plots (see below) to overcome this limitation.

In summary, no phase transfer processes involving (local) equilibrium between two phases substantial isotope effects are observed. With the exception of gas-phase diffusion no clear evidence for the relevance of phase transfer and transport processes on observed isotope fractionation in field studies has been provided yet, but it should be kept in mind that this cannot always be excluded a priori.

Transformation and Biodegradation Processes. In the area of environmental transformation processes there are three major applications of CSIA that are discussed in some detail below:

(i) An *in-situ proof of transformation reactions* in the environment not or only in rare cases possible with other methods.
(ii) A *quantification of the fraction of a compound transformed* by applying previously determined isotope enrichment factors (see Chapter 2 for a fundamental introduction to enrichment factors).
(iii) A *direct probe for prevalent reaction mechanisms* allowing in-depth insights into natural (and engineered) systems.

In the first application area (*in-situ proof of transformation reactions*), CSIA has been extensively used in contaminant hydrology and site management, either to meet requirements set for monitored natural attenuation (MNA) at contaminated sites or to provide a qualitative indicator of efficacy of an active remediation (for a discussion of the latter see below). The US EPA defined three criteria that must be met in order to accept MNA (originally formulated in ref. 220 and later refined):

(i) a proven trend of decreasing contaminant concentration and/or mass over time at appropriate monitoring points indicating a stable or retreating contaminant plume;
(ii) geochemical data providing indirect evidence of relevant potential attenuation processes at the site;
(iii) field or microcosm studies providing direct evidence of (bio)degradation processes contributing to natural attenuation at the site.

Due to the effort involved in traditional studies for criterion (iii), this is generally recommended only if the first two lines of evidence are inconclusive. However, CSIA can often provide in-situ evidence of degradation as requested in criteria (i) and (iii). An observed shift in carbon isotope composition of 2‰ towards a more positive value is deemed sufficient for this qualitative assessment if no commingling of various sources is present.[161] For many sites this will be the easiest and most cost-effective means to provide such evidence.

For other elements, there is so far an insufficient database to give similar threshold values but as for carbon a significant trend of enrichment of the heavy isotopologue along a flow path has to be observed. Several studies have pointed out that hydrogen might be better suited as a qualitative indicator of contaminant degradation (exemplified for BTEX and MTBE) due to the much larger isotope fractionation.[156,221–224] As suggested, for example, by Hunkeler *et al.*, dual isotope investigations could provide the most conclusive evidence of degradation.[156]

Quantification of degradation (second application area) based on evolution of isotope composition of a compound over space and/or time is typically done by the Rayleigh equation (see Chapter 2 for fundamentals) in its simplified form (eqn (2.60)):

$$\frac{R(^hE/^lE)_{Q_t}}{R(^hE/^lE)_{Q_0}} = f^{(\alpha-1)}$$

where $R(^hE/^lE)_{Q_0}$ and $R(^hE/^lE)_{Q_t}$ are the ratios of the heavy isotope to the light isotope in the reactant at time $t=0$ and t, respectively, f is the remaining fraction of the reactant at time t ($=c_t/c_0$), and α the fractionation factor. This form can be used in good approximation in studies at the low natural abundance level of the heavy isotopes and if the fractionation is small.[225] In particular, the Rayleigh equation has been used to determine isotope fractionation factors for biodegradation in microcosm studies. Figure 6.41 gives an example for the two important groundwater contaminants TCE and benzene (original data from Mancini et al. and Sherwood Lollar et al.).[226,227] The figure

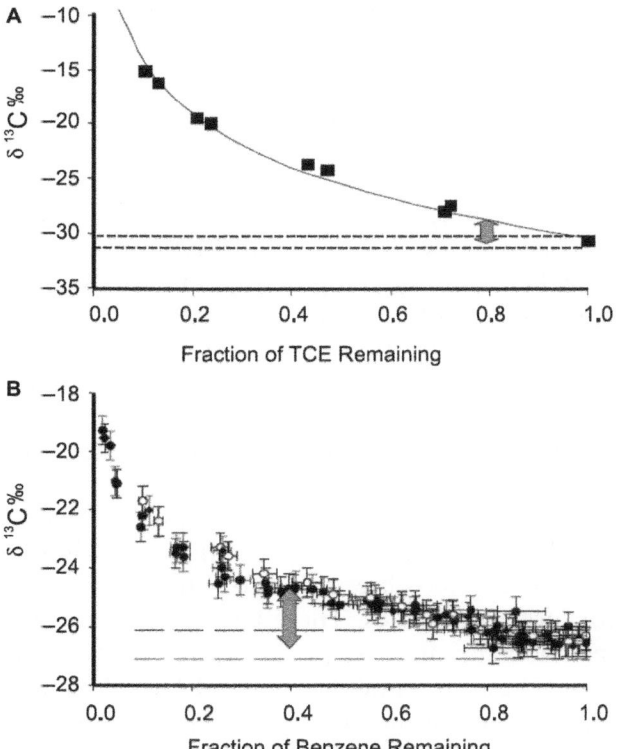

Figure 6.41 Degradation of (A) TCE and (B) benzene by enrichment cultures. The carbon isotope ratios vs. the remaining fraction of the contaminant are plotted. Dotted lines represent ±0.5‰ around the initial δ^{13}C-values of TCE and of benzene. The grey arrow represents the extent of fractionation required to observe an isotope shift of 2‰ necessary to identify transformation by isotope analysis.
(Graphic reprinted with permission from Hunkeler et al.[161])

shows clearly that (i) the more of the compound is degraded the larger is the isotope shift in the residual fraction, (ii) isotope shifts and thus enrichment factors may differ substantially among compounds (and also dominant reaction mechanisms), and (iii) therefore, a significant isotope shift (i.e., larger than 2‰) can be observed at different degrees of transformation (in this example 20% for TCE, 60% for benzene).

By use of the Rayleigh equation, fractionation factors α or enrichment factors ε have been determined for many relevant organic contaminants under specified conditions. Lists of α or ε values have been compiled in several recent reviews (e.g., refs 225, 228), a US EPA guideline[161] and a book chapter.[229] It is important to note that isotope enrichment factors should be determined in controlled laboratory experiments, not from field data.[161,230] However, field-derived values can be compared with reference data to test certain hypotheses on predominant degradation pathways.

Sometimes, rather than ε, an intrinsic enrichment factor ε_i has been reported that is simply taking into account the 'dilution' of the measured ε by non-reactive atoms of the same element in the compound, i.e., $\varepsilon_i = \varepsilon \cdot n$, where n is the number of atoms of the element in the compound of interest (see also later discussion of apparent *KIEs*). This 'dilution' comes from the fact that in the typical instrumental set-up all atoms of an element within a molecule are converted to the measuring gas, regardless of their participation in an isotope fractionating transformation. In other words, if there is only one atom out of ten in a molecule involved in the rate-limiting reaction, the measured change in isotope signature will be 10-fold smaller. Thus, degradation of compounds larger than C_{10} to C_{15} typically does not yield a measurable carbon isotope fractionation.[225,228] In addition to dilution of the isotope signal with the same element at non-reactive positions, rate-limiting transport steps partially masking the observed isotope fractionation become more important for larger molecules. There have been attempts reported on measuring position-specific isotope signatures in environmental contaminants[231] but these are rather tedious and specialized and have never been adopted for routine measurements. The decrease in measurable isotope fractionation with molecular size has been illustrated in several studies with chloroalkanes of two to four carbon atoms[232] and *n*-alkanes of three to ten carbon atoms.[233] In the latter case, a log-linear relationship between the number of carbon atoms and the isotope enrichment factor was reported. A detailed discussion of this topic can be found by Elsner.[207]

Under the assumption of the applicability of the Rayleigh equation at a site, the extent of degradation B of a compound of interest can be calculated without the need to incorporate concentration data:[228]

$$B = 1 - \left(\frac{\delta^h E_{Q_t} + 1}{\delta^h E_{Q_0} + 1}\right)^{\frac{1}{\alpha-1}} \qquad (6.1)$$

Typically, the estimates of the extent of degradation based on CSIA data and laboratory-derived fractionation factors are conservative, i.e., they tend to underestimate the true extent.

The Rayleigh equation applies in principle only to irreversible reactions in closed systems (see Chapter 2). Nevertheless, it has been extensively applied in inherently open environmental systems. Abe and Hunkeler have studied in detail under which conditions this approach may nevertheless be useful, in particular with regard to physical heterogeneity.[230] By applying an advection-dispersion transport model they found that the Rayleigh equation always leads to an underestimation of enrichment factors and degradation rates. Hence derived estimates of contaminant loss are conservative. However, the contribution of bias from application of the Rayleigh model in these estimates, in general, does not exceed that from uncertainties of other input parameters such as concentration data or travel times. Furthermore, methods not involving isotope data to quantify contaminant loss often yield much higher errors.[230] Applicability of the Rayleigh equation for quantification of (bio)degradation in the field was verified in a tracer test at a contaminated site with the injection of toluene-d_8 and toluene-d_5 as reactive tracers and bromide as conservative tracer. In this case, quantification of biodegradation was possible in two control planes downstream of the injection well by either the use of CSIA and application of the Rayleigh equation or direct quantification of the deuterated compounds. The resulting biodegraded fractions obtained by both methods agreed very well.[234]

In a study comparing various methods for quantifying degradation of o-xylene at a contaminated site it could be shown that the derived extent of degradation by integral pumping tests, carbon isotope analysis (quantification of degradation based on the Rayleigh equation) and a combination of both methods gave consistent results. This confirmed the applicability of the Rayleigh equation at the site.[235] In that context it is important to note that accurate concentration data of target contaminants are always required, both for choice of an appropriate enrichment/injection technique for subsequent CSIA and for interpretation of isotope data. It has been shown with a full error propagation that the error in concentration analysis can be more relevant for the calculated contaminant mass loss than the error in isotope analysis.[236]

Recently, van Breukelen has extended the simple Rayleigh model by a dilution term:

$$\ln\left(\frac{\delta^h E_{Q_t} + 1}{\delta^h E_{Q_0} + 1}\right) = \varepsilon \cdot \ln f = \varepsilon \cdot \ln(f_{total} \cdot F) \quad (6.2)$$

Here, F equals a dilution factor and f_{total} represents the remaining total fraction after mass reduction by dilution and degradation.[237] If ε is known, F can be calculated by rearranging eqn (6.2):

$$F = \frac{e^{\frac{\left(\frac{\delta^h E_{Q_t} + 1}{\delta^h E_{Q_0} + 1}\right)}{\varepsilon}}}{f_{total}} \quad (6.3)$$

Eqn (6.3) gives a simple means to evaluate relevance of dilution at a site. Van Breukelen substantiated this approach with the application to a benzene plume and was able to show that it allowed a more reliable prediction of degradation.

Another limitation of the simplified Rayleigh model is that it can only be applied for parent compounds, i.e., compounds that are not formed as intermediates during transformation. This is because both formation and further transformation of the intermediate can be accompanied by isotope fractionation. However, more sophisticated models allowing isotope shifts in sequential transformation reactions to be followed have recently been developed for the important sequential dechlorination of polychlorinated ethenes and have been applied with good success to describe real-world data.[212,238] These models may also be applicable for other multi-step transformations although this has not yet been demonstrated.

Finally, in a complex environment, several competing pathways for transformation may be present simultaneously which cannot be captured by the simple Rayleigh model but require a model extension[239] or application of an appropriate reactive transport model.[240]

In summary, the simple Rayleigh equation can often be sufficient for a quantification of transformation both in laboratory experiments and field sites and thus its applicability should be tested in a first step as also recommended in a US EPA guideline.[161] However, one should keep in mind the potential limitations discussed in detail above. Taking these into account it might be possible to extract useful quantitative information on contaminant behaviour even if on first sight the Rayleigh equation fails.

Under the assumption of a closed system and if all relevant transformation products are captured, the total amount of reactant and product(s) in the system will remain constant. This is true also for the total isotope composition. Thus, it is possible to set up an isotope mass balance (see also Chapter 2) that is easily calculated by summing up the molar concentrations of each species multiplied by its isotope composition divided by the total molar concentration of all species considered, as long as a substantially different extent of sorption does not bias the concentration data of some species included along a flow path. If the isotope mass balance shows an increasing total δ-value over time or space, this is an indication that transformation goes beyond the quantified transformation intermediates. Isotope mass balances have been used frequently in field studies,[241–243] most notably for carbon in chlorinated ethenes which is conserved in dechlorination products up to ethene or ethane. Aeppli et al. have recently shown that this approach successfully allowed them to distinguish contaminant plume sections where accumulation of vinyl chloride was expected and those that led to further transformation of this harmful intermediate in dechlorination.[243]

In the third application area (*direct probe of prevalent reaction mechanisms*) there have been major developments over the past few years. An early example was provided by Hirschorn et al. who investigated carbon isotope fractionation during aerobic degradation of 1,2-dichloroethane. As an unexpected result, a bimodal distribution of derived isotope enrichment factors for various microcosms, enrichment and pure cultures were found, centred around −3.9‰ and −29.2‰. Recalculation in terms of *KIEs* gave values of 1.01 and 1.06, which are typical for oxidation and substitution reactions, respectively. These different initial reaction steps among microorganisms were subsequently confirmed with dedicated pure strains of well-known enzymatic degradation pathways.[244] This example shows how powerful results of CSIA can be in unravelling primary reaction mechanisms.

Based on such results of environmental CSIA studies over the past decade and taking into account the previous knowledge on isotope effects in chemistry and biochemistry,[245] a conceptual framework of isotope fractionation to discern transformation processes in the environment has been developed by M. Elsner and others in recent years and is described in detail in a couple of landmark papers.[207,246] One of the major developments laid out in this concept is the bridging from observed isotope fractionation factors α to (apparent) kinetic isotope effects (*AKIEs*) already used in (bio)chemistry for several decades.[247] Here, the term 'apparent' indicates that *KIEs* are derived from measured data and the 'true' or 'intrinsic' *KIE* might be partially masked as discussed briefly below and much more in-depth in the mentioned papers.

Partial or total masking of *KIEs* may occur if the bond conversion process that induces isotope fractionation is preceded by a process that affects the whole molecule and is not accompanied by isotope fractionation. This includes transport to reactive sites, surface adsorption or substrate–enzyme complex formation.[246] If the reverse process is very slow essentially all molecules will be converted in the second step, thus leaving no isotope imprint on the residual substrate fraction. As an example, smaller isotope fractionation for both hydrogen and carbon has been observed in anaerobic toluene oxidation by *Geobacter metallireducens* with mineral suspensions rather than homogeneous solutions of Fe(III) as terminal electron acceptors. This was attributed to transport limitations to sessile microorganisms in the former case.[248]

Note, however, that as long as there remains a measurable *AKIE*, a Rayleigh-type model will still link concentration and isotope data.

Six important conclusions with regard to mechanistic information have been summarized by Elsner:[207]

(i) Isotope fractionation gives insight that would not be obtained from product analysis.
(ii) Evidence from isotope fractionation may elucidate degradation pathways even if no products are detected.
(iii) Isotope fractionation is mechanism-specific.
(iv) Compound-specific isotope fractionation must be evaluated in mechanistic scenarios.
(v) Observable isotope fractionation becomes smaller for larger molecules.
(vi) The need for *KIE* reference data.

We will not discuss in detail each of these points but refer the reader to this excellent in-depth overview.

The discussed work provides the basis for going beyond identifying transformation with a yes-no decision (first application area) or quantifying degradation based on enrichment factors (second application area). It shows how and under which conditions isotope information can be linked with an elucidation of reaction mechanisms without addition of artificial tracers and directly in natural systems. These opportunities make CSIA unique and so exciting for researchers working in environmental contaminant studies. It is no wonder, therefore, that the derivation of *AKIEs* has quickly become nearly routine in environmental studies.[249] We will discuss briefly how *AKIEs* are derived and for which purposes they can be used.

The *AKIE* takes into account the fact that *KIEs* are position-specific (i.e., they act only on the one reacting position where either the heavy or the light isotope of an element is present). Since in CSIA all n atoms of the same element in a compound are converted to the measuring gas, this information is 'diluted' (see also discussion of the intrinsic enrichment factor ε_i above). As an example, consider the transformation of toluene via methyl group oxidation. In this case, only one carbon out of seven is in a reacting position and thus the observed shift in isotope composition due to transformation will be seven times lower than the (true) position-specific one. In some cases, more than one atom is in a reacting position, thus the dilution is less pronounced and n needs to be corrected by the number x of atoms in the reacting position. This is the case, for example, in concerted reactions such as an epoxidation of the double bond in ethenes ($x=2$ for carbon), or in C-H bond cleavage in a methyl group ($x=3$ for hydrogen).

The *AKIE* is then calculated as follows:[207,246,249]

$$\ln\left(\frac{\delta^h E_{Q_0} + \frac{n}{x}\Delta\delta^h E_{Q_t} + 1}{\delta^h E_{Q_0} + 1}\right) = \varepsilon_{reactive\ position} \cdot \ln f \quad (6.4)$$

where n is the number of atoms of the element considered that are present in the molecule, and x of these are located at the reacting position.

For all elements except hydrogen this can, to good approximation, be simplified to:

$$\varepsilon_{reactive\ position} = \frac{n}{x}\varepsilon_{bulk} \quad (6.5)$$

where ε_{bulk} is the enrichment factor derived from measured CSIA data (denoted previously with ε).

AKIE is then rather simply derived by:

$$AKIE = \frac{1}{z \cdot \varepsilon_{reactive\ position} + 1} \quad (6.6)$$

where z is the number of atoms in a reactive position that are in intramolecular isotope competition. Note that the inverse relation takes into account that *KIE* values in the (bio)chemical literature are always reported as the ratio of the rate constants of the light to the heavy isotopologue (see Chapter 2), whereas α or ε values are defined vice versa. Looking again at the examples above, in the epoxidation of the ethene double bond with $n = x = 2$, z will be 1 because in the concerted reaction there is no intramolecular competition. In contrast, in the methyl group oxidation of toluene with $n = 7$, $x = 3$, there will be intramolecular competition between the three equivalent hydrogen atoms, thus $z = 3$. In contrast to n, correct designation of x and z requires a mechanistic hypothesis since, depending on the reaction mechanism, values will be different. Always, though, $n \geq x \geq z$. From current knowledge, x and z differ only in cases where (i) concerted reactions take place as in the double bond epoxidation or (ii) secondary isotope effects are involved (most importantly for hydrogen).

AKIEs calculated in this way can be compared to literature values of *KIEs* for certain reaction mechanisms or with approximated maximum *KIEs* ('semiclassical Streitwieser limits') which allows for a semi-quantitative comparison of measured and expected *KIEs*.[246] The latter is only semiquantitative since several simplifying assumptions have to be made, including an infinitely late transition state that resembles complete bond cleavage. Since in reality the transition state will be somewhere between reactant and product, i.e., bond cleavage is not complete, the Streitwieser limits provide maximum *KIEs*. Thus, these limits should not be overinterpreted as a physical constant.

Streitwieser limits can be calculated for any bond according to:

$$KIE = \frac{1}{e^{-100\frac{hc\tilde{v}}{2kT}\left(1 - \sqrt{\frac{\mu_L}{\mu_H}}\right)}} \tag{6.7}$$

where h, c and k are physical constants (see Appendix), $\tilde{v}$ is the stretch vibrational wavenumber in cm^{-1} as measured by infrared spectroscopy, μ_L and μ_H are the reduced masses of the light isotope and the heavy isotope, respectively. Table 6.8 gives such limits calculated with eqn (6.7) for some common bonds in organic molecules. The Streitwieser limits can be useful for a first plausibility test of experimental enrichment factors and of hypothesized reaction mechanisms in addition to *KIEs* known from literature for specific reactions.

Figure 6.42 shows ranges of *AKIEs* for hydrogen and carbon for various reaction mechanisms. Clearly, primary *KIEs* for hydrogen can be orders of magnitude higher than both secondary hydrogen and primary carbon *KIEs*. For carbon, highest *KIEs* are observed for S_N2 reactions which can, therefore, often be easily distinguished from other reaction mechanisms. Although many important chemical reactions are

Table 6.8 Calculated semiclassical Streitwieser limits at 25 °C for comparison with experimental values. For use and limitations of indicated *KIEs* see text.

Bond	Typical stretch vibrational wavenumbers in cm^{-1}	Isotope	Kinetic isotope effect (KIE)
C-H	2900	$^{12}C/^{13}C$	1.021
C-H	2900	$^{1}H/^{2}H$	6.447
C-C	1000	$^{12}C/^{13}C$	1.048
C-Cl	750	$^{12}C/^{13}C$	1.054
C-Cl	750	$^{35}Cl/^{37}Cl$	1.013
C-N	1150	$^{12}C/^{13}C$	1.060
C-N	1150	$^{14}N/^{15}N$	1.044
C-O	1100	$^{12}C/^{13}C$	1.061
C-O	1100	$^{16}O/^{18}O$	1.066
C-S	650	$^{12}C/^{13}C$	1.046
C-S	650	$^{32}S/^{34}S$	1.013

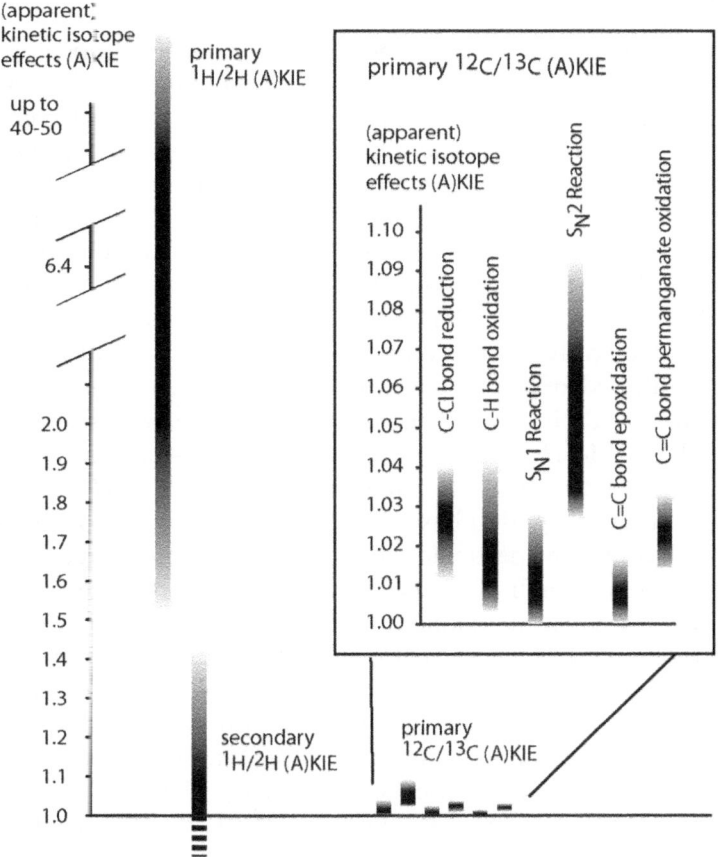

Figure 6.42 Expected *AKIEs* for hydrogen and carbon depending on reaction mechanism.
(Graphic reprinted with permission from Aelion et al.[253])

indicated already in Figure 6.42, clearly there is still a lack of data for many others and hence a vastly open research field, including *AKIEs* for elements other than carbon and hydrogen. For example, Hofstetter *et al.* have reported *KIEs* for nitrogen (and carbon) in the transformation of nitrobenzene by two different pathways. One involved, as the rate-limiting step, dioxygenation of the aromatic ring with a pronounced carbon *KIE* (1.0241) but essentially no nitrogen *KIE* (1.0008), the other involved partial reduction of the nitro group with a large nitrogen *KIE* (1.0273) but only a very small value for carbon (1.0034).[250] In addition, substituent effects on nitrogen *KIEs* have been studied in the same group.[251] Such work is extremely useful to expand the reference database for *KIEs* in environmental transformation reactions. Finally, in the future computational chemistry will help to better constrain or even predict *KIEs* in transformation reactions as has been shown by a few examples recently.[252]

Recently, the power of two-dimensional (or dual) isotope approaches has been recognized not only for better resolving sources of target compounds (see Sections 6.6.2 and 6.6.3) but also for better characterization of transformation reactions. By plotting isotope enrichment in one element vs. another, transformation mechanisms can be better characterized by the resulting slope since influences on residual concentrations by dilution, and masking effects, for example, by commitment to catalysis, that affect the apparent fractionation in each element should be eliminated.[207,246] This was exemplified in a recent study by Tobler *et al.* already discussed above, where the extent of isotope fractionation differed depending on the nature of Fe(III) as the terminal electron acceptor. However, almost identical slopes were found in a dual isotope plot of $\Delta\delta^2 H$ vs. $\Delta\delta^{13}C$, supporting the assumption of an identical reaction mechanism and partial masking of the *KIE*.[248]

Indeed, a growing database exists, so far mainly for fuel-related compounds and some pesticides, and demonstrates the huge potential of this approach although of course measurement efforts are substantially higher. Table 6.9 lists available dual isotope studies, Figure 6.43 shows how such plots can be used to distinguish different reaction mechanisms. The list of target compounds certainly needs to be extended considerably before one can judge if the dual isotope approach is really generally applicable and which limitations exist. A profound and detailed account of the current state of knowledge is given by Elsner.[207] Furthermore, although suggested as an approximation,[239] the derivation of the slope in a dual isotope plot of $\delta^h E_1$ (e.g., $\delta^2 H$) vs. $\delta^h E_2$ (e.g., $\delta^{13}C$) does not necessarily yield the same result as the approximation by ratios of enrichment factors.

Table 6.9 Overview of dual isotope correlations from degradation studies.

Compound(s)	Measured isotopes	Correlation	Reference
MTBE	H, C	Slope δ^2H vs. δ^{13}C: 1.3 (anaerobic from field data) 1.6 (anaerobic enrichment cultures)	158
MTBE	H, C	Slope δ^2H vs. δ^{13}C: 1.8 (predominately anaerobic from field data)[b]	254
MTBE, TAME	H, C	Slope δ^2H vs. δ^{13}C: MTBE: 1.3, TAME: 2.4 (presumably anaerobic from field data)	255
MTBE	H, C	Slope δ^2H vs. δ^{13}C: 14 to 18 (aerobic enrichment culture *M. petroleiphilum* PM1 and field data)[a]	256
MTBE	H, C	Slope δ^2H vs. δ^{13}C: 22 (aerobic from field bioremediation study)[b]	257
MTBE, ETBE	H, C	Slope δ^2H vs. δ^{13}C: MTBE: 15 (aerobic enrichment culture strain R8)[c] ETBE: 10 to 13 (aerobic enrichment culture strains L108 and IFP2001, R8 was not able to degrade ETBE)	258
MTBE, ETBE, TAME	H, C	Slope δ^2H vs. δ^{13}C: 45 to 49 (aerobic enrichment culture *Pseudonocardia* K1)	259
MTBE	H, C	Slope δ^2H vs. δ^{13}C: 11 (acid hydrolysis via S_N1)	260
Benzene	H, C	Slope δ^2H vs. δ^{13}C: 23 to 29 (sulfate reducing conditions) 28 to 39 (methanogenic conditions) 8 to 19 (nitrate reducing conditions)	261, 262, 226
Benzene	H, C	Slope δ^2H vs. δ^{13}C: 3.7 to 7.3 (aerobic conditions) 15 to 22 (anaerobic conditions)	263
Toluene	H, C	Slope δ^2H vs. δ^{13}C: 62 ± 3 (aerobic degradation by *P. putida*)	264
Toluene	H, C	Slope δ^2H vs. δ^{13}C: 1 to 68 (oxic conditions) 27 to 33 (sulfate reducing conditions) 11 to 15 (nitrate reducing conditions)	265
Toluene	H, C	Slope δ^2H vs. δ^{13}C: 25 to 27 (anaerobic degradation by *G. metallireducens*)	248
VC, cDCE	C, Cl	Slope δ^{13}C vs. δ^{37}Cl: 0.031 to 0.039 (oxidation) 0.073 to 0.088 (reductive dechlorination)	266
cDCE	C, Cl	Slope δ^{13}C vs. δ^{37}Cl: 0.48 (field data)	267
Nitrobenzene	C, H, N	Slope δ^{15}N vs. δ^{13}C: 52 (reduction) 0.2 (oxidation) Slope δ^2H vs. δ^{13}C: 1.6 (oxidation)	250, 249

Table 6.9 (*Continued*).

Compound(s)	Measured isotopes	Correlation	Reference
RDX	N, O[d]	Slope $\delta^{15}N$ vs. $\delta^{18}CO$: 1.2 (aerobic) 0.94 (anaerobic)	268
Atrazine	C, H, N	Photooxidation: Slope $\delta^{15}N$ vs. $\delta^{13}C$: 0.24 to 0.4 Slope $\delta^{2}H$ vs. $\delta^{13}C$: 27 to 31 Slope $\delta^{2}H$ vs. $\delta^{15}N$: 72 to 93 Direct photolysis: Slope $\delta^{15}N$ vs. $\delta^{13}C$: 1.1	269
Atrazine	C, N	Slope $\delta^{15}N$ vs. $\delta^{13}C$: 0.22, 0.26 (abiotic hydrolysis pH 12) −0.52 (abiotic hydrolysis pH 3)	270, 271
Atrazine	C, N	Slope $\delta^{15}N$ vs. $\delta^{13}C$: −0.65, −0.61, −0.32 (aerobic conditions, various strains)	271
Isoproturon	C, N	Slope $\delta^{15}N$ vs. $\delta^{13}C$: 3.0 to 3.8 (abiotic hydrolysis)	272
Isoproturon	C, H, N	Slope $\delta^{15}N$ vs. $\delta^{13}C$: 3.2 (abiotic hydrolysis) 0.84 (biotic hydrolysis) 0.45 (hydroxylation by fungus) Slope $\delta^{2}H$ vs. $\delta^{13}C$: 0.51 (abiotic hydrolysis) 0.85 (biotic hydrolysis) 21, 22, 98 (hydroxylation by various fungi species)	273

If data compilations have been provided in the given references, these have been used without referencing each individual study here.
[a]One exceptionally high value with a slope of 48 was excluded.
[b]Calculated based on given isotope data for MTBE.
[c]Two strains (L108 and IFP2001 (resting cells)) did not show any significant ^{2}H enrichment.
[d]Measured with EA-IRMS after thin layer chromatography clean-up. Abbreviations: MTBE: methyl *tert*-butyl ether; ETBE: ethyl *tert*-butyl ether; TAME: *tert*-amyl methyl ether; VC: chloroethene (vinyl chloride); *cis*DCE: *cis*-dichloroethene; RDX: research development explosive, common name for 1,3,5-trinitroperhydro-1,3,5-triazine.

Vogt *et al.* have shown that dual isotope information can much better distinguish between initial steps in microbial toluene degradation induced by either ring dioxygenase, ring mono-oxygenase, or methyl mono-oxygenase.[265]

One of the hitherto best-investigated organic compounds is MTBE. As a result of multiple studies, the isotope fractionation in both hydrogen and carbon has been well constrained for three different primary reaction mechanisms. A dual isotope plot of hydrogen vs. carbon is shown in Figure 6.43. It is evident that one can distinguish the three reaction mechanisms well if the extent of transformation (and

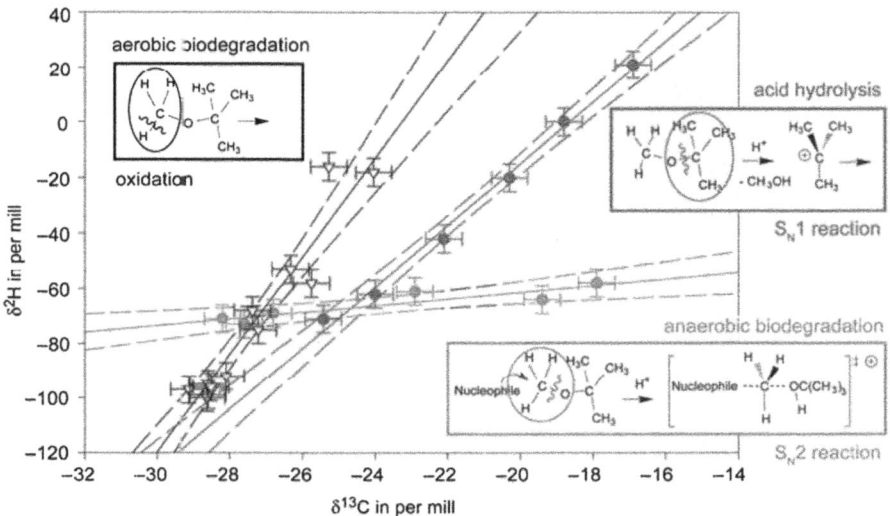

Figure 6.43 Dual isotope plot of δ^2H vs. δ^{13}C for MTBE degraded by three different reaction mechanisms.
(Graphic reprinted with permission from Elsner.[207])

accordingly the corresponding isotope shift $\Delta\delta^{13}$C and $\Delta\delta^2$H) is sufficiently large. During aerobic degradation a cleavage of the C-H bond in the methyl group is the primary step. This is accompanied by a large primary *KIE* for hydrogen but a rather small primary *KIE* for carbon (see also Figure 6.42). In contrast, a rather weak secondary hydrogen *KIE* is observed in the S_N2 reaction that represents the primary step in anaerobic biodegradation, whereas carbon isotope fractionation in that step is very strong. The abiotic hydrolysis via an S_N1 reaction is intermediate in both *KIEs*.

Dual isotope data for MTBE from a contaminated site by Zwank *et al.*[254] have been further scrutinized by van Breukelen who could show by further modelling that degradation could not have resulted exclusively from anaerobic biodegradation as proposed in the original publication (Figure 6.44) (see van Breukelen for details).[239]

The approach of van Breukelen allows us to semiquantitatively define contributions to degradation from two reaction mechanisms occurring at a site. It has also been used by Hofstetter *et al.* to represent contributions to nitrobenzene degradation by either ring oxidation (under aerobic conditions) or reduction of the nitro group (under anaerobic conditions) in microcosm experiments with two bacteria strains.[250] Since both strains utilize one pathway exclusively, no evaluation of a mixed scenario has

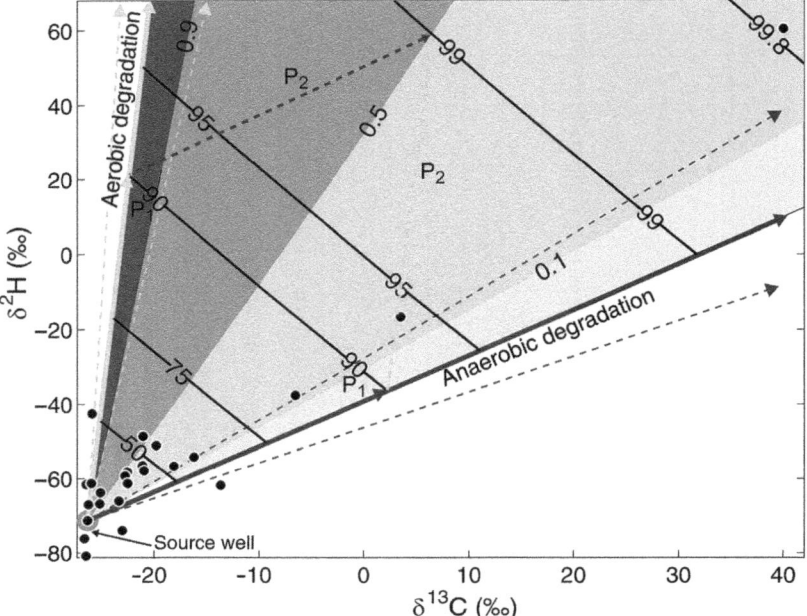

Figure 6.44 Dual isotope plot of δ^2H vs. δ^{13}C for MTBE at an industrial landfill site. Data taken from ref. 254. Black circles are observations, arrows indicate the δ^2H vs. δ^{13}C slopes (average, full lines; minimal and maximal, dashed lines) as expected based on literature values for aerobic and anaerobic degradation, shaded areas indicate the distribution, F, of aerobic and anaerobic degradation, and also the calculated extent of biodegradation in % is shown as black lines.
(Graphic reprinted with permission from van Breukelen.[239])

been carried out yet. However, this can be observed in the field with varying geochemical conditions as in the MTBE example above.

Probably the most thorough compound-specific multi-isotope study to date has been on biotic and abiotic transformation of the herbicide isoproturon. Initially, a method had to be developed that allowed for the GC-IRMS measurement of δ^2H, δ^{13}C and δ^{15}N of isoproturon. It turned out that the molecule is thermally labile and decomposes in the injector, which at first sight makes a reliable isotope measurement impossible. However, a careful investigation has shown that the decomposition can be controlled well and highly reproducible isotope data are obtained for the two fragments formed, i.e., 4-isopropylphenylisocyanate and dimethylamine.[274] The reaction is shown in Figure 6.45.

Using this method and triple isotope analysis of isoproturon via the fragments, a clear differentiation of initial reaction steps in (i) abiotic

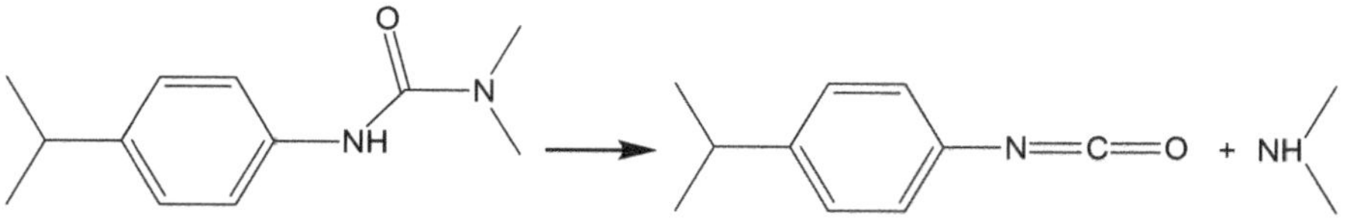

Figure 6.45 Thermal decomposition of isoproturon to 4-isopropylphenylisocyanate and dimethylamine.

hydrolysis, (ii) biotic hydrolysis leading to 4-isopropylaniline, and (iii) primarily hydroxylation of the isopropyl group by various fungi was possible (see Figure 6.46).[273]

In contrast to microbial degradation, isotope effects in photolytic transformation processes so far have hardly been studied in environmental systems. This can partly be explained by the difficult mechanistic understanding hampered by the fact that mass-independent fractionation might not be negligible in such processes.[207] For anthracene, a significant isotope shift during the first two hours of sunlight irradiation have been reported, but the results were not reproducible.[166] More recently, a comprehensive study of photo-oxidative transformation of atrazine by OH radicals, 4-carboxybenzophenone and direct photolysis has been carried out by Hartenbach et al.[269] They observed moderate H fractionation, and small isotope fractionation for C and N in the first two transformation reactions, but in the latter completely different results with no discernable H fractionation and significant inverse fractionation for C and N.

Use of CSIA in Assessment of Environmental Technical Processes. Finally, it is interesting to note that CSIA may provide further insight into environmental technical processes, including active attenuation measures for groundwater at contaminated sites (in contrast to confirming natural attenuation as has been amply discussed above) or used in drinking water purification and waste water treatment. In the former area, indeed, several studies have been carried out whereas in the latter area so far hardly any application is documented despite the great potential of CSIA.

Using carbon isotope analysis, McKelvie et al. have shown success of stimulated biodegradation based on ethanol addition to an aquifer contaminated with MTBE.[275] At the site, two lanes were directly compared, one without amendment as a field control, one with release of ethanol. By $\delta^{13}C$ analysis, an unequivocal proof of MTBE biodegradation in the ethanol release lane was possible after a relatively short

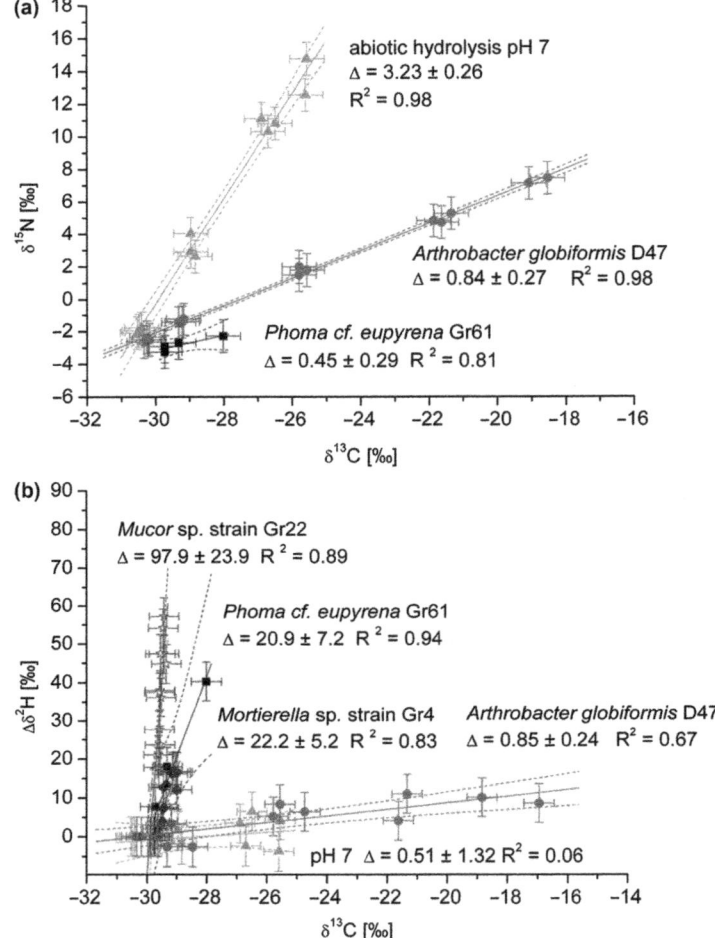

Figure 6.46 Dual isotope plots of (a) $\delta^{15}N$ vs. $\delta^{13}C$ and (b) $\Delta\delta^2H$ of the 4-isopropylphenylisocyanate fragment vs. $\delta^{13}C$ for isoproturon degraded by different fungi (*Phoma cf. eupyrena* Gr61, *Muco* sp. strain Gr22, *Mortirella* sp. strain Gr4, biotic hydrolysis by *Arthrobacter globiformis* D47, and abiotic hydrolysis. For $\delta^{15}N$ and $\delta^{13}C$, whole molecule isotope values were calculated as weighted average of the 4-isopropylphenylisocyanate and dimethylamine fragments. (Graphic reprinted with permission from Penning et al.[273])

period of time. It must be mentioned, though, that ethanol presence retards degradation of often co-occurring BTEX compounds in such contaminated aquifers.[276]

In another biostimulation study, an emulsion of soybean oil in water and lactate was injected into an aquifer with a mixed contamination plume of 1,2-dichloroethane and TCE. By supplying these additional electron donors, it was expected that microbial growth would be stimulated, consuming residual oxygen and eventually leading to anoxic conditions favourable for contaminant transformation by dehalogenation.[277] By carbon isotope analysis over a monitoring period of 183 days in just a single monitoring well it was possible to obtain a clearer picture of relevant in-situ processes. Here, dichloroelimination of 1,2-dichloroethane led to ethene, whereas reductive dehalogenation of TCE step-wise led to formation of vinyl chloride (VC) that was not further degraded. Transformation was observed within the first 90 days of monitoring. Afterwards, isotope signatures returning to their original values indicated depletion of the electron donors (corroborated by analysis of volatile fatty acids and sum parameters) and inflow of nondegraded or previously sorbed contaminants. Such information was a key for a successful up-scaling of the pilot test to a full-site treatment. Similar conclusions were drawn at other sites with active biostimulation of TCE degradation in groundwater by addition of lactate[278] and bioaugmentation of TCE degradation by injection of specific bacterial cultures at two sites.[279,280]

For groundwater contaminated with chlorinated hydrocarbons, active remediation efforts based on dehalogenation induced by zero-valent iron (ZVI) either in permeable reactive barriers (PRBs) or via slurry injection of ZVI particles in the nm to μm size range have been implemented in the past two decades at field scale. As early as 1999, a large carbon isotope fractionation during dehalogenation of various chlorinated ethenes was reported and it has been suggested that CSIA could be used in monitoring remediation efficacy.[281]

In order to predict formation of harmful intermediates such as VC one would like to be able to distinguish between biotic reduction which, in the case of TCE or PCE transformation, may lead to accumulation of VC, and abiotic reduction where this is typically not the case.[282] To this end, a more detailed investigation of isotope signature evolution in the degradation intermediates which allows a distinction between abiotic and biotic transformation is required, as depicted in Figure 6.47.

This distinction is even possible in the field as has been shown by Elsner et al.[283] Based on several studies using ZVI and other metals, VanStone et al. concluded that for reduction of chlorinated ethenes and ethanes the prevalent reaction mechanism is governing the observed isotope fractionation rather than the type of reductant.[284] In contrast, Zwank et al. have reported a bimodal distribution of isotope enrichment factors among different iron(II) minerals but concluded that in the case

(a) Isotope Patterns from Laboratory Experiments

Biotransformation (Slater et al. 2001, Bloom et al. 2000, Van Breukelen et al. 2005)

Abiotic Transformation by ZVI (Elsner et al. 2008)

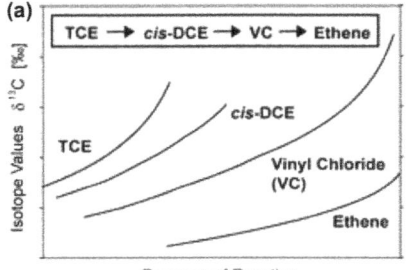
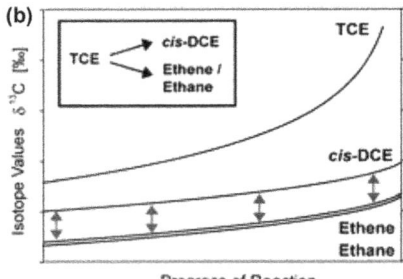

Figure 6.47 Schematic of $\delta^{13}C$ evolution during proceeding degradation of TCE and subsequently its transformation products: (a) in biodegradation leading to sequential reductive dechlorination, (b) in abiotic transformation by ZVI leading to several products.
(Graphic reprinted with permission from Elsner et al.[283])

of the lower observed enrichment a partial masking of the intrinsic fractionation occurred.[285]

The usefulness of carbon isotope data in in-situ probing of the success of PRBs as an active remediation approach has been shown by several studies to date (e.g., VanStone et al.).[286] The advantage of isotope data in addition to concentration data is not only evidence of compound transformation (which might be limited due to residual concentrations in a PRB below the method detection limit in CSIA) but also a better process understanding.

Another active groundwater remediation option utilizes single or multiple injection of permanganate in order to oxidize contaminants abiotically. CSIA use for demonstrating efficacy of this approach has been shown at a field site with TCE. In accompanying laboratory studies large enrichment factors of -25 to $-27‰$ have been found for this oxidation reaction and, indeed, after injection of permanganate, strong isotope shifts in TCE were observed. However, this shift was only transient and the original isotope composition was restored, which was explained by the presence of DNAPL and dissolution of TCE from this pool.[287]

In summary, one major advantage of using CSIA as a control measure in active site remediation aiming at a degradation of contaminants is the potential to recognize rather early success or failure of the remediation strategy. In the latter cases, results from isotope analysis might help to save on useless expenditures. In contrast, mere concentration analyses are often inconclusive.[161]

As mentioned above, in drinking water treatment, so far there have been hardly any studies reported although there seems to be great potential for a further characterization of elimination processes for micropollutants and formation of unwanted by-products. This has been demonstrated for the formation of haloforms during drinking water chlorination. The hypothesis of a common precursor molecule for all NOM-derived chloroform could be dismissed by contrary evidence based on *KIEs* during chlorination of various potential precursors. Furthermore, for real lake water processed to drinking water, phenols seemed the most likely source of chloroform based on *KIE* data.[288]

6.6.5 Further Reading

In the past couple of years, several reviews (refs 207, 225, 228, 249, 289–292), a book (253) and a guideline issued by the US Environmental Protection Agency (161) have been published devoted to the use of CSIA in environmental science that may aid the interested reader with further information.

6.7 EXTRATERRESTRIAL MATTER

One of the most fascinating areas of SIA is to find in a sample isotope ratios that are not known in the materials of our planet. Most prominent examples are primitive meteorites originating in the asteroid belt and finally arriving at the earth's surface. Of course, when analysing such meteorite samples it has to be proven that the material has not been terrestrially contaminated. Extraterrestrial isotopic composition of the bulk sample as well as constituents thereof (fractions or compounds) may be one of the strongest arguments, also pointing to the importance of CSIA in this respect. To illustrate this situation, in Figure 6.48 the ranges of stable isotope data reported in this chapter are compared with the ones known for terrestrial samples.[293–295]

Earlier reviews covering the topic of this section have been published.[296–302] Essential instrumental analytical developments in cosmochemistry were (i) static mass spectrometers capable of measuring a few ng C as CO_2,[303] and (ii) advances in isotopic techniques allowing CSIA.[304,305]

Besides interplanetary dust particles (IPDs) and ordinary chondrites, in the following text isotopic data are mainly reported for meteorites classified as carbonaceous chondrites (CCs). These meteorites are thought to represent heterogeneous aggregates of some of the most primitive early solar system materials containing chondrules,

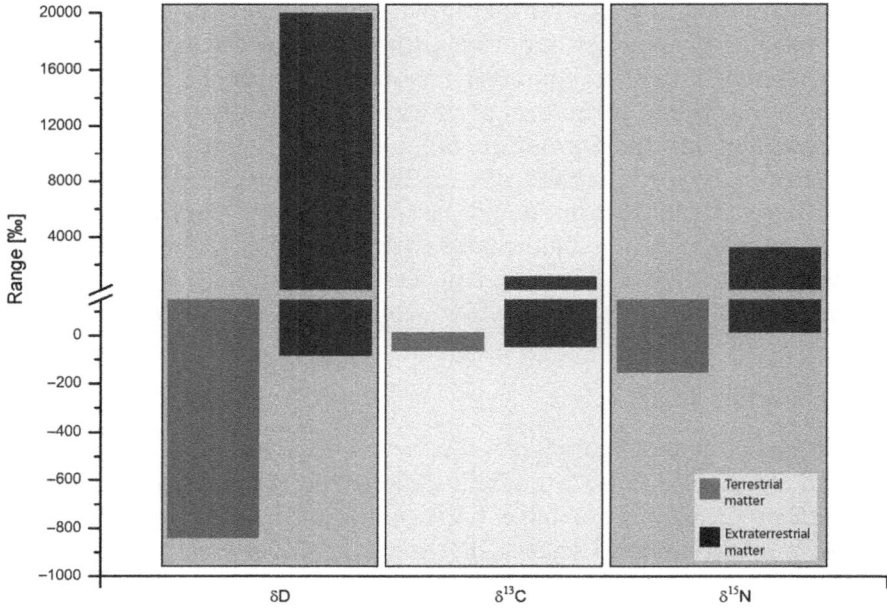

Figure 6.48 Ranges of stable isotope ratios in extraterrestrial and terrestrial materials.

calcium-aluminium-rich inclusions, presolar grains embedded in a matrix consisting of carbonates, oxides, clay minerals, sheet silicates and organics that formed through low-temperature aqueous alteration of meteorite parent bodies. The classification of CCs is based on bulk chemistry and mineralogy (CI, CM, CR, CO, CV, CK, CH, CB according to the type of specimen, e.g., CI = Ivuna or CR = Renazzo) or on carbon content (C1 to C4 with 3.6 to 0.2 wt-% C on average). While the chemistry of our sun is best matched by CI chondrites, CM and CR show only slight depletions in volatile elements relative to CI.[297] When focusing on mineralogical and textural aspects, however, a few CR chondrites appear to be the most primitive. According to present knowledge, even CI, CM and CR chondrites will have experienced aqueous and thermal processing on their parent asteroids.[297]

The CM2 meteorite Murchison fell in Western Australia in 1969, and provided approximately 100 kg sample material in a relatively good state of preservation. This meteorite has been demonstrated to contain the most extensive series of extraterrestrial organic compounds such as aliphatic and aromatic hydrocarbons as well as amino, carboxylic, sulfonic and phosphonic acids, alcohols, aldehydes, ketones, sugars, amides or N and S heterocycles together with abundant macromolecular, water-insoluble organic material:[296] thus, the Murchison organic inventory has become a valuable reference to which all other extraterrestrial matter can be compared.[300]

CR (Renazzo type) chondrites found in Antarctica and the Sahara desert have a petrology closely similar to that of the Renazzo meteorite, which fell in 1824. The unclassified CC Tagish Lake (composition between CM and CI) may be the best preserved meteorite, and was collected from frozen ice shortly after falling to Earth on January 18, 2000.[306] This chondrite, resembling one of the most primitive solar system materials, contains up to 5.8 wt-% C, approximately 2.6 wt-% of which is organic; >99% of organic matter occurs in acid-insoluble form (IOM = insoluble organic matter) of highly aromatic character.[307]

Stable Hydrogen Isotopes. It is generally agreed that deuterium existing in the universe was produced largely during the Big Bang.[308] Deuterated molecules were detected both in interstellar translucent clouds and in cold, dark clouds, as well as star forming regions. For example, up to triply deuterated ammonia has been reported in cold, dark clouds.[309] As a consequence, primordial matter from cool interplanetary regions should be highly enriched in ^{2}H. This is long known for primitive meteorites, for example, methanol extracts of selected CCs show $+300‰ < \delta^2H < +500‰$, and thus are significantly enriched in ^{2}H compared to terrestrial samples.[294]

Ion–molecule reactions which proceed essentially independent of temperature seem to be the only feasible route to such major fractionations. In dark clouds the greatest ^{2}H-enrichments (10^3 to 10^4 compared to SMOW) are found.[310] Yang and Epstein have used data from acid-resistant residues of the three most ^{2}H-enriched meteorites to create a mixing model which implies a δ^2H value of $+10 000‰$ for the end member.[311]

For the meteorite Murchison, published data for various organic extracts, IOM, and sulfonic acids significantly point to extraterrestrial origin ($+100‰ < \delta^2H < +1000‰$).[312–314] While this trend is even more pronounced in the case of amino acids ($+150‰ < \delta^2H < +3600‰$),[300,315] several of the monocarboxylic acids measured by Huang et al.[316] could originate from terrestrial contamination (Figure 6.49). There is no apparent easy explanation for terrestrial values of Murchison bulk samples, also because no general trend seem to be associated with the total CC group which shows a broad isotopic range of $-25‰ < \delta^2H < 0‰$.[317]

Although the extraterrestrial values found in Murchison materials as reported in Figure 6.49 are very impressive, numerous similar results were obtained with other samples: IOM of ordinary chondrites revealed

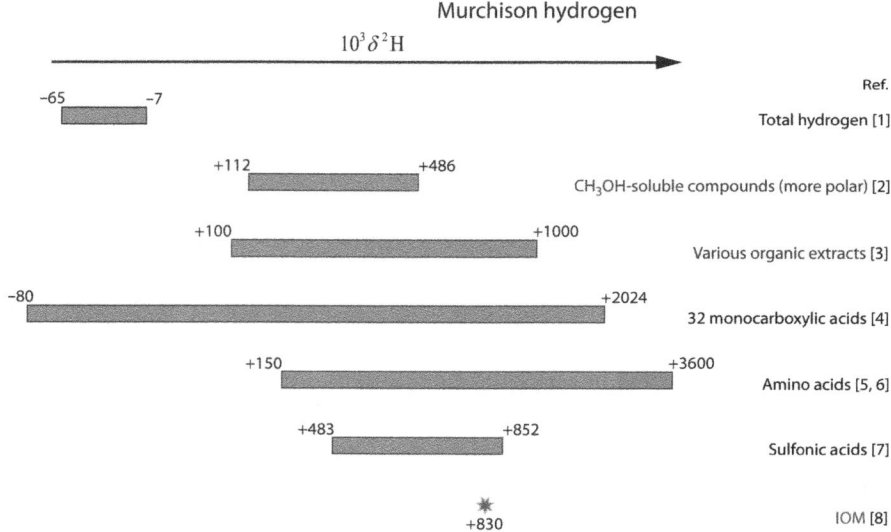

Figure 6.49 Summary of stable hydrogen isotope ratios measured in constituents of Murchison meteorite.
Data sources: [1][317], [2][312], [3][314], [4][316], [5][300], [6][315], [7][313], [8][294].

δ^2H up to 12 000‰; CSIA was performed on pyrolysates δ^2H of individual aliphatic and aromatic compounds and resulted in $+200‰ < \delta^2H < 4500‰$, and typical values found in IPDs are in the order of 1000‰.[299,318,319] IOM of Tagish Lake meteorite is characterized by δ^2H 596 ± 4‰, in ≤ μm-sized 'hotspots' of this IOM even up to δ^2H 15 000‰.[307] Measurements by (nano) secondary ion mass spectrometry resulted in δ^2H up to 20 000‰ for IOM hotspots in CCs. In particular, for Murchison IOM up to δ^2H 1740‰ have been detected in hotspots, and 712‰ in the bulk.[320] Pizzarello and Holmes analysed amino acids, amines and aldehydes in two CRs from Antarctica (species more abundant than in CMs by factors 4–8): $700‰ < \delta^2H < 8000‰$.[321] Protein amino acids (such as threonine, tryosine and phenylalanine) were identified which had never been found in meteorites before. Ultracarbonaceous micrometeorites from central Antarctic snow contain organics with extreme ^{2}H excess (10 to 30 times terrestrial values), also indicating an origin from the cold regions of the protoplanetary disc.[322]

Recent results demonstrate that there are not only isotopic but also chemical effects differentiating between extraterrestrial and terrestrial material: by EPR it could be shown that Orgueil IOM is dominated by ^{2}H rich biradicaloids (S = 0 or 1) and biradicals (S = 1), while terrestrial kerogen almost exclusively contains $S = \frac{1}{2}$ radicals.[323,324] It was also

Applications of Compound-specific Stable Isotope Analysis

observed that clustering of radicals is a common feature of synthetic and extraterrestrial carbonaceous material, while radicals are homogeneously distributed in biomass.[325]

Stable Nitrogen Isotopes. With respect to extraterrestrial fingerprints, nitrogen isotope composition shows some parallels to that of hydrogen: the heavy isotope is enriched, for example in organics and amino acids of Murchison (Figure 6.50) or IOM globules thereof ($+250‰ < \delta^{15}N < +500‰$), organic nanoglobules in IPDs were reported as $\delta^{15}N$ 1120‰, and in IOM of Tagish Lake CC was $\delta^{15}N$ $73 \pm 1.5‰$ in the bulk, and up to $\delta^{15}N$ 1000‰ in $\leq \mu$m-sized 'hotspots'; by (nano) SIMS up to $\delta^{15}N$ 3200‰ have been found in hotspots of IOM in CCs.[307,320,326]

Individual amino acid enantiomers from Murchison are enriched in ^{15}N relative to their terrestrial counterparts, confirming an extraterrestrial source for a slight but significant L-excess of alanine, glutamic and aspartic acid.[327,328] Pizarello and Holmes analysed amino acids, amines and aldehydes in two CRs from Antarctica and found $+70‰ < \delta^{15}N < +130‰$.[321]

The overall isotopic anomalies for extraterrestrial nitrogen (as well as carbon, see below), however, are not as clear as described above for hydrogen. For example, the variability of $\delta^{15}N$ in primitive meteorites, coupled with the wide range of stable nitrogen composition of lunar samples has discouraged most authors from speculating on the isotopic composition of nitrogen in the solar system, or that of the bulk earth.

Stable Carbon Isotopes. ^{12}C (like ^{14}N) is formed in stars as a primary isotope during He burning, while ^{13}C (like ^{15}N) is a secondary isotope

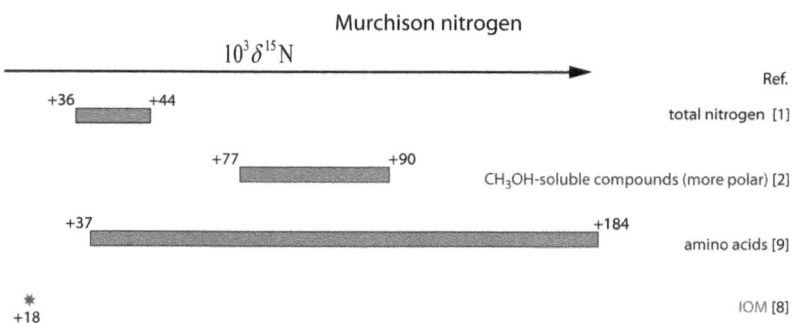

Figure 6.50 Summary of stable nitrogen isotope ratios measured in constituents of Murchison meteorite.
Data sources: [1][317], [2][312], [8][294], [9][327].

built up from ^{12}C (resp. ^{14}N) in the course of the CNO cycle; both isotopes enter interstellar space via stellar envelopes. Based on these processes the ratios of ^{12}C/^{13}C were calculated as 2.8 for novae, 20 to 30 for envelopes of Red Giants as well as 89 at the time of the formation of our solar system and 66 at present.[329–331]

Presolar grains are tagged by exotic noble gases, and can be separated from meteorites in essentially pure form by chemical processing; they are considered as resembling grains of stardust that originated from stellar outflows and supernova ejecta prior to the formation of the solar system.[332,333] For example, presolar SiC grains are believed to originate from Red Giants, in particular asymptotic giant branch stars. Isotopic measurements are performed by ion microprobe of individual grains from chemically resistant residues. Most single-grain studies have been carried out on Murchison. Approximately 90% of all grains have ^{12}C/^{13}C ratios between 15 and 100.[332] Up to 5 ppm carbon enriched in ^{13}C up to 1100‰ were found to be associated with neon-22 and anomalous krypton and xenon showing the signature of the s-process

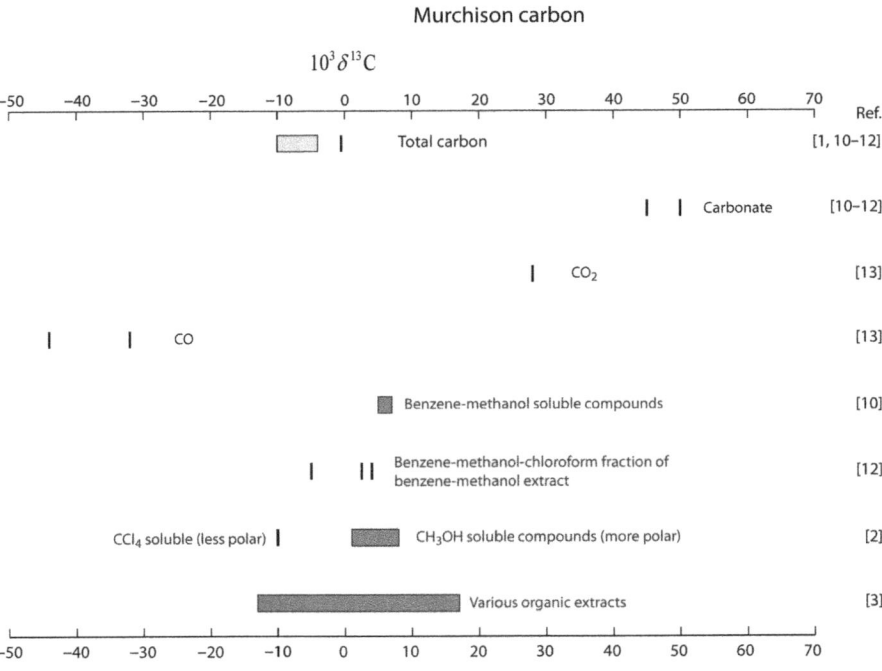

Figure 6.51 Summary of stable carbon isotope ratios measured in bulk as well as inorganic and organic fractions of Murchison meteorite.
Data sources: [1][317], [2][312], [3][314], [10][335], [11][334], [12][336], [13][337].

(neutron capture on a slow time scale) typical for interstellar grains ejected from late-type stars.[334]

The stable carbon isotopic compositions of inorganic components and organic extracts of the Murchison meteorite are compiled in Figure 6.51. Although the reported bulk isotopic composition is identical with the one of our earth and other solar system bodies, at least part of the extractable organics is significantly enriched in ^{13}C compared to known terrestrial organic material. However carbonates exhibit the highest enrichment, leading to isotopic fractionations between reduced and oxidized carbon in the meteorite which are similar to those obtained experimentally in thermodynamic equilibrium or by the Fischer-Tropsch process.[336,338–340]

The most serious problem when analysing extraterrestrial components is terrestrial contamination (time between fall and find; transport, handling, and storage of samples). Applying stepped combustion, Swart et al.[341] could demonstrate that even for the most valuable samples, such as Apollo 11 lunar soil, contamination by terrestrial carbon could not be avoided. Krishnamurthy et al. compared two different samples of Murchison, the so-called Chicago and ASU stones, and obtained $\delta^{13}C$-values of -11.5‰ and -5‰, respectively.[314] Because the content of n-alkanes is higher in the Chicago stone, they found the latter to be heavier contaminated than the ASU stone. When comparing Murchison samples stored in a museum for different times, Hirner found ^{12}C enrichments of roughly 1‰ per year in the benzene-methanol fraction.[336] The observation that within the group of monocarboxylic acids the branched acids are substantially enriched in both ^{2}H and ^{13}C relative to the straight chain acids, was interpreted by Huang et al.[316] as terrestrial contamination mainly by straight chain acids.

From Figure 6.52 it can be clearly seen that larger n-alkanes and aromatic hydrocarbons, in particular, are prone to terrestrial contamination, while there is apparently no detectable contamination for C_1–C_3 hydrocarbons, nucleobases as well as C_2–C_5 monocarboxylic, sulfonic and amino acids. More than 80 amino acids have been identified in Murchison; most of these are not protein-forming amino acids and are rare in the terrestrial biosphere.[344] Individual amino acid enantiomers from Murchison are enriched in ^{13}C relative to their terrestrial counterparts, confirming an extraterrestrial source for a slight but significant L-excess of alanine, glutamic and aspartic acid.[327,328] The largest L-enantiomeric excess ever reported to date was 18.5% found in isovaline of Murchison.[345] Pizarello and Holmes analysed amino acids, amines and aldehydes also in two CRs from Antarctica (species more abundant than in CMs by factors 4–8) showing -6‰ $<\delta^{13}C<+27$‰;

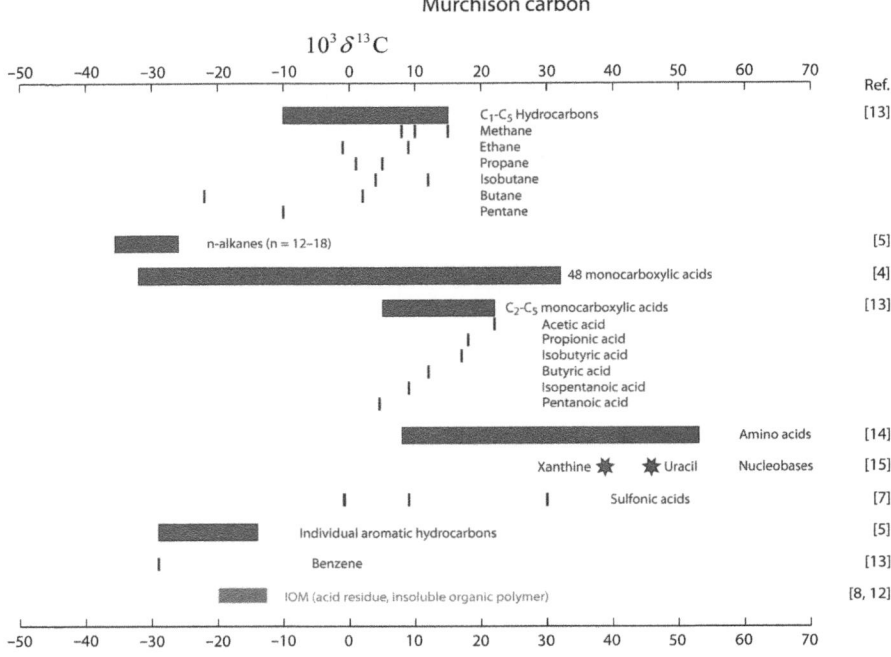

Figure 6.52 Summary of stable carbon isotope ratios measured in individual and classes of organic compounds of Murchison meteorite.
Data sources: [4][316], [5][300], [7][313], [8][294], [12][336], [13][337], [14][342], [15][343].

as has already been mentioned for hydrogen above, several of the protein amino acids (such as threonine, tyrosine and phenylalanine) identified had never been found in meteorites before.[321] Comparable carbon isotope ratios were obtained for glycine in Murchison ($+22‰ < \delta^{13}C < +41‰$), in the CI meteorite Orgueil ($\delta^{13}C \; +22‰$) as well as in cometary dust collected by the stardust mission ($\delta^{13}C \; +29‰$).[344,346,347] Furthermore, the mineralogy as well as the chemical and oxygen isotope compositions of cometary dust are very similar to those of CCs.[348]

The situation described is completely different from the carbon isotope effects associated with terrestrial amino acid biosynthesis.[349] Also in the context of discussing the origin of extraterrestrial chiral asymmetry, as manifested presently in eight amino acids and one hydroxyl acid, isotopic enrichment of stable isotopes plays a predominant role;[350] furthermore, processes such as attainment of enantiomeric excesses via phase changes (crystallization or sublimation) on meteorite parent bodies have been postulated in this respect. When discussing hypotheses on the origin of life it is also very important to be aware of the discovery of isotopically heavy nucleobases such as uracil and xanthine, to be added to the list of extraterrestrial prebiotic precursor molecules[343] (Figure 6.52). Such recent observations are corroborating older hypotheses stating that CCs may have provided the first prebiotic building blocks of life on our earth (e.g., see Chyba and Sagan).[351]

Stable isotope distributions are a valuable tool in evaluating possible pathways for experimentally investigating possible mechanisms for extraterrestrial formation of organic matter: although Lancet and Anders[340] demonstrated that the observed C isotopic fractionations in Murchison could be paralleled by experiments of Fischer-Tropsch type, a Miller-Urey type synthesis produced a fractionation of just -0.4 ± 0.2‰.[339] To shed further light on this problem, it is necessary to extend such investigations from the bulk to the molecular level (i.e., CSIA):

- $\delta^{13}C$ decreases with increasing C number of hydrocarbons, monocarboxylic acids and α-amino acids in a roughly parallel manner, pointing towards kinetically controlled synthesis of higher homologues from lower ones[337,342] (see also Figure 6.52).
- However, higher molecular weight alkanes seem to follow the opposite trend implying they were produced by cracking.[299]
- The marked structural relationship between α-amino acids and α-hydroxy acids suggest their formation during Strecker-cyanohydrin reactions on the meteorite parent body.[300]

Other Stable Isotopes. Beside the light elements already discussed, in stable isotope geochemistry also oxygen (^{16}O, ^{17}O, ^{18}O) and sulfur (^{32}S, ^{33}S, ^{34}S, ^{36}S) are of interest. In the context of possible extraterrestrial fingerprints, two relevant observations are to be mentioned:

- Plotting $\delta^{17}O$ vs. $\delta^{18}O$ in Al-rich inclusions of meteorites (esp. C3 meteorite Allende) is interpreted as an addition of almost pure ^{16}O indicative of explosive carbon burning.[352–354]
- Sulfur isotope composition in meteorites is very uniform, and usually there exist linear $\delta^{34}S$ relationships between meteorites and terrestrial samples. However, specific anomalies have been identified, such as spallogenically produced ^{33}S and ^{36}S in the metallic phase of iron meteorites or 1‰ excess in ^{33}S in Allende.[355,356]

No CSIA data are available for these elements. Only Briggs[357] has quoted $\delta^{34}S$-values for 'organic' sulfur. However, due to the extraction method applied, also elemental sulfur is separated in the same fraction and thus CSIA is not possible when using this procedure.

6.8 GENERAL CONCLUSIONS

We will close this chapter with some general conclusions that can be drawn from the many applications of CSIA described here and elsewhere. These should provide readers with some guidance on key points to be considered when planning to utilize CSIA for their own applications. The order of appearance is no indicator of relevance.

- *One isotope analysis is no isotope analysis.* All examples discussed in Chapter 6 show that if one wants to utilise CSIA for source tracking it is of the utmost importance to have a reliable database at hand to compare the samples/compounds under question against. Nowadays, in most applications a limited set of samples of known origin is analysed along with the target samples in order to create an internal database. Otherwise, only some general conclusions may be drawn (e.g., merely distinguishing C_3 and C_4 plant-derived compounds). A prerequisite for utilising external isotope databases is comparability of data. Therefore, QA measures ensuring trueness of results are mandatory. In order to make use of small isotopic differences, measures to maximise precision of measurements are also required. In the case of transformation monitoring these requirements can be less strict since in most cases the change of the isotope composition over time or space provides the information aimed for. Of course, in these cases a number of isotope analyses is inherently necessary to observe changes in isotope composition, thus the first-sentence paradigm applies here as well.
- *CSIA is a target analysis.* There have been some reports on advantages of a combined identification, quantification and isotope analysis typically by splitting the effluent of the GC column to a regular mass spectrometric detector (quadrupole, ion trap, TOF) and to the IRMS. Split ratios are typically in the order of 1 : 10 to 1 : 20, with the larger flow going to the IRMS due to its poorer sensitivity. This approach allows for the isotope analysis of (previously) unknown compounds in complex samples. The major advantage though is the control of potential co-elution of target analytes with unknowns that may falsify the obtained isotope composition. However, in practice samples that are subjected to isotope analysis are typically well characterized prior to CSIA. This is even a prerequisite for samples of unknown composition. Thus in

many cases the regular MS detector would not be frequently required, at least not in carbon isotope analysis. In our own stable isotope lab we, therefore, run two independent instruments, i.e., a GC-MS and a GC-IRMS with the same peripherals that allow for easy control measurements and method transfer but can be used simultaneously for different samples. It needs to be emphasized, though, that for the most rigorous demands for correct compound identification along with isotope composition, in particular in forensic investigations, a combined GC-MS/IRMS instrument might be a prerequisite, preferably with an MS detector enabling confirmatory MS^n experiments. In LC-IRMS, no split approaches have been described so far but several regular LC detectors can be installed in line (UV, RI, FL) before the chemical oxidation interface. None of these detectors is universal for all organic compounds though and they deliver only very limited structural information on the analytes detected. Since co-elution in LC is more likely than in GC due to the much lower peak capacity, further developments in that area are urgently needed.
- Often *multiple isotope approaches may be required* to help resolve the problem at hand. This is the case in source discrimination as has been demonstrated in various sections of this chapter. Furthermore, it has been shown that limitations of isotope investigations in characterizing transformation reactions can be circumvented by using dual isotope or 2D isotope approaches, such as effects by dilution or commitment to catalysis, as has been discussed above.
- *At best, multiple components can be analysed in single runs* (one major advantage of CSIA over EA-IRMS). In case of multiple component analysis it is often advantageous to use an internal reference compound (also called i-IST or ERC in doping analysis) to eliminate external effects shifting all isotope values. Thus, rather than plotting measured isotope compositions, frequently $\Delta\delta$-values are presented, which may be useful for creating isotopic fingerprints and/or authenticity ranges for certain products or groups of samples. So far, this approach has been most frequently used in flavour analysis, fatty acid analysis in archaeology, and steroid analysis in doping control and often allows a more sensitive detection of various sources for individual compounds than relying on an expected range of δ-values alone. Furthermore, it provides the opportunity of visualizing isotope signatures or isotopic differences in two or more compounds in isotope crossplots or radar plots as exemplified above.

- In many applications, combining information from CSIA and other parameters (concentration data, bulk isotope data, etc.) or evaluating multiple component isotope data is useful or even mandatory in order to provide a clear sample classification. As discussed above for several examples as a tool to process such multiple data *statistical treatment and chemometric approaches are required*. The latter involves, in particular, PCA. A protocol for carrying out such evaluations including PCA followed by multiple analysis of variance (MANOVA) and hierarchical cluster analysis has been provided recently for environmental source discrimination[203] but can be adapted for other applications.
- CSIA can provide one important line of evidence in many application areas. However, a word of caution is required to avoid over-enthusiasm: with a few exceptions *CSIA alone is insufficient* to resolve unequivocally the question at hand or problem to be addressed. Thus, the full power of CSIA is only exploited when accompanied by other suitable methods, including structural and concentration analysis of target analytes and diagnostic metabolites (also called signature metabolites or metabolic biomarkers), historical evidence, geochemical parameters, molecular biological analysis of genes involved in degradation, statistical/chemometric data treatment and appropriate modelling at environmental sites, other evidence of illegal conduit in doping and forensic analysis, and in food adulteration etc. Several case studies highlighting this point have been reported including environmental source discrimination[199] and biodegradation.[359]
- Only *differences* in isotope profiles of samples can be used to unequivocally discriminate between different sources or prove degradation. If *similar* isotope profiles are found among samples, this does not exclude different sources of similar isotope signatures or transformation reactions without (discernable) isotope fractionation. Since with the exception of methane-derived organic compounds, the isotopic range observed is quite limited for carbon, this is of particular relevance for carbon isotope analysis and gives further support for the recommended 2D and multicomponent approach. For carbon isotope analysis, differences of measured $\delta^{13}C$-values should exceed at least 1‰, better 2‰, to warrant further interpretation.[161,290]
- The analysis of minor components (i.e., compounds present in a sample in trace concentrations (nmol L^{-1} to µmol L^{-1})) is most common in environmental applications of CSIA. However, the sample preparation techniques developed for such applications

could be transferred to other areas, thus widening the concentration range of analytes amenable to CSIA. So far, these possibilities have rarely been utilized.
- Other instrumental analytical opportunities described in literature for years but not widely used include enantiomer separation in GC-IRMS, and multidimensional GC or even comprehensive GC × GC coupled to IRMS. The latter, though, may require further developments with regard to speed of data acquisition and processing and is ultimately limited by the ion source design (see Chapters 3 and 7).
- As, hopefully, is evident from this chapter there are numerous further opportunities for CSIA applications that surely will be tackled in the future. At best, the interested reader has already obtained an idea of possibilities in his or her own research. Recent and future instrumental advances as, for example, discussed in Chapter 7 will provide the basis for an ever-increasing number of potential applications. Many well-equipped research laboratories and a few commercial ones are available; thus researchers without access to such facilities themselves should be able to find collaborators with ease.

6.9 WEB FINDINGS

6.9.1 General Links

- World Anti-Doping Agency: http://www.wada-ama.org/en/
- Forensic Isotope Mass Spectrometry Network: http://www.forensic-isotopes.org
- Global Network of Isotopes in Precipitation (GNIP) run by IAEA and World Metereological Organization: http://isohis.iaea.org/
- Working Group Stable Isotopes Analytics within GDCh, Society of Food Chemistry Division: http://www.gdch.de/strukturen/fg/lm/ag/stabilisotopen_e.htm
- German Assocation for Stable Isotope Research (GASIR) (only in German): http://www.gasir.de/ (Publishing society of 'Isotopes in Environmental and Health Studies')
- Benelux Association of Stable Isotope Scientists (BASIS): http://www.basis-online.eu
- British Mass Spectrometry Society, Special Interest Group Stable Isotopes: http://www.bmss.org.uk/SIG_stable-iso.shtml
- US Environmental Protection Agency: http://www.epa.gov; specifically: D Hunkeler, R. U. Meckenstock, B. Sherwood Lollar, T. C.

Schmidt and J. T. Wilson, *A Guide for Assessing Biodegradation and Source Identification of Organic Ground Water Contaminants using Compound Specific Isotope Analysis (CSIA)*, US EPA Office of Research and Development, Report EPA 600/R-08/148, 2008; available free of charge at http://www.epa.gov/nrmrl/pubs/600r08148/600r08148.pdf
- US Geological Survey: http://www.usgs.gov; specifically: Reston Stable Isotope Laboratory (Tyler B. Coplen and co-workers): http://www.isotopes.usgs.gov
- Isotopically characterized working standards by Arndt Schimmelmann: http://mypage.iu.edu/~aschimme

6.9.2 European Isotope Application Projects

- ISOSTER: Determination of the origins of hormones in cattle (no existing homepage, 2002–2005)
- ISONET: 400 years of Annual Reconstruction of European Climate Variability using a High Resolution Isotopic Network (http://www.isonet-online.de, 2002–2005)
- TRACE: Tracing the origin of food (http://trace.eu.org, 2005–2009)
- ISOSOIL: Contaminant-specific isotope analyses as sharp environmental-forensics tools for site characterization, monitoring and source apportionment of pollutants in soil (http://isosoil.eu, 2009–2012)
- Marie-Curie Initial Training Network CSI: Environment: Isotope forensics meets biogeochemistry – linking sources and sinks of organic contaminants by compound-specific isotope investigation – CSI: ENVIRONMENT (http://www.csi-environment.ufz.de/, 2010–2014)

REFERENCES

1. S. E. Woodbury, R. P. Evershed and J. B. Rossell, *J. Am. Oil Chem. Soc.*, 1998, **75**, 371–379.
2. H. Förstel, *Anal. Bioanal. Chem.*, 2007, **388**, 541–544.
3. G. Calderone and C. Guillou, *Food Chem.*, 2008, **106**, 1399–1405.
4. I. D. Clark and P. Fritz, *Environmental Isotopes in Hydrogeology*, Lewis Publishers, Boca Raton, FL, 1997.
5. R. E. Criss, *Principles of Stable Isotope Distribution*, Oxford University Press, New York, 1999.
6. B. Fry, *Stable Isotope Ecology*, Springer, New York, [London], 2006.

7. J. Hoefs, *Stable Isotope Geochemistry*, 6th edn., Springer, Berlin, 2009.
8. C. Kendall and J. J. McDonnell, *Isotope Tracers in Catchment Hydrology*, Elsevier, Amsterdam, Oxford, 1998.
9. Z. Sharp, *Principles of Stable Isotope Geochemistry*, 1st edn., Pearson/Prentice Hall, Upper Saddle River, NJ, London, 2007.
10. C. Cordella, I. Moussa, A. C. Martel, N. Sbirrazzouli and L. Lizzani-Cuvelier, *J. Agric. Food. Chem.*, 2002, **50**, 1751–1764.
11. A. Gonzalvez, S. Armenta and M. de la Guardia, *Trends Anal. Chem.*, 2009, **28**, 1295–1311.
12. S. Kelly, K. Heaton and J. Hoogewerff, *Trends Food Sci. Technol.*, 2005, **16**, 555–567.
13. A. Rossmann, *Food Rev. Int.*, 2001, **17**, 347–381.
14. S. Nogués, I. Aranjuelo, A. Pardo and J. Azcón-Bieto, *Rapid Commun. Mass Spectrom.*, 2008, **22**, 1017–1022.
15. T. E. Cerling, J. M. Harris, B. J. MacFadden, M. G. Leakey, J. Quade, V. Eisenmann and J. R. Ehleringer, *Nature*, 1997, **389**, 153–158.
16. R. Braunsdorf, U. Hener, S. Stein and A. Mosandl, *Z. Lebensm. Unters. F. A.*, 1993, **197**, 137–141.
17. C. Frank, A. Dietrich, U. Kremer and A. Mosandl, *J. Agric. Food Chem.*, 1995, **43**, 1634–1637.
18. L. Schipilliti, P. Q. Tranchida, D. Sciarrone, M. Russo, P. Dugo, G. Dugo and L. Mondello, *J. Sep. Sci.*, 2010, **33**, 617–625.
19. A. Kaunzinger, D. Juchelka and A. Mosandl, *J. Agric. Food Chem.*, 1997, **45**, 1752–1757.
20. S. Bilke and A. Mosandl, *Eur. Food Res. Technol.*, 2002, **214**, 532–535.
21. S. Sewenig, U. Hener and A. Mosandl, *Eur. Food Res. Technol.*, 2003, **217**, 444–448.
22. D. Juchelka and A. Mosandl, *Pharmazie*, 1996, **51**, 417–422.
23. B. Weckerle, R. Bastl-Borrmann, E. Richling, K. Hör, C. Ruff and P. Schreier, *Flavour Fragr. J.*, 2001, **16**, 360–363.
24. J. Jung, S. Sewenig, U. Hener and A. Mosandl, *Eur. Food Res. Technol.*, 2005, **220**, 232–237.
25. S. Faulhaber, U. Hener and A. Mosandl, *J. Agric. Food Chem.*, 1997, **45**, 4719–4725.
26. S. Faulhaber, U. Hener and A. Mosandl, *J. Agric. Food Chem.*, 1997, **45**, 2579–2583.
27. K. Fink, E. Richling, F. Heckel and P. Schreier, *J. Agric. Food Chem.*, 2004, **52**, 3065–3068.
28. H. Tamura, M. Appel, E. Richling and P. Schreier, *J. Agric. Food Chem.*, 2005, **53**, 5397–5401.

29. C. Preston, E. Richling, S. Elss, M. Appel, F. Heckel, A. Hartlieb and P. Schreier, *J. Agric. Food Chem.*, 2003, **51**, 8027–8031.
30. S. Sewenig, D. Bullinger, U. Hener and A. Mosandl, *J. Agric. Food Chem.*, 2005, **53**, 838–844.
31. M. Greule, L. D. Tumino, T. Kronewald, U. Hener, J. Schleucher, A. Mosandl and F. Keppler, *Eur. Food Res. Technol.*, 2010, **231**, 933–941.
32. A. Mosandl, U. Hener, H. G. Schmarr and M. Rautenschlein, *HRC J. High Resolut. Chromatogr.*, 1990, **13**, 528–531.
33. V. Karl, A. Dietrich and A. Mosandl, *Phytochem. Anal.*, 1994, **5**, 32–37.
34. D. Juchelka, T. Beck, U. Hener, F. Dettmar and A. Mosandl, *HRC J. High Resolut. Chromatogr.*, 1998, **21** , 145–151.
35. F. Keppler, D. B. Harper, R. M. Kalin, W. Meier-Augenstein, N. Farmer, S. Davis, H. L. Schmidt, D. M. Brown and J. T. G. Hamilton, *New Phytologist*, 2007, **176**, 600–609.
36. H. L. Schmidt, *Naturwissenschaften*, 2003, **90**, 537–552.
37. M. A. Jochmann, D. Steinmann, M. Stephan and T. C. Schmidt, *J. Agric. Food Chem.*, 2009, **57**, 10489–10496.
38. G. Calderone, N. Naulet, C. Guillou, F. Reniero and A. I. B. Cortes, *Rapid Commun. Mass Spectrom.*, 2005, **19**, 701–705.
39. J. Jung, T. Jaufmann, U. Hener, A. Münch, M. Kreck, H. Dietrich and A. Mosandl, *Eur. Food Res. Technol.*, 2006, **223**, 811–820.
40. U. Hener, A. Mosandl, A. Hilkert, J. Bahrs-Windsberger, M. Großmann and W. R. Sponholz, *Vitic. Enol. Sci.*, 1998, **53**, 49–53.
41. U. Hener, A. Mosandl, U. Hagenauer-Hener and H. Dietrich, *Vitic. Enol. Sci.*, 1995, **50**, 113–117.
42. A. Cabañero, J. L. Recio and M. Ruperez, *Rapid Commun. Mass Spectrom.*, 2008, **22**, 3111–3118.
43. A. I. Cabañero, J. L. Recio and M. Ruperez, *J. Agric. Food Chem.*, 2010, **58**, 722–728.
44. M. E. Spitzke and C. Fauhl-Hassek, *Eur. Food Res. Technol.*, 2010, **231**, 247–257.
45. B. O. Aguilar-Cisneros, M. G. Lopez, E. Richling, F. Heckel and P. Schreier, *J. Agric. Food Chem.*, 2002, **50**, 7520–7523.
46. I. G. Parker, S. D. Kelly, M. Sharman, M. J. Dennis and D. Howie, *Food Chem.*, 1998, **63**, 423–428.
47. K. Schumacher, U. Hener, C. Patz, H. Dietrich and A. Mosandl, *Eur. Food Res. Technol.*, 1999, **209**, 12–15.
48. M. L. Fogel and N. Tuross, *J. Archaeol. Sci.*, 2003, **30**, 535–545.

49. A. I. Cabañero, J. L. Recio and M. Ruperez, *J. Agric. Food Chem.*, 2006, **54**, 9719–9727.
50. L. Elflein and K. P. Raezke, *Apidologie*, 2008, **39**, 574–587.
51. J. Bricout and J. Koziet, *J. Agric. Food Chem.*, 1987, **35**, 758–760.
52. S. D. Kelly, C. Rhodes, J. H. Lofthouse, D. Anderson, C. E. Burwood, M. J. Dennis and P. Brereton, *J. Agric. Food Chem.*, 2003, **51**, 1801–1806.
53. M. Gensler, A. Rossmann and H. L. Schmidt, *J. Agric. Food Chem.*, 1995, **43**, 2662–2666.
54. W. Meier-Augenstein, *Anal. Chim. Acta*, 2002, **465**, 63–79.
55. F. Angerosa, L. Camera, S. Cumitini, G. Gleixner and F. Reniero, *J. Agric. Food Chem.*, 1997, **45**, 3044–3048.
56. G. Bianchi, F. Angerosa, L. Camera, F. Reniero and C. Anglani, *J. Agric. Food Chem.*, 1993, **41**, 1936–1940.
57. F. Angerosa, O. Breas, S. Contento, C. Guillou, F. Reniero and E. Sada, *J. Agric. Food Chem.*, 1999, **47**, 1013–1017.
58. J. Jung, B. Puff, T. Eberts, U. Hener and A. Mosandl, *Eur. Food Res. Technol.*, 2007, **225**, 191–197.
59. J. E. Spangenberg, S. A. Macko and J. Hunziker, *J. Agric. Food Chem.*, 1998, **46**, 4179–4184.
60. J. E. Spangenberg and N. Ogrinc, *J. Agric. Food Chem.*, 2001, **49**, 1534–1540.
61. S. E. Woodbury, R. P. Evershed, J. B. Rossell, R. E. Griffith and P. Farnell, *Anal. Chem.*, 1995, **67**, 2685–2690.
62. E. K. Richter, J. E. Spangenberg, M. Kreuzer and F. Leiber, *J. Agric. Food Chem.*, 2010, **58**, 8048–8055.
63. R. Hrastar, M. G. Petrisic, N. Ogrinc and I. J. Kosir, *J. Agric. Food Chem.*, 2009, **57**, 579–585.
64. L. X. Guo, X. M. Xu, J. P. Yuan, C. F. Wu and J. H. Wang, *J. Am. Oil Chem. Soc.*, 2010, **87**, 839–848.
65. B. Weinert, M. Ulrich and A. Mosandl, *Eur. Food Res. Technol.*, 1999, **208**, 277–281.
66. J. E. Spangenberg and F. Dionisi, *J. Agric. Food Chem.*, 2001, **49**, 4271–4277.
67. L. J. Zhang, D. M. Kujawinski, E. Federherr, T. C. Schmidt and M. A. Jochmann, *Anal. Chem.*, 2012, **84**, 2805–2810.
68. J. F. Carter, E. L. Titterton, H. Grant and R. Sleeman, *Chem. Commun.*, 2002, **8**, 2590–2591.
69. S. Benson, C. Lennard, P. Maynard and C. Roux, *Forensic Sci. Int.*, 2006, **157**, 1–22.
70. W. Meier-Augenstein, *Stable Isotope Forensics*, Wiley, 2010.

71. N. Gentile, R. T. W. Siegwolf and O. Delémont, *Rapid Commun. Mass Spectrom.*, 2009, **23**, 2559–2567.
72. W. Meier-Augenstein, W. Brand, G. F. Hoffmann and D. Rating, *Biol. Mass Spectrom.*, 1994, **23**, 376–378.
73. M. Desage, R. Guilluy, J. L. Brazier, H. Chaudron, J. Girard, H. Cherpin and J. Jumeau, *Anal. Chim. Acta*, 1991, **247**, 249–254.
74. J. R. Ehleringer, J. F. Casale, M. J. Lott and V. L. Ford, *Nature*, 2000, **408**, 311–312.
75. F. Besacier, R. Guilluy, J. L. Brazier, H. Chaudron-Thozet, J. Girard and A. Lamotte, *J. Forensic Sci.*, 1997, **42**, 429–433.
76. J. F. Casale, J. R. Ehleringer, D. R. Morello and M. J. Lott, *J. Forensic Sci.*, 2005, **50**, 1315–1321.
77. J. Casale, E. Casale, M. Collins, D. Morello, S. Cathapermal and S. Panicker, *J. Forensic Sci.*, 2006, **51**, 603–606.
78. S. Dautraix, R. Guilluy, H. Chaudron-Thozet, J. L. Brazier and A. Lamotte, *J. Chromatogr. A*, 1996, **756**, 203–210.
79. F. A. Idoine, J. F. Carter and R. Sleeman, *Rapid Commun. Mass Spectrom.*, 2005, **19**, 3207–3215.
80. H. A. S. Buchanan, N. N. Daéid, W. J. Kerr, J. F. Carter and J. C. Hill, *Anal. Chem.*, 2010, **82**, 5484–5489.
81. J. F. Carter, E. L. Titterton, M. Murray and R. Sleeman, *Analyst*, 2002, **127**, 830–833.
82. F. Mas, B. Beemsterboer, A. C. Veltkamp and A. M. A. Verweij, *Forensic Sci. Int.*, 1995, **71**, 225–231.
83. F. Palhol, C. Lamoureux and N. Naulet, *Anal. Bioanal. Chem.*, 2003, **376**, 486–490.
84. S. Schneiders, T. Holdermann and R. Dahlenburg, *Sci. Justice*, 2009, **49**, 94–101.
85. C. Saudan, M. Augsburger, P. Mangin and M. Saugy, *Rapid Commun. Mass Spectrom.*, 2007, **21**, 3956–3962.
86. F. Marclay, D. Pazos, O. Delémont, P. Esseiva and C. Saudan, *Forensic Sci. Int.*, 2010, **198**, 46–52.
87. B. J. Smallwood, R. P. Philp and J. D. Allen, *Org. Geochem.*, 2002, **33**, 149–159.
88. I. D. Bull, R. Berstan, A. Vass and R. P. Evershed, *Sci. Justice*, 2009, **49**, 142–149.
89. J. A. Lee-Thorp, *Archaeometry*, 2008, **50**, 925–950.
90. H. Friedli, H. Lötscher, H. Oeschger, U. Siegenthaler and B. Stauffer, *Nature*, 1986, **324**, 237–238.
91. E. A. Reber and R. P. Evershed, *Archaeometry*, 2004, **46**, 19–33.

92. G. Docherty, V. Jones and R. P. Evershed, *Rapid Commun. Mass Spectrom.*, 2001, **15**, 730–738.
93. M. A. Sephton, W. Meredith, C. G. Sun and C. E. Snape, *Rapid Commun. Mass Spectrom.*, 2005, **19**, 3339–3342.
94. S. N. Dudd and R. P. Evershed, *Science*, 1998, **282**, 1478–1481.
95. H. R. Mottram, S. N. Dudd, G. J. Lawrence, A. W. Stott and R. P. Evershed, *J. Chromatogr. A*, 1999, **833**, 209–221.
96. R. P. Evershed, S. N. Dudd, M. S. Copley, R. Berstan, A. W. Stott, H. Mottram, S. A. Buckley and Z. Crossman, *Acc. Chem. Res.*, 2002, **35**, 660–668.
97. J. E. Spangenberg, S. Jacomet and J. Schibler, *J. Archaeol. Sci.*, 2006, **33**, 1–13.
98. M. S. Copley, R. Berstan, S. N. Dudd, G. Docherty, A. J. Mukherjee, V. Straker, S. Payne and R. P. Evershed, *Proc. Natl. Acad. Sci. USA*, 2003, **100**, 1524–1529.
99. J. E. Spangenberg, I. Matuschik, S. Jacomet and J. Schibler, *Isot. Environ. Health Stud.*, 2008, **44**, 189–200.
100. R. P. Evershed, S. Payne, A. G. Sherratt, M. S. Copley, J. Coolidge, D. Urem-Kotsu, K. Kotsakis, M. Özdogan, A. E. Özdogan, O. Nieuwenhuyse, P. M. M. G. Akkermans, D. Bailey, R. R. Andeescu, S. Campbell, S. Farid, I. Hodder, N. Yalman, M. Özbasaran, E. Bicakci, Y. Garfinkel, T. Levy and M. M. Burton, *Nature*, 2008, **455**, 528–531.
101. M. S. Copley, H. A. Bland, P. Rose, M. Horton and R. P. Evershed, *Analyst*, 2005, **130**, 860–871.
102. R. P. Evershed, *Archaeometry*, 2008, **50**, 895–924.
103. A. K. Outram, N. A. Stear, R. Bendrey, S. Olsen, A. Kasparov, V. Zaibert, N. Thorpe and R. P. Evershed, *Science*, 2009, **323**, 1332–1335.
104. R. Berstan, S. N. Dudd, M. S. Copley, E. D. Morgan, A. Quye and R. P. Evershed, *Analyst*, 2004, **129**, 270–275.
105. M. W. Gregg and G. F. Slater, *Archaeometry*, 2010, **52**, 833–854.
106. K. Romanus, J. Poblome, K. Verbeke, A. Luypaerts, P. Jacobs, D. De Vos and M. Waelkens, *Archaeometry*, 2007, **49**, 729–747.
107. L. T. Corr, J. C. Sealy, M. C. Horton and R. P. Evershed, *J. Archaeol. Sci.*, 2005, **32**, 321–330.
108. J. A. Silfer, M. H. Engel, S. A. Macko and E. J. Jumeau, *Anal. Chem.*, 1991, **63**, 370–374.
109. S. Jim, V. Jones, M. S. Copley, S. H. Ambrose and R. P. Evershed, *Rapid Commun. Mass Spectrom.*, 2003, **17**, 2283–2289.
110. A. Balzer, G. Gleixner, G. Grupe, H. L. Schmidt, S. Schramm and S. Turban-Just, *Archaeometry*, 1997, **39**, 415–429.

111. W. Meier-Augenstein, in *Handbook of Stable Isotope Analytical Techniques*, ed. P. A. de Groot, Elsevier B.V., Amsterdam, 2004, vol. I, pp. 153–176.
112. L. T. Corr, R. Berstan and R. P. Evershed, *Rapid Commun. Mass Spectrom.*, 2007, **21**, 3759–3771.
113. Y. I. Naito, N. V. Honch, Y. Chikaraishi, N. Ohkouchi and M. Yoneda, *Am. J. Phys. Anthropol.*, 2010, **143**, 31–40.
114. E. P. Hare, M. L. Fogel, T. W. Stafford Jr, A. D. Mitchell and T. C. Hoering, *J. Archaeol. Sci.*, 1991, **18**, 277–292.
115. J. S. O. McCullagh, D. Juchelka and R. E. M. Hedges, *Rapid Commun. Mass Spectrom.*, 2006, **20**, 2761–2768.
116. K. Choy, C. I. Smith, B. T. Fuller and M. P. Richards, *Geochim. Cosmochim. Acta*, 2010, **74**, 6093–6111.
117. M. Raghavan, J. S. O. McCullagh, N. Lynnerup and R. E. M. Hedges, *Rapid Commun. Mass Spectrom.*, 2010, **24**, 541–548.
118. O. A. Sherwood, M. F. Lehmann, C. J. Schubert, D. B. Scott and M. D. McCarthy, *Proc. Natl. Acad. Sci. USA*, 2011, **108**, 1011–1015.
119. A. W. Stott and R. P. Evershed, *Anal. Chem.*, 1996, **68**, 4402–4408.
120. M. Tsivou, N. Kioukia-Fougia, E. Lyris, Y. Aggelis, A. Fragkaki, X. Kiousi, P. Simitsek, H. Dimopoulou, I. P. Leontiou, M. Stamou, M. H. Spyridaki and C. Georgakopoulos, *Anal. Chim. Acta*, 2006, **555**, 1–13.
121. M. Thevis, *Mass Spectrometry in Sports Drug Testing: Characterization of Prohibited Substances and Doping Control Analytical Assays*, John Wiley & Sons, Hoboken, 2010.
122. P. E. Sottas, C. Saudan, C. Schweizer, N. Baume, P. Mangin and M. Saugy, *Forensic Sci. Int.*, 2008, **174**, 166–172.
123. R. Aguilera, C. K. Hatton and D. H. Catlin, *Clin. Chem.*, 2002, **48**, 629–636.
124. M. Ueki and M. Okano, *Rapid Commun. Mass Spectrom.*, 1999, **13**, 2237–2243.
125. A. T. Cawley, G. J. Trout, R. Kazlauskas, C. J. Howe and A. V. George, *Steroids*, 2009, **74**, 379–392.
126. A. T. Cawley, E. R. Hine, G. J. Trout, A. V. George and R. Kazlauskas, *Forensic Sci. Int.*, 2004, **143**, 103–114.
127. R. Aguilera, T. E. Chapman, H. Pereira, G. C. Oliveira, R. P. Illanes, T. F. Fernandes, D. A. Azevedo and F. A. Neto, *J. Steroid Biochem. Mol. Biol.*, 2009, **115**, 107–114.
128. R. Aguilera, D. H. Catlin, M. Becchi, A. Phillips, C. Wang, R. S. Swerdloff, H. G. Pope and C. K. Hatton, *J. Chromatogr. B*, 1999, **727**, 95–105.
129. R. D. Blackledge, *Clin. Chim. Acta*, 2009, **406**, 8–13.

130. E. Bichon, F. Kieken, N. Cesbron, F. Monteau, S. Prévost, F. André and B. Le Bizec, *Rapid Commun. Mass Spectrom.*, 2007, **21**, 2613–2620.
131. A. T. Cawley and U. Flenker, *J. Mass Spectrom.*, 2008, **43**, 854–864.
132. C. Saudan, C. Emery, F. Marclay, E. Strahm, P. Mangin and M. Saugy, *J. Chromatogr. B*, 2009, **877**, 2321–2329.
133. Y. Zhang, H. J. Tobias and J. T. Brenna, *Steroids*, 2009, **74**, 369–378.
134. G. L. Sacks, Y. Zhang and J. T. Brenna, *Anal. Chem.*, 2007, **79**, 6348–6358.
135. H. J. Tobias, G. L. Sacks, Y. Zhang and J. T. Brenna, *Anal. Chem.*, 2008, **80**, 8613–8621.
136. U. Flenker, M. Hebestreit, T. Piper, F. Häsemann and W. Schänzer, *Anal. Chem.*, 2007, **79**, 4162–4168.
137. T. Piper, M. Thevis, U. Flenker and W. Schänzer, *Rapid Commun. Mass Spectrom.*, 2009, **23**, 1917–1926.
138. E. M. van Warmerdam, S. K. Frape, R. Aravena, R. J. Drimmie, H. Flatt and J. A. Cherry, *Appl. Geochem.*, 1995, **10**, 547–552.
139. K. M. Beneteau, R. Aravena and S. K. Frape, *Org. Geochem.*, 1999, **30**, 739–753.
140. B. D. Holt, N. C. Sturchio, T. A. Abrajano and L. J. Heraty, *Anal. Chem.*, 1997, **69**, 2727–2733.
141. N. Jendrzejewski, H. G. M. Eggenkamp and M. L. Coleman, *Anal. Chem.*, 1997, **69**, 4259–4266.
142. N. Jendrzejewski, H. G. M. Eggenkamp and M. L. Coleman, *Appl. Geochem.*, 2001, **16**, 1021–1031.
143. O. Shouakar-Stash, S. K. Frape and R. J. Drimmie, *J. Contam. Hydrol.*, 2003, **60**, 211–228.
144. O. Shouakar-Stash, R. J. Drimmie, M. Zhang and S. K. Frape, *Appl. Geochem.*, 2006, **21**, 766–781.
145. M. M. G. Chartrand, S. K. Hirschorn, G. Lacrampe-Couloume and B. Sherwood Lollar, *Rapid Commun. Mass Spectrom.*, 2007, **21**, 1841–1847.
146. W. M. Jarman, A. Hilkert, C. E. Bacon, J. W. Collister, K. Ballschmiter and R. W. Risebrough, *Environ. Sci. Technol.*, 1998, **32**, 833–836.
147. P. J. Yanik, T. H. O'Donnell, S. A. Macko, Y. Qian and M. C. Kennicutt II, *Org. Geochem.*, 2003, **34**, 239–251.
148. Y. Horii, K. Kannan, G. Petrick, T. Gamo, J. Falandysz and N. Yamashita, *Environ. Sci. Technol.*, 2005, **39**, 4206–4212.
149. M. Mandalakis, H. Holmstrand, P. Andersson and Ö. Gustafsson, *Chemosphere*, 2008, **71**, 299–305.

150. N. J. Drenzek, C. H. Tarr, T. I. Eglinton, L. J. Heraty, N. C. Sturchio, V. J. Shiner and C. M. Reddy, *Org. Geochem.*, 2002, **33**, 437–444.
151. W. Vetter, W. Armbruster, T. R. Betson, J. Schleucher, T. Kapp and K. Lehnert, *Anal. Chim. Acta*, 2006, **577**, 250–256.
152. H. Holmstrand, M. Mandalakis, Z. Zencak, P. Andersson and Ö. Gustafsson, *Chemosphere*, 2007, **69**, 1533–1539.
153. W. Vetter, S. Gaul and W. Armbruster, *Environ. Int.*, 2008, **34**, 357–362.
154. H. S. Dempster, B. Sherwood Lollar and S. Feenstra, *Environ. Sci. Technol.*, 1997, **31**, 3193–3197.
155. R. R. Harrington, S. R. Poulson, J. I. Drever, P. J. S. Colberg and E. F. Kelly, *Org. Geochem.*, 1999, **30**, 765–775.
156. D. Hunkeler, N. Anderson, R. Aravena, S. M. Bernasconi and B. J. Butler, *Environ. Sci. Technol.*, 2001, **35**, 3462–3467.
157. B. J. Smallwood, R. P. Philp, T. W. Burgoyne and J. D. Allen, *Environ. Forensics*, 2001, **2**, 215–221.
158. T. Kuder, J. T. Wilson, P. Kaiser, R. Kolhatkar, P. Philp and J. Allen, *Environ. Sci. Technol.*, 2005, **39**, 213–220.
159. G. O'Sullivan and R. M. Kalin, *Environ. Forensics*, 2008, **9**, 166–176.
160. R. B. Coffin, P. H. Miyares, C. A. Kelley, L. A. Cifuentes and C. M. Reynolds, *Environ. Toxicol. Chem.*, 2001, **20**, 2676–2680.
161. D. Hunkeler, R. U. Meckenstock, T. C. Schmidt, B. Sherwood Lollar and J. T. Wilson, *A Guide for Assessing Biodegradation and Source Identification of Organic Ground Water Contaminants using Compound Specific Isotope Analysis (CSIA)* EPA 600/R-08/148, US Environmental Protection Agency, Office of Research and Development, Ada, OK, 2008.
162. S. E. Walker, R. M. Dickhut, C. Chisholm-Brause, S. Sylva and C. M. Reddy, *Org. Geochem.*, 2005, **36**, 619–632.
163. L. Mazeas and H. Budzinski, *J. Chromatogr. A*, 2001, **923**, 165–176.
164. V. P. O'Malley, T. A. Abrajano Jr and I. Hellou, *Environ. Sci. Technol.*, 1996, **30**, 634–639.
165. B. T. Hammer, C. A. Kelley, R. B. Coffin, L. A. Cifuentes and J. G. Mueller, *Chem. Geol.*, 1998, **152**, 43–58.
166. V. P. O'Malley, T. A. Abrajano and J. Hellou, *Org. Geochem.*, 1994, **21**, 809–822.
167. A. Stark, T. Abrajano, J. Hellou and J. L. Metcalf-Smith, *Org. Geochem.*, 2003, **34**, 225–237.
168. T. Okuda, H. Kumata, H. Naraoka and H. Takada, *Org. Geochem.*, 2002, **33**, 1737–1745.

169. C. McRae, C. Sun, C. F. McMillan, C. E. Snape and A. E. Fallick, *Polycyclic Aromat. Compd.*, 2000, **20**, 97–109.
170. C. McRae, C. G. Sun, C. E. Snape, A. E. Fallick and D. Taylor, *Org. Geochem.*, 1999, **30**, 881–889.
171. D. Saber, D. Mauro and T. Sirivedhin, *Environ. Forensics*, 2006, **7**, 65–75.
172. C. McRae, C. E. Snape and A. E. Fallick, *Analyst*, 1998, **123**, 1519–1523.
173. T. Okuda, H. Kumata, M. P. Zakaria, H. Naraoka, R. Ishiwatari and H. Takada, *Atmos. Environ.*, 2002, **36**, 611–618.
174. T. Okuda, H. Kumata, H. Naraoka and H. Takada, *Geochem. J.*, 2004, **38**, 89–100.
175. V. P. O'Malley, R. A. Burke and W. S. Schlotzhauer, *Org. Geochem.*, 1997, **27**, 567–581.
176. D. C. Ballentine, S. A. Macko, V. C. Turekian, W. P. Gilhooly and B. Martincigh, *Org. Geochem.*, 1996, **25**, 97–104.
177. C. McRae, C. E. Snape, C. G. Sun, D. Fabbri, D. Tartari, C. Trombini and A. E. Fallick, *Environ. Sci. Technol.*, 2000, **34**, 4684–4686.
178. M. Blessing, PhD Thesis, Eberhard-Karls-University, 2008.
179. A. Smirnov, T. A. Abrajano, A. Smirnov and A. Stark, *Org. Geochem.*, 1998, **29**, 1813–1128.
180. B. Yan, T. A. Abrajano, R. F. Bopp, L. A. Benedict, D. A. Chaky, E. Perry, J. Song and D. P. Keane, *Org. Geochem.*, 2006, **37**, 674–687.
181. L. Mazeas and H. Budzinski, *Org. Geochem.*, 2002, **33**, 1253–1258.
182. D. Fabbri, I. Vassura, C. G. Sun, C. E. Snape, C. McRae and A. E. Fallick, *Mar. Chem.*, 2003, **84**, 123–135.
183. C. McRae, C.-G. Sun, C. E. Snape and A. E. Fallick, *Am. Lab.*, 1999, **31**.
184. L. Zhang, Z. Bai, Y. You, J. Wu, Y. Feng and T. Zhu, *Chemosphere*, 2009, **75**, 453–461.
185. J. E. Silliman, P. A. Meyers, P. H. Ostrom, N. E. Ostrom and B. J. Eadie, *Org. Geochem.*, 2000, **31**, 1133–1142.
186. W. Wilcke, M. Krauss and W. Amelung, *Environ. Sci. Technol.*, 2002, **36**, 3530–3535.
187. T. Gocht, J. A. C. Barth, M. Epp, M. Jochmann, M. Blessing, T. C. Schmidt and P. Grathwohl, *Appl. Geochem.*, 2007, **22**, 2652–2663.
188. R. L. Hough, M. Whittaker, A. E. Fallick, T. Preston, J. G. Farmer and S. J. T. Pollard, *Environ. Pollut.*, 2006, **143**, 489–498.
189. L. Mansuy, R. P. Philp and J. Allen, *Environ. Sci. Technol.*, 1997, **31**, 3417–3425.

190. K. L. Pond, Y. S. Huang, Y. Wang and C. F. Kulpa, *Environ. Sci. Technol.*, 2002, **36**, 724–728.
191. L. M. Dowling, C. J. Boreham, J. M. Hope, A. P. Murray and R. E. Summons, *Org. Geochem.*, 1995, **23**, 729–737.
192. A. Vieth and H. Wilkes, *Geochim. Cosmochim. Acta*, 2006, **70**, 651–665.
193. R. P. Philp, J. Allen and T. Kuder, *Environ. Forensics*, 2002, **3**, 341–348.
194. K. M. Rogers and M. M. Savard, *Org. Geochem.*, 1999, **30**, 1559–1569.
195. L. Mazeas and H. Budzinski, *Environ. Sci. Technol.*, 2002, **36**, 130–137.
196. B. R. T. Simoneit, *Atmos. Environ.*, 1997, **31**, 2225–2233.
197. D. Hunkeler, N. Chollet, X. Pittet, R. Aravena, J. A. Cherry and B. L. Parker, *J. Contam. Hydrol.*, 2004, **74**, 265–282.
198. S. M. Eberts, C. Braun and S. Jones, *Environ. Forensics*, 2008, **9**, 85–95.
199. M. Blessing, T. C. Schmidt, R. Dinkel and S. B. Haderlein, *Environ. Sci. Technol.*, 2009, **43**, 2701–2707.
200. Y. Wang, Y. Huang, J. N. Huckins and J. D. Petty, *Environ. Sci. Technol.*, 2004, **38**, 3689–3697.
201. S. A. Mancini, G. Lacrampe-Couloume and B. Sherwood Lollar, *Environ. Forensics*, 2008, **9**, 177–186.
202. S. M. Mudge, W. Meier-Augenstein, C. Eadsforth and P. DeLeo, *J. Environ. Monit.*, 2010, **12**, 1846–1856.
203. T. J. Boyd, C. L. Osburn, K. J. Johnson, K. B. Birgl and R. B. Coffin, *Environ. Sci. Technol.*, 2006, **40**, 1916–1924.
204. G. F. Slater, H. S. Dempster, B. Sherwood Lollar and J. Ahad, *Environ. Sci. Technol.*, 1999, **33**, 190–194.
205. S. R. Poulson and J. I. Drever, *Environ. Sci. Technol.*, 1999, **33**, 3689–3694.
206. Y. Wang and Y. Huang, *Appl. Geochem.*, 2003, **18**, 1641–1651.
207. M. Elsner, *J. Environ. Monit.*, 2010, **12**, 2005–2031.
208. C. Aeppli, M. Berg, O. A. Cirpka, C. Holliger, R. P. Schwarzenbach and T. B. Hofstetter, *Environ. Sci. Technol.*, 2009, **43**, 8813–8820.
209. C. Schüth, H. Taubald, N. Bolano and K. Maciejczyk, *J. Contam. Hydrol.*, 2003, **64**, 269–281.
210. G. F. Slater, J. M. E. Ahad, B. Sherwood Lollar, R. Allen-King and B. Sleep, *Anal. Chem.*, 2000, **72**, 5669–5672.
211. F. D. Kopinke, A. Georgi, M. Voskamp and H. H. Richnow, *Environ. Sci. Technol.*, 2005, **39**, 6052–6062.
212. B. M. van Breukelen, D. Hunkeler and F. Volkering, *Environ. Sci. Technol.*, 2005, **39**, 4189–4197.

213. B. M. van Breukelen and H. Prommer, *Environ. Sci. Technol.*, 2008, **42**, 2457–2463.
214. E. M. LaBolle, G. E. Fogg, J. B. Eweis, J. Gravner and D. G. Leaist, *Water Resour. Res.*, 2008, **44**.
215. M. Rolle, G. Chiogna, R. Bauer, C. Griebler and P. Grathwohl, *Environ. Sci. Technol.*, 2010, **44**, 6167–6173.
216. A. de Visscher, I. de Pourcq and J. Chanton, *J. Geophys. Res. - Atmos.*, 2004, **109**.
217. D. Bouchard, P. Höhener and D. Hunkeler, *Environ. Sci. Technol.*, 2008, **42**, 7801–7806.
218. D. Bouchard, D. Hunkeler, P. Gaganis, R. Aravena, P. Höhener, M. M. Broholm and P. Kieldsen, *Environ. Sci. Technol.*, 2008, **42**, 596–601.
219. T. Kuder, P. Philp and J. Allen, *Environ. Sci. Technol.*, 2009, **43**, 1763–1768.
220. *Use of Monitored Natural Attenuation at Superfund, RCRA Corrective Action, and Underground Storage Tank Sites* 9200.4–17P, US Environmental Protection Agency, Washington, DC, 1999.
221. J. R. Gray, G. Lacrampe-Couloume, D. Gandhi, K. M. Scow, R. D. Wilson, D. M. Mackay and B. Sherwood Lollar, *Environ. Sci. Technol.*, 2002, **36**, 1931–1938.
222. S. A. Mancini, G. Lacrampe-Couloume, H. Jonker, B. M. van Breukelen, J. Groen, F. Volkering and B. Sherwood Lollar, *Environ. Sci. Technol.*, 2002, **36**, 2464–2470.
223. B. Morasch, H. H. Richnow, B. Schink and R. U. Meckenstock, *Appl. Environ. Microbiol.*, 2002, **68**, 5191–5194.
224. J. A. M Ward, J. M. E. Ahad, G. Lacrampe-Couloume, G. F. Slater, E. A. Edwards and B. Sherwood Lollar, *Environ. Sci. Technol.*, 2000, **34**, 4577–4581.
225. T. C. Schmidt, L. Zwank, M. Elsner, M. Berg, R. U. Meckenstock and S. B. Haderlein, *Anal. Bioanal. Chem.*, 2004, **378**, 283–300.
226. S. A. Mancini, A. C. Ulrich, G. Lacrampe-Couloume, B. E. Sleep, E. A. Edwards and B. Sherwood Lollar, *Appl. Environ. Microbiol.*, 2003, **69**, 191–198.
227. B. Sherwood Lollar, G. F. Slater, J. Ahad, B. Sleep, J. Spivack, M. Brennan and P. MacKenzie, *Org. Geochem.*, 1999, **30**, 813–820.
228. R. U. Meckenstock, B. Morasch, C. Griebler and H. H. Richnow, *J. Contam. Hydrol.*, 2004, **75**, 215–255.
229. D. Hunkeler and B. Morasch, in *Environmental Isotopes in Biodegradation and Bioremediation*, eds. C. M. Aelion, P. Höhener, D. Hunkeler and R. Aravena, CRC Press, Boca Raton, 2010, pp. 79–125.

230. Y. Abe and D. Hunkeler, *Environ. Sci. Technol.*, 2006, **40**, 1588–1596.
231. C. M. Reddy, N. J. Drenzek, T. I. Eglinton, L. J. Heraty, N. C. Sturchio and V. J. Shiner, *Environ. Sci. Pollut. Res.*, 2002, **9**, 183–186.
232. Y. Abe, J. Zopfi and D. Hunkeler, *Isot. Environ. Health Stud.*, 2009, **45**, 18–26.
233. D. Bouchard, D. Hunkeler and P. Höhener, *Org. Geochem.*, 2008, **39**, 23–33.
234. A. Fischer, J. Bauer, R. U. Meckenstock, W. Stichler, C. Griebler, P. Maloszewski, M. Kästner and H. H. Richnow, *Environ. Sci. Technol.*, 2006, **40**, 4245–4252.
235. A. Peter, A. Steinbach, R. Liedl, T. Ptak, W. Michaelis and G. Teutsch, *J. Contam. Hydrol.*, 2004, **71**, 127–154.
236. C. Griebler, M. Safinowski, A. Vieth, H. H. Richnow and R. U. Meckenstock, *Environ. Sci. Technol.*, 2004, **38**, 617–631.
237. B. M. van Breukelen, *Environ. Sci. Technol.*, 2007, **41**, 4980–4985.
238. P. L. Morrill, B. E. Sleep, G. F. Slater, E. A. Edwards and B. Sherwood Lollar, *Environ. Sci. Technol.*, 2006, **40**, 3886–3892.
239. B. M. van Breukelen, *Environ. Sci. Technol.*, 2007, **41**, 4004–4010.
240. K. E. Pooley, M. Blessing, T. C. Schmidt, S. B. Haderlein, K. T. B. Macquarrie and H. Prommer, *Environ. Sci. Technol.*, 2009, **43**, 7458–7464.
241. D. Hunkeler, R. Aravena, K. Berry-Spark and E. Cox, *Environ. Sci. Technol.*, 2005, **39**, 5975–5981.
242. D. Hunkeler, R. Aravena and B. J. Butler, *Environ. Sci. Technol.*, 1999, **33**, 2733–2738.
243. C. Aeppli, T. B. Hofstetter, H. I. F. Amaral, R. Kipfer, R. P. Schwarzenbach and M. Berg, *Environ. Sci. Technol.*, 2010, **44**, 3705–3711.
244. S. K. Hirschorn, M. J. Dinglasan, M. Elsner, S. A. Mancini, G. Lacrampe-Couloume, E. A. Edwards and B. Sherwood Lollar, *Environ. Sci. Technol.*, 2004, **38**, 4775–4781.
245. A. Kohen and H.-H. Limbach, *Isotope Effects in Chemistry and Biology*, Taylor & Francis, Boca Raton, FL, 2006.
246. M. Elsner, L. Zwank, D. Hunkeler and R. P. Schwarzenbach, *Environ. Sci. Technol.*, 2005, **39**, 6896–6916.
247. L. Melander and W. H. Saunders, Jr, *Reaction Rates of Isotopic Molecules*, Wiley, Chichester, 1980.
248. N. B. Tobler, T. B. Hofstetter and R. P. Schwarzenbach, *Environ. Sci. Technol.*, 2008, **42**, 7786–7792.

249. T. B. Hofstetter, R. P. Schwarzenbach and S. M. Bernasconi, *Environ. Sci. Technol.*, 2008, **42**, 7737–7743.
250. T. B. Hofstetter, J. C. Spain, S. F. Nishino, J. Bolotin and R. P. Schwarzenbach, *Environ. Sci. Technol.*, 2008, **42**, 4764–4770.
251. T. B. Hofstetter, A. Neumann, W. A. Arnold, A. E. Hartenbach, J. Bolotin, C. J. Cramer and R. P. Schwarzenbach, *Environ. Sci. Technol.*, 2008, **42**, 1997–2003.
252. A. Dybala-Defratyka, L. Szatkowski, R. Kaminski, M. Wujec, A. Siwek and P. Paneth, *Environ. Sci. Technol.*, 2008, **42**, 7744–7750.
253. C. M. Aelion, P. Höhener, D. Hunkeler and R. Aravena, eds., *Environmental Isotopes in Biodegradation and Bioremediation*, CRC Press, Boca Raton, FL, 2010.
254. L. Zwank, M. Berg, M. Elsner, T. C. Schmidt, S. B. Haderlein and R. P. Schwarzenbach, *Environ. Sci. Technol.*, 2005, **39**, 1018–1029.
255. D. M. Kujawinski, M. Stephan, M. A. Jochmann, K. Krajenke, J. Haas and T. C. Schmidt, *J. Environ. Monit.*, 2010, **12**, 347–354.
256. J. R. Gray, G. Lacrampe-Couloume, D. Gandhi, K. M. Scow, R. D. Wilson, D. M. Mackay and B. Sherwood Lollar, *Environ. Sci. Technol.*, 2002, **36**, 1931–1938.
257. L. E. Lesser, P. C. Johnson, R. Aravena, G. E. Spinnler, C. L. Bruce and J. P. Salanitro, *Environ. Sci. Technol.*, 2008, **42**, 6637–6643.
258. M. Rosell, D. Barcelo, T. Rohwerder, U. Breuer, M. Gehre and H. H. Richnow, *Environ. Sci. Technol.*, 2007, **41**, 2036–2043.
259. J. R. McKelvie, M. R. Hyman, M. Elsner, C. Smith, D. M. Aslett, G. Lacrampe-Couloume and B. Sherwood Lollar, *Environ. Sci. Technol.*, 2009, **43**, 2793–2799.
260. M. Elsner, J. McKelvie, G. L. Couloume and B. Sherwood Lollar, *Environ. Sci. Technol.*, 2007, **41**, 5693–5700.
261. A. Fischer, M. Gehre, J. Breitfeld, H. H. Richnow and C. Vogt, *Rapid Commun. Mass Spectrom.*, 2009, **23**, 2439–2447.
262. S. A. Mancini, C. E. Devine, M. Elsner, M. E. Nandi, A. C. Ulrich, E. A. Edwards and B. Sherwood Lollar, *Environ. Sci. Technol.*, 2008, **42**, 8290–8296.
263. A. Fischer, K. Theuerkorn, N. Stelzer, M. Gehre, M. Thullner and H. H. Richnow, *Environ. Sci. Technol.*, 2007, **41**, 3689–3696.
264. S. A. Mancini, S. K. Hirschorn, M. Elsner, G. Lacrampe-Coulome, E. A. Edwards and B. Sherwood Lollar, *Environ. Sci. Technol.*, 2006, **40**, 7675–7681.
265. C. Vogt, E. Cyrus, I. Herklotz, D. Schlosser, A. Bahr, S. Herrmann, H. H. Richnow and A. Fischer, *Environ. Sci. Technol.*, 2008, **42**, 7793–7800.

266. Y. Abe, R. Aravena, J. Zopfi, O. Shouakar-Stash, E. Cox, J. D. Roberts and D. Hunkeler, *Environ. Sci. Technol.*, 2009, **43**, 101–107.
267. D. Hunkeler, Y. Abe, M. M. Broholm, S. Jeannottat, C. Westergaard, C. S. Jacobsen, R. Aravena and P. L. Bjerg, *J. Contam. Hydrol.*, 2011, **119**, 69–79.
268. A. Bernstein, Z. Ronen, E. Adar, R. Nativ, H. Lowag, W. Stichler and R. U. Meckenstock, *Environ. Sci. Technol.*, 2008, **42**, 7772–7777.
269. A. E. Hartenbach, T. B. Hofstetter, P. R. Tentscher, S. Canonica, M. Berg and R. P. Schwarzenbach, *Environ. Sci. Technol.*, 2008, **42**, 7751–7756.
270. A. H. Meyer, H. Penning, H. Lowag and M. Elsner, *Environ. Sci. Technol.*, 2008, **42**, 7757–7763.
271. A. H. Meyer, H. Penning and M. Elsner, *Environ. Sci. Technol.*, 2009, **43**, 8079–8085.
272. H. Penning, C. J. Cramer and M. Elsner, *Environ. Sci. Technol.*, 2008, **42**, 7764–7771.
273. H. Penning, S. R. Sörensen, A. H. Meyer, J. Aamand and M. Elsner, *Environ. Sci. Technol.*, 2010, **44**, 2372–2378.
274. H. Penning and M. Elsner, *Anal. Chem.*, 2007, **79**, 8399–8405.
275. J. R. McKelvie, D. M. Mackay, N. R. de Sieyes, G. Lacrampe-Couloume and B. Sherwood Lollar, *J. Contam. Hydrol.*, 2007, **94**, 157–165.
276. D. M. Mackay, N. R. De Sieyes, M. D. Einarson, K. P. Feris, A. A. Pappas, I. A. Wood, L. Jacobson, L. G. Justice, M. N. Noske, K. M. Scow and J. T. Wilson, *Environ. Sci. Technol.*, 2006, **40**, 6123–6130.
277. S. K. Hirschorn, A. Grostern, G. Lacrampe-Couloume, E. A. Edwards, L. MacKinnon, C. Repta, D. W. Major and B. Sherwood Lollar, *J. Contam. Hydrol.*, 2007, **94**, 249–260.
278. D. L. Song, M. E. Conrad, K. S. Sorenson and L. Alvarez-Cohen, *Environ. Sci. Technol.*, 2002, **36**, 2262–2268.
279. M. M. G. Chartrand, P. L. Morrill, G. Lacrampe-Couloume and B. Sherwood Lollar, *Environ. Sci. Technol.*, 2005, **39**, 4848–4856.
280. P. L. Morrill, G. Lacrampe-Coulome, G. F. Slater, B. Sleep, E. A. Edwards, M. L. McMaster, D. W. Major and B. Sherwood Lollar, *J. Contam. Hydrol.*, 2005, **76**, 279–293.
281. H. Dayan, T. Abrajano, N. C. Sturchio and L. Winsor, *Org. Geochem.*, 1999, **30**, 755–763.
282. M. Elsner, M. Chartrand, N. Vanstone, G. L. Couloume and B. Sherwood Lollar, *Environ. Sci. Technol.*, 2008, **42**, 5963–5970.

283. M. Elsner, G. L. Couloume, S. Mancini, L. Burns and B. Sherwood Lollar, *Ground Water Monit. Rem.*, 2010, **30**, 79–95.
284. N. VanStone, M. Elsner, G. Lacrampe-Couloume, S. Mabury and B. Sherwood Lollar, *Environ. Sci. Technol.*, 2008, **42**, 126–132.
285. L. Zwank, M. Elsner, A. Aeberhard, R. P. Schwarzenbach and S. B. Haderlein, *Environ. Sci. Technol.*, 2005, **39**, 5634–5641.
286. N. VanStone, A. Przepiora, J. Vogan, G. Lacrampe-Couloume, B. Powers, E. Perez, S. Mabury and B. Sherwood Lollar, *J. Contam. Hydrol.*, 2005, **78**, 313–325.
287. D. Hunkeler, R. Aravena, B. L. Parker, J. A. Cherry and X. Diao, *Environ. Sci. Technol.*, 2003, **37**, 798–804.
288. W. A. Arnold, J. Bolotin, U. Von Gunten and T. B. Hofstetter, *Environ. Sci. Technol.*, 2008, **42**, 7778–7785.
289. T. B. Hofstetter and M. Berg, *Trends Anal. Chem.*, 2011, **30**, 618–627.
290. G. F. Slater, *Environ. Forensics*, 2003, **4**, 13–23.
291. T. C. Schmidt and M. A. Jochmann, *Annu. Rev. Anal. Chem.*, 2012, **5**, 133–155.
292. M. Thullner, F. Centler, H. H. Richnow and A. Fischer, *Org. Geochem.*, 2012, **42**, 1440–1460.
293. T. B. Coplen, J. K. Böhlke, P. De Bièvre, T. Ding, N. E. Holden, J. A. Hopple, H. R. Krouse, A. Lamberty, H. S. Peiser, K. Révész, S. E. Rieder, K. J. R. Rosman, E. Roth, P. D. P. Taylor, R. D. Vocke and Y. K. Xiao, *Pure Appl. Chem.*, 2002, **74**, 1987–2017.
294. F. Robert and S. Epstein, *Geochim. Cosmochim. Acta*, 1982, **46**, 81–95.
295. M. Wahlen, in *Stable Isotopes in Ecology and Environmental Science*, ed. K. Lajtha and R. H. Michener, Blackwell, Oxford, 1994.
296. J. R. Cronin and S. Chang, in *The Chemistry of Life's Origins*, ed. J. M. Greenberg, C. X. Mendoza-Gomez and V. Pirronello, Kluwer, 1993, pp. 209–225.
297. Z. Martins, *Elements*, 2011, **7**, 35–40.
298. F. Mullie and J. Reisse, in *Top. Curr. Chem.*, ed. M. J. S. Dewar, J. D. Dunitz and K. Hafner, Springer, Berlin, 1987, vol. 139, pp. 83–117.
299. C. T. Pillinger, *Geochim. Cosmochim. Acta*, 1984, **48**, 2739–2766.
300. M. A. Sephton, *Natural Product Reports*, 2002, **19**, 292–311.
301. M. A. Sephton and I. Gilmour, *Mass Spectrom. Rev.*, 2001, **20**, 111–120.
302. I. Gilmour, in *Treatise on Geochemistry*, ed. A. M. Davis, Elsevier, Amsterdam, 2005, vol. 1, pp. 269–290.
303. R. H. Carr, I. P. Wright, C. T. Pillinger, R. S. Lewis and E. Anders, *Meteoritics*, 1983, **18**, 277.

304. J. M. Hayes, K. H. Freeman, B. N. Popp and C. H. Hoham, *Org. Geochem.*, 1990, **16**, 1115–1128.
305. A. W. Hilkert, C. B. Douthitt, H. J. Schlüter and W. A. Brand, *Rapid Commun. Mass Spectrom.*, 1999, **13**, 1226–1230.
306. P. G. Brown, A. R. Hildebrand, M. E. Zolensky, M. Grady, R. N. Clayton, T. K. Mayeda, E. Tagliaferri, R. Spalding, N. D. MacRae, E. L. Hoffman, D. W. Mittlefehldt, J. F. Wacker, J. A. Bird, M. D. Campbell, R. Carpenter, H. Gingerich, M. Glatiotis, E. Greiner, M. J. Mazur, P. J. A. McCausland, H. Plotkin and T. R. Mazur, *Science*, 2000, **290**, 320–325.
307. T. J. Zega, C. M. O. Alexander, H. Busemann, L. R. Nittler, P. Hoppe, R. M. Stroud and A. F. Young, *Geochim. Cosmochim. Acta*, 2010, **74**, 5966–5983.
308. H. Reeves, *Annu. Rev. Astron. Astrophys.*, 1974, **12**, 437–469.
309. E. Roueff and M. Gerin, *Space Sci. Rev.*, 2003, **106**, 61–72.
310. J. Geiss and H. Reeves, *Astron. Astrophys.*, 1981, **93**, 189–199.
311. J. Yang and S. Epstein, *Lunar Planet Sci.*, 1983, **XIII**, 885–886.
312. R. H. Becker and S. Epstein, *Geochim. Cosmochim. Acta*, 1982, **46**, 97–103.
313. G. W. Cooper, M. H. Thiemens, T. L. Jackson and S. Chang, *Science*, 1997, **277**, 1072–1074.
314. R. V. Krishnamurthy, S. Epstein, J. R. Cronin, S. Pizzarello and G. U. Yuen, *Geochim. Cosmochim. Acta*, 1992, **56**, 4045–4058.
315. S. Pizzarello and Y. S. Huang, *Geochim. Cosmochim. Acta*, 2005, **69**, 599–605.
316. Y. S. Huang, Y. Wang, M. R. Alexandre, T. Lee, C. Rose-Petruck, M. Fuller and S. Pizzarello, *Geochim. Cosmochim. Acta*, 2005, **69**, 1073–1084.
317. J. F. Kerridge, *Geochim. Cosmochim. Acta*, 1985, **49**, 1707–1714.
318. S. Messenger, *Nature*, 2000, **404**, 968–971.
319. S. A. Sandford, M. P. Bernstein and J. P. Dworkin, *Meteorit. Planet. Sci.*, 2001, **36**, 1117–1133.
320. H. Busemann, A. F. Young, C. M. O. Alexander, P. Hoppe, S. Mukhopadhyay and L. R. Nittler, *Science*, 2006, **312**, 727–730.
321. S. Pizzarello and W. Holmes, *Geochim. Cosmochim. Acta*, 2009, **73**, 2150–2162.
322. J. Duprat, E. Dobrica, C. Engrand, J. Aleon, Y. Marrocchi, S. Mostefaoui, A. Meibom, H. Leroux, J. N. Rouzaud, M. Gounelle and F. Robert, *Science*, 2010, **328**, 742–745.
323. O. Delpoux, D. Gourier, H. Vezin, L. Binet, S. Derenne and F. Robert, *Geochim. Cosmochim. Acta*, 2011, **75**, 326–336.

324. D. Gourier, F. Robert, O. Delpoux, L. Binet, H. Vezin, A. Moissette and S. Derenne, *Geochim. Cosmochim. Acta*, 2008, **72**, 1914–1923.
325. L. Binet and D. Gourier, *Appl. Magn. Reson.*, 2006, **30**, 207–231.
326. B. T. De Gregorio, R. M. Stroud, L. R. Nittler, C. M. O. Alexander, A. L. D. Kilcoyne and T. J. Zega, *Geochim. Cosmochim. Acta*, 2010, **74**, 4454–4470.
327. M. H. Engel and S. A. Macko, *Nature*, 1997, **389**, 265–268.
328. M. H. Engel, S. A. Macko and J. A. Silfer, *Nature*, 1990, **348**, 47–49.
329. D. S. P. Dearborn, in *CNO Isotopes in Astrophysics*, ed. J. Audouze, D Reidel Publ. Comp., Dordrecht, 1977, pp. 39–44.
330. E. R. Wollman, *Astrophys. J.*, 1973, **184**, 773–785.
331. L. Vigroux, J. Audouze and J. Lequeux, *Astron. Astrophys.*, 1976, **52**, 1–9.
332. E. K. Zinner, *Annu. Rev. Earth Planet Sci.*, 1998, **26**, 147–188.
333. E. K. Zinner, in *Treatise on Geochemistry*, ed. A. M. Davis, Elsevier, Amsterdam, 2005, vol. 1, pp. 16–40.
334. P. K. Swart, M. M. Grady, C. T. Pillinger, R. S. Lewis and E. Anders, *Science*, 1983, **220**, 406–410.
335. K. Kvenvolden, J. Lawless, K. Pering, E. Peterson, J. Flores, C. Ponnampe, I. R. Kaplan and C. Moore, *Nature*, 1970, **228**, 923–926.
336. A. V. Hirner, *Bull. Ver. schweiz. Petroleum-Geol. u Ing.*, 1979, **45**, 47–55.
337. G. U. Yuen, N. Blair, D. J. Des Marais and S. Chang, *Nature*, 1984, **307**, 252–254.
338. Y. Bottinga, *Geochim. Cosmochim. Acta*, 1969, **33**, 49–64.
339. M. S. Lancet, PhD Thesis, University of Chicago, 1972.
340. M. S. Lancet and E. Anders, *Science*, 1970, **170**, 980–982.
341. P. K. Swart, M. M. Grady and C. T. Pillinger, *Meteoritics*, 1983, **18**, 137–153.
342. S. Pizzarello, Y. S. Huang and M. Fuller, *Geochim. Cosmochim. Acta*, 2004, **68**, 4963–4969.
343. Z. Martins, O. Botta, M. L. Fogel, M. A. Sephton, D. P. Glavin, J. S. Watson, J. P. Dworkin, A. W. Schwartz and P. Ehrenfreund, *Earth. Planet. Sci. Lett.*, 2008, **270**, 130–136.
344. Z. Martins and M. A. Sephton, in *Amino Acids, Peptides and Proteins in Organic Chemistry*, ed. A. B. Hughes, Wiley VCH, Weinheim, 2009, pp. 2–42.
345. D. P. Glavin and J. P. Dworkin, *Proc. Natl. Acad. Sci. USA*, 2009, **106**, 5487–5492.

346. J. E. Elsila, D. P. Glavin and J. P. Dworkin, *Meteorit. Planet. Sci.*, 2009, **44**, 1323–1330.
347. S. Pizzarello, G. W. Cooper and G. J. Flynn, in *Meteorites and the Early Solar System*, ed. D. S. Lauretta and H. Y. J. McSween, University of Arizona Press, Tuscon, 2006, vol. II, pp. 625–651.
348. M. Gounelle, *Elements*, 2011, **7**, 29–34.
349. J. H. Scott, D. M. O'Brien, D. Emerson, H. Sun, G. D. McDonald, A. Salgado and M. L. Fogel, *Astrobiol.*, 2006, **6**, 867–880.
350. S. Pizzarello and T. L. Groy, *Geochim. Cosmochim. Acta*, 2011, **75**, 645–656.
351. C. Chyba and C. Sagan, *Nature*, 1992, **355**, 125–132.
352. A. G. W. Cameron and J. W. Truran, *Icarus*, 1977, **30**, 447–461.
353. R. N. Clayton, L. Grossman and T. K. Mayeda, *Science*, 1973, **182**, 485–488.
354. R. N. Clayton, in *Treatise on Geochemistry*, ed. A. M. Davis, Elsevier, Amsterdam, 2005, vol. 1, pp. 129–142.
355. J. R. Hulston and H. G. Thode, *J. Geophys. Res.*, 1965, **70**, 4435–4442.
356. C. E. Rees and H. G. Thode, *Geochim. Cosmochim. Acta*, 1977, **41**, 1679–1682.
357. M. H. Briggs, *Nature*, 1963, **197**, 1290.
358. A. Mikolajczuk, E. P. Przyk, B. Geypens, M. Berglund and P. Taylor, *Isot. Environ. Health Stud.*, 2010, **46**, 2–12.
359. J. R. McKelvie, S. K. Hirschorn, G. Lacrampe-Couloume, J. Lindstrom, J. Braddock, K. Finneran, D. Trego and B. Sherwood Lollar, *Ground Water Monit. Rem.*, 2007, **27**, 63–73.

CHAPTER 7
Further Developments in Compound-Specific Isotope Analysis

7.1 SCOPE

In this chapter topics will be covered that go beyond the normal instrumental background of CSIA by GC- or LC-IRMS. In Section 7.2, the first attempts to use multidimensional, comprehensive GC will be introduced. In Section 7.3 the application of multicollector inductively coupled plasma mass spectrometry will be introduced for analysis of sulfur isotope ratio determination. In Section 7.4 recent developments in the measurement of chlorine and bromine isotopes by GC with quadrupole MS as detector will be discussed. In Section 7.5 the fundamentals of position-specific isotope ratio (PSIA) analysis are introduced.

7.2 MULTIDIMENSIONAL GC (GC-GC), COMPREHENSIVE GC (GC × GC)

The maximum number of resolvable peaks or 'peak capacity' n in capillary GC is high and can exceed 150 peaks under optimal conditions within one run.[1] In spite of this high number of totally separated compounds, the separation of complex mixtures can be limited, even after sophisticated and laborious clean-up steps. For such complex samples, different paths can be taken in order to enhance the sample capacity and resolution of a chromatographic system.

Compound-specific Stable Isotope Analysis
By Maik A. Jochmann and Torsten C. Schmidt
© The Royal Society of Chemistry 2012
Published by the Royal Society of Chemistry, www.rsc.org

The sample capacity q_{cc} of a capillary column is defined by:[2]

$$q_{cc} = c_M d_c^{\frac{5}{2}}(1+k)(LH)^{\frac{1}{2}} \tag{7.1}$$

where c_M is the concentration of a compound in the mobile phase, d_c the internal column diameter, k the retention factor, L the column length and H the height equivalent of a theoretical plate. As pointed out in Chapter 3, sample capacity can be problematic in GC-IRMS, especially when using large volume injection techniques where sample capacity might be exceeded due to a large amount of matrix components being co-injected. As seen in eqn (7.1) the sample capacity decreases rapidly with a decreasing internal column diameter d_c. To overcome sample capacity problems, either wider internal column diameters or removal of interfering matrix is possible.[2]

Resolution can be enhanced, (i) by changing the selectivity of the column by changing the stationary phase material (however, depending on the compounds, this may only shift the problem from one critical peak pair to another), (ii) by coupling two independent columns (column 1 and 2) with different retention mechanisms (orthogonal separation) via an interface. Orthogonality in this context means that the elution times in the two columns or dimensions can be treated as statistically independent, which is fulfilled if the two columns possess totally different retention mechanisms (polarity, chirality). In this context one can distinguish between two techniques: (i) multidimensional GC (MD-GC, GC-GC or 2D-GC) and (ii) comprehensive GC (GC × GC).

In MD-GC, only specific fractions or heart-cuts from the first dimension (column 1) are introduced into a second dimension (column 2) for further additional separation. This approach was introduced by McEwen in 1964.[3] The principles of MD-GC and GC × GC are illustrated in Figure 7.1.

A heart-cut window is set for one or more pre-selected peaks (selected bands). These peaks will be transferred to the second dimension where a further separation of the heart-cut section of the first dimension takes place by another separation interaction mechanism. Preliminary one-dimensional separations have to be carried out to select the heart-cut section.[4] In this case, the peak capacity of the two dimensions is additive ($n_1 + n_2$). The 'heart-cutting' in MD-GC can be carried out by a switching valve, a 'Deans switch' or a 'moving capillary stream switch' (MCSS) device such as shown in Figure 7.2. Perhaps the most significant additional modification that can be made is refocusing of analytes prior to the secondary separation. As highlighted earlier, the peak widths from the primary column separation fundamentally limit the resolving power of the second column, with refocusing being the key step in

Further Developments in Compound-Specific Isotope Analysis

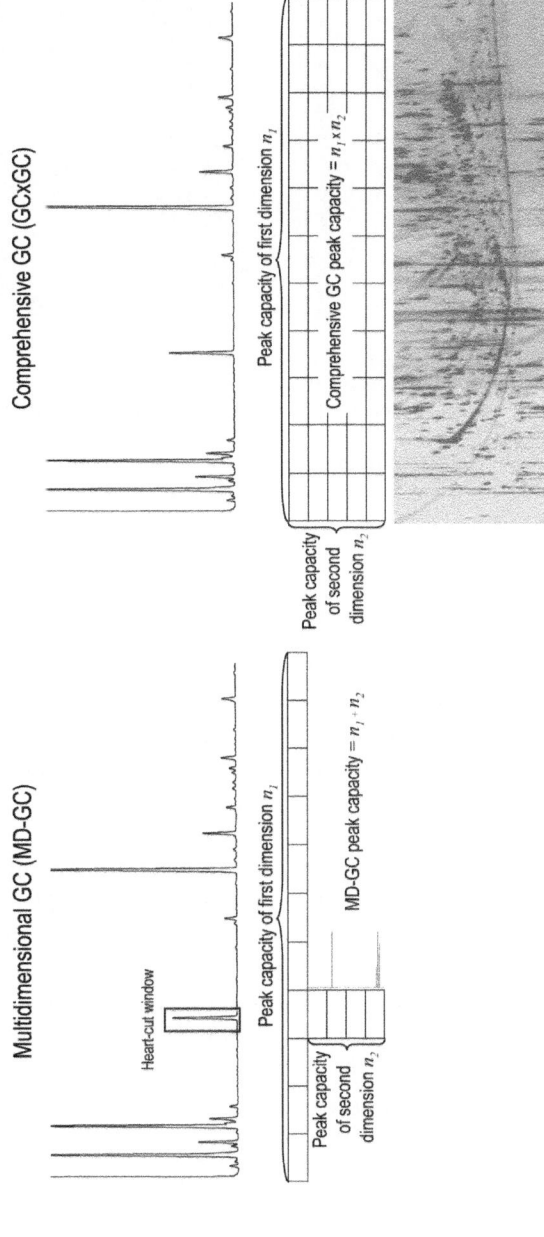

Figure 7.1 Differentiation between multidimensional gas chromatography (MD-GC, GC-GC) and comprehensive gas chromatography (GC × GC). For description see text.

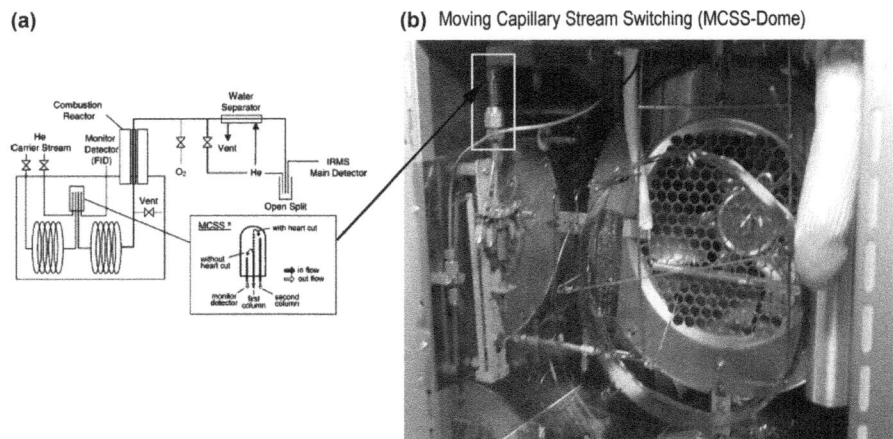

Figure 7.2 Moving capillary stream switching. a) Schematic view of the MD-GC-IRMS with an MCSS. (Graphic from Horii et al.[5] with permission of the American Chemical Society.) b) Picture of an MD-GC-IRMS system with an MCSS.
(Picture by Maik Jochmann.)

reducing this dispersion effect by using a cryogenic trap, cooled typically with CO_2 or liquid nitrogen.[2]

In contrast to MD-GC, in comprehensive GC (GC × GC) the whole chromatogram of the first dimension (column 1) is transferred to the second dimension (column 2). The transfer between the two columns is obtained by a retention modulation device.

Nitz et al. published the first work on MD-GC-IRMS for carbon isotope ratio analysis of flavours. In this study a Siemens Sichromat double-oven MD-GC (Siemens, Karlsruhe, Germany) that employs a 'Deans switching system' for heart-cutting was used.

Juchelka et al. used a similar system for the carbon isotope ratio analysis of the synthetic enantiomers $(R)/(S)$-γ-octalactone.[6] However, apart from the fact that the Siemens system is out of production, the employed system was not suitable for the determination of hydrogen isotope ratios when a temperature-programmed column switching is necessary because no constant carrier gas flow could be obtained, which is necessary for reproducible residence times in the high-temperature conversion oven.

Sewenig et al. overcame these carrier gas flow problems by using a Multi Column Switching System MCS 2 (Gerstel, Mühlheim, Germany) for authentication of (E)-$\alpha(\beta)$-Ionone from raspberries.[7] Horii et al. used MD-GC-IRMS to obtain better resolved peaks for the analysis of PCBs and polychlorinated naphthalenes (PCNs).[5,8] In these studies an MCSS (CE Instruments, Mainz-Kastel, Germany) was used.

Due to the pre-determination of the heart-cut window the MD-GC-IRMS approach is only useful in target analysis of specific compounds that are not well separated from other compounds or matrix.

When screening of an entire sample is required, MD-GC is not applicable and a comprehensive GC (GC × GC) approach such as that shown in the right panel of Figure 7.1 has to be used. In comprehensive GC a so-called modulator is used as interface between the two GC columns. Its function is to accumulate, refocus, and rapidly release adjacent fractions of the first-dimension column (see upper part of Figure 7.3).

The modulator, commonly using cryogenics, between the first and second column allows a plug of eluting components from the first column to be trapped and continuously transferred as a narrow band onto the second column within 2–10 s, generating an entire secondary chromatogram every 2–10 s.[9] The first column is usually a conventional capillary column that separates components in the order of minutes, whereas the second column is rather short (1–3 m), of small diameter (0.1–0.15 mm) and with a thin coating (0.1–0.15 µm). Fast isothermal or nearly isothermal separations can be achieved in the order of seconds within each modulation interval.[9] The method is 'comprehensive' because all of the first column eluent is subjected to the second independent separation mechanism and the peak capacity of the two dimensions in an ideal case is multiplicative ($n_1 \times n_2$).

Tobias et al. reported the first coupling of GC × GC to an IRMS (see Figure 7.3). However, the implementation into a commercial GC-IRMS system requires instrument and software modifications: (i) low flow rates are necessary for the narrow GC columns of the second dimension for which commercial combustion reactor tubes are too wide. To overcome this problem, Tobias et al. constructed a customized microreactor.[10] (ii) Sacks et al. developed an algorithm to reconstruct isotope ratios of target compounds from the slices of several fractions supporting full width at half-maximum (fwhm) peaks of > 250 ms.[11]

Although GC × GC in combination with IRMS is sophisticated and not yet commercially available, the method is promising for better sensitivity and resolution enhancement. Tobias et al. have shown the potential of GC × GC-IRMS for the detection of synthetic testosterone misuse in urine[12] (see also Chapter 6 for further details on doping analysis).

7.3 CSIA BY MULTICOLLECTOR INDUCTIVELY COUPLED PLASMA MASS SPECTROMETRY

Multicollector inductively coupled plasma mass spectrometers (MC-ICP-MS) are generally used for the determination of isotope

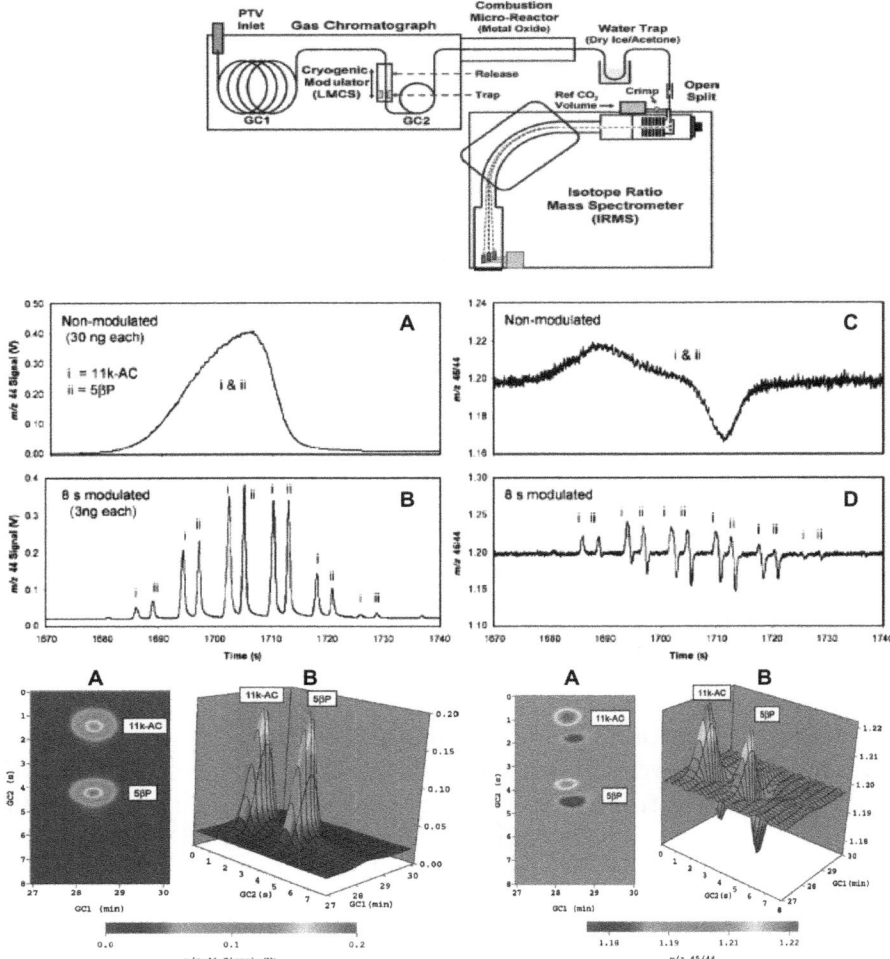

Figure 7.3 GC × GC-IRMS system. Top: schematic of the GC × GC-IRMS system used by Tobias et al.[9] Middle: GC × GC-IRMS chromatograms, where the steroids 5androstan-3α-ol-11,17-dione acetate (11k-AC) and 5-pregnan-3α-20β-diol (5βP) co-eluted in GC1, but separated in GC2. (A) nonmodulated m/z 44, (B) 8 s modulated m/z 44, (C) nonmodulated m/z 45/44, and (D) 8 s modulated m/z 45/44 chromatograms. Bottom left: an m/z 44 2D contour and 3D surface plot of baseline separated 11k-AC and 5βP. Bottom right: an m/z 45/44 ratio trace 2D contour and 3D surface plot of the same compounds.
(Graphics adapted from Tobias et al.[9] with permission of the American Chemical Society.)

abundances of metals although more recently also isotope ratio determinations of elements such as carbon,[13] sulfur,[14] chlorine and bromine have been reported.

An MC-ICP-MS consists of four main parts:

(i) a sample introduction system via which the sample is introduced into the instrument, either as liquid (LC), gas (GC), or solid (e.g., laser ablation) sample;
(ii) an inductively-coupled argon plasma in which the sample is vaporized, atomized and ionized;
(iii) an MS interface that separates the atmospheric pressure of the plasma from the vacuum of the analyser; and
(iv) a mass analyser with a multicollector array for isotope ratio determination. In the following we will discuss MC-ICP-MS use for sulfur isotope ratio analysis. In Section 7.4 various approaches for chlorine and bromine stable isotope analysis are presented that also include MC-ICP-MS. For a more thorough introduction into the field of MC-ICP-MS we refer to several reviews.[15–19]

Sulfur possesses the four stable isotopes ^{32}S, ^{33}S, ^{34}S and ^{36}S with abundances of 0.9499(26), 0.0075(2), 0.0425(24) and 0.0001(1), respectively. Sulfur isotope ratios are widely applied to studying biological and geochemical processes. Sulfur isotope ratio determination by EA-IRMS (^{34}S/^{32}S) or dual-inlet IRMS is applied routinely by conversion of sulfur-containing compounds either into SO_2 or SF_6. Amrani et al. recently reported on the first CSIA method for sulfur by hyphenation of a GC with a MC-ICP-MS (see Figure 7.4).[14]

The ionized sulfur species ^{32}S$^+$, ^{33}S$^+$ and ^{34}S$^+$ are detected in the collector cups whereas the mass of the ion ^{36}S$^+$ cannot be measured because of an interference with ^{36}Ar$^+$.

The measurement of monoatomic ions (^{32}S$^+$) rather than molecular ions (SO_2^+) has the advantage that it avoids the need for ^{18}O or ^{17}O corrections in the isotopic ratio of SO_2.[14] The main problem in measuring S$^+$ isotopes by ICP-MS is the existence of significant isobaric interferences between S and O_2 ions (i.e., ^{32}S$^+$ and ^{16}O^{16}O$^+$, ^{34}S$^+$ and ^{18}O^{16}O$^+$). The isobaric interference from O_2^+ is minimized by employing dry plasma conditions and is fully resolved at all masses using medium resolution ($m/\Delta m$) of 5000.[14] As reference gas 2% SF_6 in helium was employed. The precision of measured δ^{34}S-values is around 0.1‰ for analytes containing >40 pmol S and is still better than 0.5‰ for those containing as little as 6 pmol S. The external accuracy was

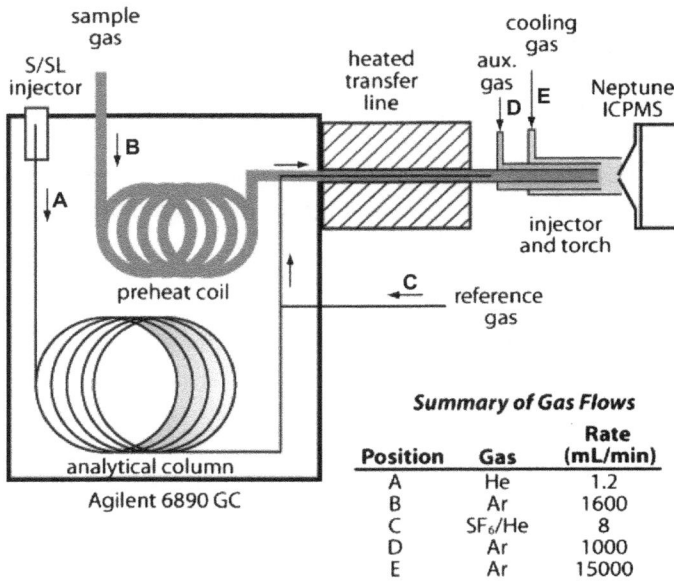

Figure 7.4 Scheme of the GC-MC-ICP-MS coupling used by Amrani et al. for CSIA of sulfur.
(Graphic from Amrani et al.[14] with permission of the American Chemical Society.)

reported to be better than 0.3‰. The integration of only the centre of chromatographic peaks, rather than the entire peak, offers a significant gain in precision and chromatographic resolution with minimal effect on the accuracy.[14]

7.4 CSIA OF CHLORINE- AND BROMINE-CONTAINING COMPOUNDS

Chlorine is the major anion in surface- and mantle-derived fluids[20] (e.g., 1.9% in seawater[21]) which possesses two stable isotopes ^{35}Cl and ^{37}Cl with abundances of 0.7576(10) and 0.2424(10), respectively.[14] $\delta^{37}Cl$-values of naturally occurring samples cover a range of about 16‰.[22] The determination of chlorine isotope ratios is of growing interest by using it as additional isotope (dual isotope plots) for the determination of sources and degradation processes of chlorinated solvents in groundwater[23] and other environmental systems[24] (see also Chapter 6). Chlorinated solvents have a range of $\delta^{37}Cl$-values between –6‰ for methyl chloride and +4.4‰ for trichloroethylene relative to SMOC.[25–28]

Comparable to chlorine, seawater is the largest natural pool of bromine on earth (6.5×10^{-3}%[21]). Bromine has two stable isotopes, ^{79}Br

and ^{81}Br, with abundances of 0.5069(7) and 0.4931(7), respectively.[14] The natural variation of δ^{81}Br has rarely been reported. Only in one study results for oil field formation waters have been reported, which range from +0.08 to +1.27‰ relative to SMOB.[21] Shouakar-Stash reported an overall δ^{81}Br range between −0.64‰ and +1.80‰.[21] Bromine is nowadays extensively used in brominated flame retardants for textiles and polymers, biocides for industrial water treatment and intermediates for pesticides and pharmaceuticals.[21] As for chlorinated compounds the distribution of such bromine-containing compounds as well as their possible degradation pathways can be traced and determined by compound-specific isotope ratio measurements. For chlorine and bromine isotopic reference materials we refer to Chapter 5.

The high potential in applying chlorine and bromine stable isotopes in environmental sciences and other fields is, however, hampered by analytical limitations.[29] The applied methods involve labour-intensive steps in which organochlorine compounds have to be separated off-line, enriched and converted into a measurable species[29–31] such as methyl chloride (CH_3Cl) for dual-inlet IRMS (DI-IRMS),[26,28,32–34] silver chloride (AgCl) for fast atom bombardment IRMS (FAB-IRMS)[35] or caesium chloride (CsCl) for thermal ionization mass spectrometry (TIMS).[36–38]

Over recent years several attempts to measure on-line chlorine isotope ratios have been reported. These attempts include:

(i) the determination of organohalogen compounds by GC-MC-ICP-MS;[39]
(ii) on-line high-temperature conversion of organic chlorine into gaseous HCl and GC-quadrupole MS;[30] and
(iii) measurement of isotopologues containing multiple chlorine substituents by GC-IRMS[40] and GC-quadrupole MS.[29,41]

(i) CSIA of Chlorine- and Bromine-containing Compounds by GC-MC-ICP-MS. The measurement of chlorine isotope ratios ($^{37}Cl/^{35}Cl$) is possible by inductively coupled plasma combined with a multicollector MS. Van Acker *et al.* presented a method for the determination of stable chlorine isotope ratios of PCE and TCE by GC-MC-ICP-MS.[39] This method is very precise (1σ = 0.06‰) and universally applicable, but it suffers from high instrument costs, low ionization efficiency to Cl^+ and interference of ArH^+ ions from the inductively coupled plasma. The required sample size for a precise isotope analysis is therefore relatively high (several μmol of Cl).[31]

Similar to the chlorine determination with GC-MC-ICP-MS, Sylva et al. determined the δ^{81}Br of brominated organic compounds (BOCs). The analysis of three brominated benzenes showed a precision of 0.3‰ for injected Br amounts of 0.3 nmol.[42]

Gelman et al. determined the bromine isotope ratio of individual organic compounds based on the simultaneous introduction of BOC and strontium as an external spike.[43] External precision as low as 0.1‰ (2σ) has been attained.[43] Carrizo et al. determined δ^{81}Br-values for six industrially synthesized and one natural BOC by GC-MC-ICP-MS. The δ^{81}Br-values for industrial BOCs ranged from 0.4 to 4.3‰, showing that the δ^{81}Br-values of the six analysed industrial compounds overlap to some extent with the δ^{81}Br-values of the natural samples.[44]

(ii) High-temperature Conversion of Organic Chlorine into Gaseous HCl and GC-quadrupole MS. Hitzfeld et al.[30] presented an on-line high-temperature conversion (HTC) method that converts organic chlorine to gaseous hydrochloric acid HCl under hydrogen gas flow according to:

$$C_xH_yCl_2 + H_2 \xrightarrow{T = 1300\,°C} xC_{(s)} + zHCl \qquad (7.2)$$

After conversion the ratio of the HCl isotopologues with m/z 38 (^{1}H^{35}Cl) and 36 (^{1}H^{37}Cl) was determined by quadrupole MS. The precision of the method is between 0.5 and 1‰ (1σ).[30] In principle, the method can also be applied to sulfur and bromine isotope ratio determination by HTC to HBr and H$_2$S isotopologues.[30]

(iii) CSIA by measurement of isotopologues containing multiple chlorine substituents by GC-IRMS and GC-quadrupole MS. Instead of species that contain only one chlorine atom such as ^{37}Cl/^{35}Cl or H^{37}Cl/H^{35}Cl, isotopologues containing multiple chlorine substituents are analysed, for example C$_2$H$_2$^{37}Cl^{35}Cl/C$_2$H$_2$^{35}Cl^{35}Cl. Evaluations either involve molecular ions,[24,29,45] fragment ions[24,40] or a weighted combination of both.[41]

Shouakar-Stash et al. devoloped a method for determining compound-specific chlorine isotope compositions for tetrachloroethene (PCE), trichloroethene (TCE), *cis*-dichloroethene (*cis*-DCE), *trans*-dichloroethene (*trans*-DCE) and 1,1-dichloroethene (1,1-DCE).[21] To that end, they used an instrument equipped with nine collector cups (see Figure 7.5).

The obtained precisions of this analytical technique were ± 0.12‰ (1σ, n = 30), ± 0.06‰ (1σ, n = 30), and ± 0.08‰ (1σ, n = 15) for PCE, TCE and DCE isomers, respectively, with limits of quantification for

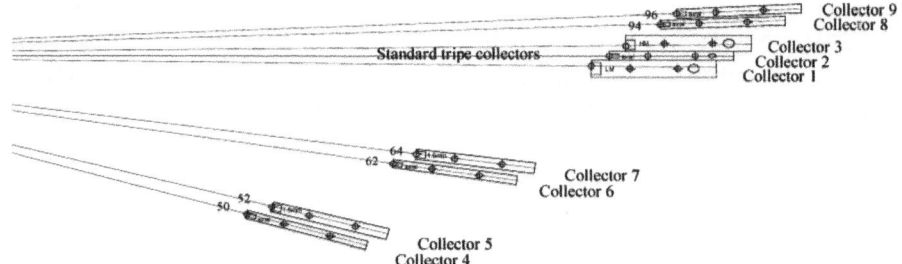

Figure 7.5 Scheme of the multicollector array arrangement (Isoprime GV Instruments) used by Shouakar-Stash et al. with indication of the m/z ratios (collectors 4–9) for the specific mass fragments used for $\delta^{37}Cl$ determination of PCE, TCE, cis-DCE, trans-DCE and 1,1-DCE.[21]
(Graphic reprinted with permission of the American Chemical Society.)

analysing chlorine isotopic composition in aqueous solutions of 5–20 µg L^{-1} per compound, which corresponds to 6–9 nmol chlorine.[21]

However, the Faraday-cup configuration in the IRMS has to be (manually) adjusted for each compound, in practice limiting this method to a narrow set of compounds.

Sakaguchi-Söder et al. and Aeppli et al. showed that the determination of isotope ratios for chlorine-containing organic compounds such as perchloroethylene (PCE), trichloroethylene (TCE), p,p′-dichlorodiphenyltrichloroethane (DDT), and pentachlorophenol (PCP) by GC-quadrupole MS with sufficient precision for characterization of degradation processes is possible.[29,41] Jin et al. investigated the method in terms of instrumental settings[45] and a validation of the method was carried out in an inter-laboratory comparison study by Bernstein et al.[24]

7.5 POSITION-SPECIFIC ISOTOPE RATIO ANALYSIS

In the case of position-specific isotope analysis (PSIA) the natural isotope ratio of an isotope at a specific atomic position within a molecule is determined (intramolecular isotope ratios).[46] Thus, position-specific measurements offer an insight into isotopic fractionation processes without dilution of the isotopic signature by non-reactive positions of the element in the whole molecule. In particular, fractionation at fundamental biochemical synthesis steps or rate-limiting steps in metabolic studies can be investigated by PSIA.[46,47] PSIA can be carried out by different techniques. One of the most prominent is site-specific natural isotope fractionation measured by nuclear magnetic resonance (SNIF-NMR). Especially, the intramolecular $^2H/^1H$ distribution of caffeine, ethanol in wine and flavour compounds has been investigated by SNIF-NMR for food authenticity control.[48–51] More recently, the $^{13}C/^{12}C$ distribution within molecules has been determined.[52–54] However, due to the fact that it has not so far been hyphenated with separation techniques it will not be discussed within this book. For a detailed discussion of SNIF-NMR we refer to the literature.[55,56]

In the same manner as compound-specific measurements were carried out before the advent of continuous-flow IRMS, single compounds were isolated and specific positions were measured by isolation of specific molecular groups and adjacent off-line determination by dual-inlet-IRMS. As an example, in 1961 Abelson and Hoering isolated individual amino acids from algal and bacterial cultures and found after decarboxylation with ninhydrin that the carboxyl groups are enriched in ^{13}C relative to the whole molecule.[57] Other works in the field are reviewed by Lichtfouse[58] and in various PSIA-related publications.[46,59] With deep respect to these early works, which show a high and sometimes forgotten level of chemical arts, we will nevertheless restrict the further discussion to continuous-flow IRMS methods.

Corso and Brenna introduced an on-line method using a tandem gas chromatograph with a pyrolysis furnace (see Figure 7.6).[60] In a first GC oven (GC I) mixtures of precursor compounds were separated and subsequently pyrolysed at around 508 °C in a pyrolysis furnace. The developed fragments passed through a heated transfer line to the second GC (GC II). Here, the fragments were cryofocused at −40 °C before separation.

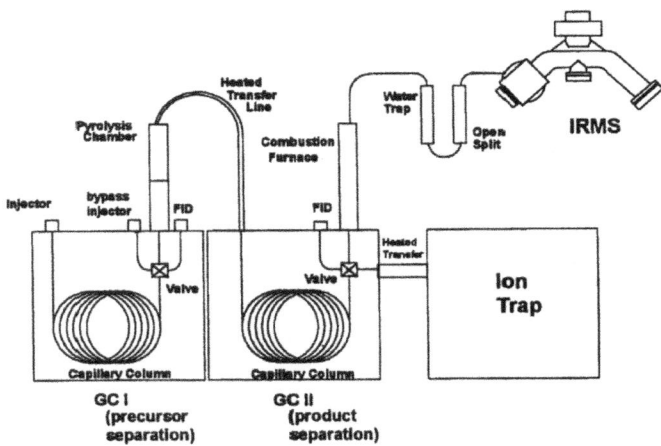

Figure 7.6 GC-Py-GCC-IRMS system for on-line PSIA.
(Graphic from Corso and Brenna[60] reprinted with permission of National Academy of Sciences of the United States of America.)

The separated fragments were converted to CO_2 in a combustion furnace, and the isotopic signature could be determined by IRMS.[60]

Several compound classes, including FAMEs, fatty alcohols,[60] leucine and methionine analogues,[46] and n-alkanes have been investigated.[60] Although these examples demonstrate the possibility of CF-IRMS of individual fragments it is not a wide-spread approach and it cannot yet be clearly decided for which problems the additional effort might be useful.

REFERENCES

1. M. Adahchour, J. Beens and U. A. T. Brinkmann, *J. Chromatogr. A*, 2008, **1186**, 67–108.
2. L. Mondello, A. C. Lewis and K. D. Bartle, *Multidimensional Chromatography*, Wiley, West Sussex, England; New York, 2002.
3. D. J. McEwen, *Anal. Chem.*, 1964, **36**, 279–282.
4. L. Mondello, P. Quinto Tranchida, P. Dugo and G. Dugo, *Mass Spectrom. Rev.*, 2008, **27**, 101–124.
5. Y. Horii, G. Petrick, M. Okada, K. Amano, T. Katase, T. Gamo and N. Yamashita, *Bunseki Kagaku*, 2005, **54**, 361–372.
6. D. Juchelka, T. Beck, U. Hener, F. Dettmar and A. Mosandl, *J. High. Resolut. Chromatogr.*, 1998, **21**, 145–151.
7. S. Sewenig, D. Bullinger, U. Hener and A. Mosandl, *J. Agric. Food Chem.*, 2005, **53**, 838–844.
8. Y. Horii, K. Kannan, G. Petrick, T. Gamo, J. Falandysz and N. Yamashita, *Environ. Sci. Technol.*, 2005, **39**, 4206–4212.
9. H. J. Tobias, G. L. Sacks, Y. Zhang and J. T. Brenna, *Anal. Chem.*, 2008, **80**, 8613–8621.
10. H. J. Tobias and J. T. Brenna, *Microfluid. Nanofluid.*, 2010, **9**, 461–470.
11. G. L. Sacks, Y. Zhang and J. T. Brenna, *Anal. Chem.*, 2007, **79**, 6348–6358.
12. H. J. Tobias, Y. Zhang, R. J. Auchus and J. T. Brenna, *Anal. Chem.*, 2011, **83**, 7158–7165.
13. R. Santamaria-Fernandez, D. Carter and R. Hearn, *Anal. Chem.*, 2008, **80**, 5963–5969.
14. A. Amrani, A. Sessions and J. Adkins, *Anal. Chem.*, 2009, **81**, 9027–9034.
15. M. Moldovan, E. M. Krupp, A. E. Holliday and O. F. X. Donard, *J. Anal. At. Spectrom.*, 2004, **19**, 815–822.
16. M. E. Wieser and J. B. Schwieters, *Int. J. Mass Spectrom.*, 2005, **242**, 97–115.
17. F. Vanhaecke, L. Balcaen and D. Malinovsky, *J. Anal. At. Spectrom.*, 2009, **24**, 863–886.

18. C. B. Douthitt, *Anal. Bioanal. Chem.*, 2008, **390**, 437–440.
19. C. B. Douthitt, *J. Anal. At. Spectrom.*, 2008, **23**, 685–689.
20. J. Hoefs, *Stable Isotope Geochemistry*, 6th edn., Springer, Berlin, 2009.
21. O. Shouakar-Stash, S. K. Frape and R. J. Drimmie, *Anal. Chem.*, 2005, **77**, 4027–4033.
22. T. B. Coplen, J. K. Böhlke, P. De Bièvre, T. Ding, N. E. Holden, J. A. Hopple, H. R. Krouse, A. Lamberty, H. S. Peiser, K. Révész, S. E. Rieder, K. J. R. Rosman, E. Roth, P. D. P. Taylor, R. D. Vocke Jr and Y. K. Xiao, *Pure Appl. Chem.*, 2002, **74**, 1987–2017.
23. D. Hunkeler, B. M. Van Breukelen and M. Elsner, *Environ. Sci. Technol.*, 2009, **43**, 6750–6756.
24. A. Bernstein, O. Shouakar-Stash, K. Ebert, C. Laskov, D. Hunkeler, S. Jeannottat, K. Sakaguchi-Söder, J. Laaks, M. A. Jochmann, S. Cretnik, J. Jager, S. B. Haderlein, T. C. Schmidt, R. Aravena and M. Elsner, *Anal. Chem.*, 2011, **83**, 7624–7634.
25. E. M. vanWarmerdam, S. K. Frape, R. Aravena, R. J. Drimmie, H. Flatt and J. A. Cherry, *Appl. Geochem.*, 1995, **10**, 547–552.
26. N. C. Sturchio, J. L. Clausen, L. J. Heraty, L. Huang, B. D. Holt and T. A. Abrajano, *Environ. Sci. Technol.*, 1998, **32**, 3037–3042.
27. K. M. Beneteau, R. Aravena and S. K. Frape, *Org. Geochem.*, 1999, **30**, 739–753.
28. O. Shouakar-Stash, S. K. Frape and R. J. Drimmie, *Journal of Contaminant Hydrology*, 2003, **60**, 211–228.
29. C. Aeppli, H. Holmstrand, P. Andersson and O. Gustafsson, *Anal. Chem.*, 2010, **82**, 420–426.
30. K. L. Hitzfeld, M. Gehre and H. H. Richnow, *Rapid Commun. Mass Spectrom.*, 2011, **25**, 3114–3122.
31. M. Elsner, M. A. Jochmann, T. B. Hofstetter, D. Hunkeler, A. Bernstein, T. C. Schmidt and A. Schimmelmann, *Anal. Bioanal. Chem.*, 2012, 1–21.
32. J. W. Hill and A. Fry, *J. Am. Chem. Soc.*, 1962, **84**, 2763–2769.
33. R. Kaufmann, A. Long, H. Bentley and S. Davis, *Nature*, 1984, **309**, 338–340.
34. B. D. Holt, N. C. Sturchio, T. A. Abrajano and L. J. Heraty, *Anal. Chem.*, 1997, **69**, 2727–2733.
35. K. C. Westaway, T. Koerner, Y. R. Fang, J. Rudzinski and P. Paneth, *Anal. Chem.*, 1998, **70**, 3548–3552.
36. M. Numata, N. Nakamura, H. Koshikawa and Y. Terashima, *Environ. Sci. Technol.*, 2002, **36**, 4389–4394.
37. H. Holmstrand, P. Andersson and Ö. Gustafsson, *Anal. Chem.*, 2004, **76**, 2336–2342.

38. M. Mandalakis, H. Holmstrand, P. Andersson and Ö. Gustafsson, *Chemosphere*, 2008, **71**, 299–305.
39. M. Van Acker, A. Shahar, E. D. Young and M. L. Coleman, *Anal. Chem.*, 2006, **78**, 4663–4667.
40. O. Shouakar-Stash, R. J. Drimmie, M. Zhang and S. K. Frape, *Appl. Geochem.*, 2006, **21**, 766–781.
41. K. Sakaguchi-Söder, J. Jäger, H. Grund, F. Matthaus and C. Schüth, *Rapid Commun. Mass Spectrom.*, 2007, **21**, 3077–3084.
42. S. P. Sylva, L. Ball, R. K. Nelson and C. M. Reddy, *Rapid Commun. Mass Spectrom.*, 2007, **21**, 3301–3305.
43. F. Gelmar and L. Halicz, *Int. J. Mass Spectrom.*, 2010, **289**, 167–169.
44. D. Carrizo, M. Unger, H. Holmstrand, P. Andersson, Ö. Gustafsson, S. P. Sylva and C. M. Reddy, *Environ. Chem.*, 2011, **8**, 127–132.
45. B. Jin, C. Laskov, M. Rolle and S. B. Haderlein, *Environ. Sci. Technol.*, 2011, **45**, 5279–5286.
46. G. L. Sacks and J. T. Brenna, *Anal. Chem.*, 2003, **75**, 5495–5503.
47. J. T. Brenna, *Rapid Commun. Mass Spectrom.*, 2001, **15**, 1252–1262.
48. N. Ogrinc, I. J. Košir, J. E. Spangenberg and J. Kidrič, *Anal. Bioanal. Chem.*, 2003, **376**, 424–430.
49. P. Lesot, Z. Serhan and I. Billault, *Anal. Bioanal. Chem.*, 2011, **399**, 1187–1200.
50. C. Aghemo, A. Albertino, R. Gobetto and F. Spanna, *J. Sci. Food Agric.*, 2011, **91**, 2088–2094.
51. W. C. Ko, J. Y. Cheng, P. Y. Chen and C. W. Hsieh, *Food Bioprocess Technol.*, 2012, 1–5.
52. F. Thomas and E. Jamin, *Anal. Chim. Acta*, 2009, **649**, 98–105.
53. F. Thomas, C. Randet, A. Gilbert, V. Silvestre, E. Jamin, S. Akoka, G. Remaud, N. Segebarth and C. Guillou, *J. Agric. Food Chem.*, 2010, **58**, 11580–11585.
54. C. Thibaudeau, G. Remaud, V. Silvestre and S. Akoka, *Anal. Chem.*, 2010, **82**, 5582–5590.
55. M. L. Martin and G. J. Martin, *Analusis*, 1999, **27**, 209–212.
56. C. Cordella, I. Moussa, A. C. Martel, N. Sbirrazzouli and L. Lizzani-Cuvelier, *J. Agric. Food Chem.*, 2002, **50**, 1751–1764.
57. P. H. Abelson and T. C. Hoering, *P. Natl. Acad. Sci. USA*, 1961, **47**, 623–632.
58. E. Lichtfouse, *Rapid Commun. Mass Spectrom.*, 2000, **14**, 1337–1344.
59. J. T. Brenna, T. N. Corso, H. J. Tobias and R. J. Caimi, *Mass Spectrom. Rev.*, 1997, **16**, 227–258.
60. T. N. Corso and J. T. Brenna, *P. Natl. Acad. Sci. USA*, 1997, **94**, 1049–1053.

Appendix

PREFIXES

Factor	Prefix	Symbol
10^{15}	peta	P
10^{12}	tera	T
10^{9}	giga	G
10^{6}	mega	M
10^{3}	kilo	k
10^{2}	hecto	h
10	deca	da
10^{-1}	deci	d
10^{-2}	centi	c
10^{-3}	milli	m
10^{-6}	micro	µ
10^{-9}	nano	n
10^{-12}	pico	p
10^{-15}	femto	f

SI-UNITS USED IN THIS BOOK

Quantity	Symbol	SI-unit	Symbol
Length	l	metre	m
Mass	m	kilogram	kg
Time	t	second	s
Electric current	I	ampere	A
Thermodynamic temperature	T	kelvin	K
Amount of substance	n	mole	mol

Compound-specific Stable Isotope Analysis
By Maik A. Jochmann and Torsten C. Schmidt
© The Royal Society of Chemistry 2012
Published by the Royal Society of Chemistry, www.rsc.org

Appendix

PHYSICAL CONSTANTS

Constant	Symbol	Value[a]
Planck constant	h	$6.62606957(29) \times 10^{-34}$ J s
Elementary charge	e	$1.602176565 \times 10^{-19}$ C
Neutron mass	m_n	$1.674927351(74) \times 10^{-27}$ kg
Electron rest mass	m_e	$9.10938291(40) \times 10^{-31}$ kg
Avogadro's constant	N_A	$6.02214129(27) \times 10^{23}$ mol^{-1}
Atomic mass unit (amu)	u	$1.660538921(73) \times 10^{-27}$ kg
Proton mass	m_p	$1.672621777(74) \times 10^{-27}$ kg
Rydberg constant	$\Re$	$10973731.568539(55)$ m^{-1}
Boltzmann's constant	k_B	$1.3806488(13) \times 10^{-23}$ J K^{-1}
Pi	π	3.14159
Universal gas constant	R	$8.3144621(75)$ J mol^{-1} K^{-1}

[a]Values according to NIST.[1]

USEFUL CONVERSIONS IN THE LAB

Pressure

1 mbar = 0.75 Torr
1 Torr = 1.33 mbar
1 bar = 14.5 PSI
1 PSI = 0.068 bar

Flow

$1 \text{ mL min}^{-1} = 0.06 \text{ L h}^{-1} = 1.44 \text{ L d}^{-1}$

Time

500 s = 8.3 min

Capillary Diameters ⌀

$1/64'' = 0.04$ mm
$1/32'' = 0.79$ mm
$1/16'' = 1.57$ mm
$1/8'' = 3.18$ mm
$3/16'' = 4.78$ mm
$1/4'' = 6.35$ mm

Energy

$1\,\text{eV} = 1.6002 \times 10^{-19}\,\text{J}$

Mesh Size Conversions

Mesh	mm	Mesh	mm
20	0.84	100	0.149
25	0.71	120	0.125
30	0.59	140	0.105
35	0.50	170	0.088
40	0.42	200	0.074
45	0.35	230	0.062
50	0.30	270	0.053
60	0.25	325	0.045
70	0.21	400	0.038
80	0.177		

IMPORTANT IONS

Ions and their Corresponding m/z

Mass to charge ratio (m/z)	Ion	Mass to charge ratio (m/z)	Ion
2	$^1H^1H^+$, $^4He^{2+}$	26	$^{12}C_2{}^1H_2{}^+$
3	$^2H^1H^+$	28	$^{14}N^{14}N^+$, $^{12}C^{16}O^+$
4	$^4He^+$	29	$^{15}N^{14}N^+$, $^{13}C^{16}O^+$
12	$^{12}C^+$	30	$^{14}N^{16}O^+$, $^{12}C^{18}O^+$
14	$^{14}N^+$	32	$^{16}O^{16}O^+$
15	$^{12}C^1H_3{}^+$	40	$^{40}Ar^+$, $^{12}C_3{}^1H_4{}^+$
16	$^{16}O^+$	43	$C_3H_7{}^+$
17	$^{16}O^1H^+$	44	$^{12}C^{16}O_2{}^+$
18	$^1H_2{}^{16}O^+$	57	$C_4H_9{}^+$
22	$^{12}C^{16}O_2{}^{2+}$		

REFERENCE

1. Physical Measurement Laboratory of NIST; *http://physics.nist.gov/cuu/index.html*.

Subject Index

accelerated solvent extraction 162
accuracy 185–7
acetaminophen 261
additive correction factor 203
adipocere 264
AKIEs *see* apparent kinetic isotope effects
alcoholic drinks 249–53
   apple-derived 251, 252
n-alkanes 162, 163, 164
   source apportionment 290–2
α-particles 15
amino acid isotope analysis 269–72
anabolic steroids, structure 273
ANSCA (automatic nitrogen, sulfur and carbon analyser) 91
apparent kinetic isotope effects (AKIEs) 303–7
apple-derived alcoholic drinks 251, 252
archaeology 264–73
   amino acid isotope analysis 269–72
   bog butters 267–8
   paleodiet studies 266–9
aromas 240–9
Aston's isotope rule 21
atomic model 14–19
atomic number 16

atrazine 310
automatic nitrogen analyser 91
Avogadro's constant 22, 37, 60, 216

backflush systems 95–8
Becker, Herbert 15
Becquerel, Henri 14
beer 249–53
benzene 309
bergamot 246
binary-encounter Bethe model 59
bit noise 76
bog butters 267–8
Bohr, Nils 14
Boltzmann constant 36, 53, 76
Bothe, Walter 15
bromine
   isotope ratio 51
   referencing materials 199–200
bromine-containing compounds 356–9
BSIA *see* bulk stable isotope ratio analysis
bulk stable isotope ratio analysis (BSIA) 4, 5–7

cactus pear fruit 246
caffeine 261
calvados 251

Calvin cycle 235
Camelina sativa oil 257
camellia seed oil 257
carbon 233
  fixation in
    photosynthesis 235–40
  ion corrections 78–81
  isotope abundance 24
  isotope ratios 28, 94–106, 189
    extraterrestrial matter 321–5
    isobaric interference 103–5
  referencing materials 192, 194–7
carbon dioxide
  alcoholic beverages 249
  fixation 235–40
    Calvin cycle 235–6
    Crassulacean acid metabolism 237–9
    Hatch-Slack cycle 236–7
    plant classification 239
carbonaceous chondrites 317–19
carrier gas 129–33
certified reference materials 164
Chadwick, James 16
changeover valve 84
chemical reaction interface 136–7
chlorinated hydrocarbons
  groundwater contamination 315
  source apportionment 292–3
  source values 282
chlorine
  isotope abundance 24
  isotope ratio 51
  referencing materials 199–200
chlorine-containing compounds 356–9
cholesterol 272–3
chromatography *see* GC-IRMS; LC-IRMS
cider 251, 252
cinnamaldehyde 242, 245
cinnamon oils 242, 245
clean-up
  semi-volatile compounds 162–4
  volatile compounds 159
cocoa 258–9
coffee 258–9
collagen 269
combustion ovens 98–103
Commission on Isotopic Abundance and Atomic Weights (CIAAW) 19–20
compound-specific isotope analysis *see* CSIA
comprehensive GC (GC×GC) 352–3, 354
continuous-flow IRMS 83, 86
  peripheral devices 91–3
coriander 246, 248
Craig correction 79–81
Crassulacean acid metabolism 237–9
CSIA 4, 5–7
  applications 230–348
    archaeology 264–73
    doping control 273–80
    environmental science 280–317
    extraterrestrial matter 317–26
    food authenticity 234–59
    forensics 259–64
  GC-IRMS *see* GC-IRMS
  historical development 7–9
  quality assurance/control 218–21
  referencing in 204–11
  sample preparation 155–84
  scope of 230–4
  see also IRMS; and specific techniques
cutting agents 261

DDT 359
Deans switch 350
decarboxylation 175
degradation studies
 dual isotope
  correlations 308–10
 quantification 298–303
delta-scale 27–30
derivatization 168–80
 decarboxylation 175
 esterification 177–8
 methylation 178–80
 silylation 175–7
 volatilization
  by oxidation 174–5
  by reduction 174
detection limits 213, 215–18, 219
deuterium detection 72
diacetylmorphine see heroin
dichloroethene 309, 358
diffusion 37, 39
doping control 273–80
 future developments 279–80
 steroid analysis 277–9
dual isotope correlations 308–10
dual viscous flow inlet system 2–3
dual-inlet systems 83–6

ecstasy 260, 261, 262, 263
efficiency of IRMS 216
effusion 37, 38
Einzel lenses 62
electromagnets 64–70
electron multipliers 72
electrons 14
electrostatic filters 114–15
elemental analyser-IRMS (EA-IRMS) 91–3
endogenous reference
 steroids 274–5

environmental science 280–317
 isotope fractionation 295–317
  environmental technical
   processes 313–17
  physical processes 295–7
  transformation/
   biodegradation
   processes 297–313
 source apportionment 283–95
  n-alkanes 290–2
  chlorinated
   hydrocarbons 292–3
  future developments 294–5
  polycyclic aromatic
   hydrocarbons 283–90
 source values 280–3
epitestosterone 274
 testosterone/epitestosterone
  ratio 276, 277
equilibrium isotope effects 30, 32
equilibrium stable isotope
 fractionation 30
errors 185–7
essential oils 240–9
esterification 177–8
ethanol 251
 see also alcoholic drinks
ethyl tert-butyl ether (ETBE) 309
Eurachem Guide 156
excipients 261
expansion factor 203
extraction
 semi-volatile compounds 161–2
 volatile compounds 159
extraterrestrial matter 317–26
 isotope ratios 319–26
  carbon 321–5
  hydrogen 319–21
  nitrogen 321
 meteorites 317–19

FAMEs *see* fatty acid methyl esters
Faraday cups 1, 2, 64, 70, 72, 74
fatty acid methyl esters (FAMEs) 168, 179
  archaeological studies 266
  forensic studies 264
  in natural stimulants 258–9
  in oils 256
feedback resistor values 74
Fick's law 38
flavours 240–9
flax seed oil 257
flow injection analysis (FIA-IRMS) 9, 210
food authenticity 234–59
  alcoholic drinks 249–53
  aromas, flavours and essential oils 240–9
  carbon fixation in photosynthesis 235–40
  cocoa, coffee and tea 258–9
  honey 253–4
  juices 254–6
  oils 256–8
forensics 259–64
free fatty acid phases 168
fruit juices 254–6
fused silica capillary 116
fusel alcohols 251

γ-butyrolactone (GBL) 262
γ-hydroxybutyric acid (GHB) 262
gas chromatography *see* GC
gas diffusion cell (GDC) 136
gas-liquid chromatography (GLC) 116
gas-solid chromatography (GSC) 116
gasoline components, source values 282–3
gauges 55–6

GC-IRMS 4, 8–9, 93–134
  carbon/nitrogen isotope ratio analysis 93–106
    backflush system 95–8
    combustion oven 98–103
    isobaric interferences 103–6
    reduction oven 103
  carrier gas, gas flow and columns 129–33
  holdup time 119
  multidetection 133–4
  peak detection 127–9
  relative atomic mass 21–2
  retention time 117, 120
  separation 116–27
  stationary phases 131–2
GC-MC-ICP-MS 8
  chlorine- and bromine-containing compounds 357–8
GC-quadrupole-MS 8, 50–1, 326
Geobacter metallireducens 304
Global Network of Isotopes in Precipitation (GNIP) 231–2
glycerol 251, 253
Graham's law of effusion 37
groundwater contaminants 221, 315, 316

Hagen-Poiseuille law 87
hair, amino acids in 272
halogenated compounds
  source values 282
  see also bromine; chlorine
Harkins, William D. 15
Hatch-Slack cycle 236–7
headspace analysis 159
height equivalent of theoretical plate (HETP) 121
Heisenberg, Werner 14
helium, effect on hydrogen isotope ratio 111
heroin 260, 261

Subject Index

high temperature, high
  performance liquid
  chromatography
  (HT-HPLC) 116
honey 253–4, 255
horse milk analysis 267
hydrogen 231–2, 233
  detection 72
  ion corrections 82–3
  isotope abundance 24
  isotope ratios 28, 106–15, 189
    extraterrestrial
      matter 319–21
    referencing materials 190–1,
      192–3

inlet systems 83–91
  continuous-flow 86
  dual-inlet 83–6
  open-split 86–9
  reference gas inlet 89–91
instrumentation 2–3, 50–154
  see also specific instruments
integer mass number 16
internal isotopic standard
  (i-IST) 241–2
International Atomic Energy
  Agency (IAEA) 188
International Union of Pure and
  Applied Chemistry (IUPAC) 21
International Union of Pure and
  Applied Physics (IUPAP) 21
interplanetary dust particles
  317
inverse isotope effect 125
ion collection 70–7
ion correction 77–83
  carbon 78–81
  hydrogen 82–3
  nitrogen 81–2
ion separation 63–70
ion sources 56–63

IRMS 1, 2–3, 50–115
  continuous-flow 83, 86
    peripheral devices 91–3
  Faraday cups 1, 2, 64, 70,
    72, 74
  GC-IRMS see GC-IRMS
  inlet systems 83–91
    continuous-flow 86
    dual-inlet 83–6
    open-split 86–9
    reference gas inlet 89–91
  ion collection and signal
    pathway 70–7
  ion corrections 77–83
  ion separation in magnet sector
    field 63–70
  ion source and focusing beam
    optics 56–63
  LC-IRMS see LC-IRMS
  single ion monitoring 51
  total ion current 51
  vacuum systems 53–6
    gauges 55–6
    molecular flow 54
    pumps 54–5
    transition (Knudsen) flow 54
  see also CSIA
isobaric interferences
  carbon 103–5
  nitrogen 106
isobars 16, 18
IsoPrime™ mass
  spectrometer 107, 114–15
isoproturon 310, 312–13, 314
isotones 18
isotope abundance 19–21, 24–30
  variations in 25–6
isotope fractionation 30–5
  during transport processes 36–9
  environmental science 295–317
    environmental technical
      processes 313–17

isotope fractionation (*continued*)
    physical processes 295–7
    transformation/
        biodegradation
        processes 297–313
    irreversible reaction
        closed system 42–5
        open system 45–6
    reversible reaction
        closed system 40–2
        open system 45
isotope mass balance 35–6
isotope radio mass spectrometry *see* IRMS
isotope ratio-monitoring GC-MS (irm-GCMS) 6
isotope ratios 26, 60
    bromine 51
    carbon 28, 94–106, 189
        extraterrestrial matter 321–5
        isobaric interference 103–5
    chlorine 51
    hydrogen 28, 106–15, 189
        extraterrestrial matter 319–21
    international scales 189
    linearity 61
    nitrogen 28, 94–106
        extraterrestrial matter 321
        isobaric interference 106
    normalization 200–4
        with reference material 202
        relative to true isotopic composition 200–2
        two-point 202–4
    oxygen 28, 115, 189
    sulfur 28, 189, 355
isotope-amount fraction 24–5
isotopic difference 32–5
isotopic fractionation factor 32–5
isotopocules 22–3

isotopologues 22–3, 30, 32, 37–9, 45, 60, 72–4, 77–82, 84, 103, 125–6, 245, 297, 298, 306, 357–9
isotopomers 22–3

Johnson noise 76
Joliot-Curie, Frédéric 15
Joliot-Curie, Irene 15
juices 254–6

KIEs *see* kinetic isotope effects
kinetic isotope effects (KIEs) 30, 32, 34, 172–3, 177, 231, 233, 270, 308
    apparent (AKIEs) 303–7
Knudsen flow 54
Knudsen number 53

large volume injection 167
laser ablation prior to GCIRMS (LA-GC-IRMS) 5
lavender 244, 246
LC-IRMS 4, 134–46
    interfaces 135–6
        chemical reaction interface 136–7
        moving belt/moving wire interface 137–9
        wet chemical combustion interface 139–41
    membrane separation unit 142–4
    oxidation reactor 141–2
    separation columns 144–6
LC-IsoLink™ 139
lemon 246
limits of detection for volatile compounds 159–61
linearity 213, 214
Liouville's theorem 62
liquid chromatography *see* LC
liquid injection 165–7

liquid-liquid chromatography (LLC) 116
liquid-liquid extraction 161
LiquiFace™ 139
lithium carbonate L-SVEC 195
Lorentz force 64

maize oil 257
mandarin 242, 246
mass spectrometry, isotope ratio *see* IRMS
mass-to-charge ratios *see* ion correction
MDMA 260, 261, 262, 263
membrane inlet devices 5
membrane permeation GC isotope ratio mass spectrometry (MP-GC-IRMS) 5
membrane separation unit 142–4
meteorites *see* extraterrestrial matter
method detection limit 218, 219
methyl *tert*-butyl ether (MTBE) 309, 310–12
molecular flow 54
morphine 260
Moseley, Henry G. J. 15
moving belt/moving wire interface 137–9
moving capillary stream switch 350, 352
MS *see* mass spectrometry
multi-point normalization 204
multicollector inductively coupled plasma mass spectrometry (MC-ICP-MS) 2, 353–6
multidetection 133–4
multidimensional GC (MD-GC) 349–52
myrcene 242

Nafion™ membranes 94–106, 108
National Institute of Standards and Technology (NIST) 188
neutrons 16
Nichrome 107
Nier, Alfred 2, 3, 21
nitrobenzene 309
nitrogen 231
  ion corrections 81–2
  isotope abundance 24
  isotope ratios 28, 94–106
    extraterrestrial matter 321
    isobaric interference 106
  referencing materials 197–8
normal phase chromatography (NPC) 116
normalization 200–4
  in CSIA 209–10
  referencing 202
  relative to true isotopic composition 200–2
  standard conversion identity 201
  two-point 202–4
nuclear drop model 21
nuclides 14–19

oils 256–8
Ölander, Arne 21
olive oil 257
open-split inlet systems 86–9
orbitals 14
organic residue analysis 266
oxidative volatilization 174–5
oxidation reactor 141–2
oxygen 231–2, 233
  isotope abundance 24
  isotope ratio 28, 115, 189
  referencing materials 190–1

paleodiet studies 266–9
palladium filter system 113

palmitic acid 259, 264, 265, 267, 268
partitioning chromatography 117
peak detection 127–9
peak shape scan 70
Penning vacuum gauges 55–6
pentachlorophenol 359
perchloroethylene 359
perilla seed oil 257
permeable reaction barriers (PRBs) 315, 316
phosphoenolpyruvic acid 236
photosynthesis, carbon fixation 235–40
pineapple 247
Pirani guages 55
Pitman estimators 45
plants
    carbon isotope composition in human diet 239–40
    photosynthesis 235–40
        C3 (Calvin cycle) 236, 237, 239
        C4 (Hatch-Slack cycle) 236, 237, 239
        Crassulacean acid metabolism 237–9
polycyclic aromatic hydrocarbons (PACs) 162, 163, 164
    source apportionment 283–90
poppy seed oil 257
porous layer open tubular columns (PLOT) 116
position-specific isotope analysis (PSIA) 4, 174, 359–61
precision 185–7
preconcentration
    semi-volatile compounds 167–8
    volatile compounds 159–61
protonation 58

protons 14, 15
Proud, William 18
Prunus fruits 247
PSIA *see* position-specific isotope analysis
pumps 54–5

quality assurance/control 218, 220–1
quantization error 76

rape seed oil 257
raspberry 247
Rayleigh equation 44–5, 298–303
    dilution factor 302
    limitations of 302–3
reductive volatilization 174
reduction ovens 103
reference gas inlet 89–91
referencing 187–200
    carbon 192, 194–7
    chlorine and bromine 199–200
    in CSIA 204–11
    hydrogen and oxygen 190–1, 192–3
    nitrogen 197–8
    normalization using 202
    sulfur 198–9
repeatability 211–12
reproducibility 212–13
retention volume 117, 120
reversed phase chromatography (RPC) 116
ribulose 1,5-bisphosphate carboxylase oxygenase (RuBisCO) 235
ribulose 1,5-bisphosphate (RuBP) 235
rotary vane oil pumps 54
Rutherford, Ernest 15

Subject Index

safrole 261, 262
sample preparation 155–84
  derivatization 168–80
    decarboxylation 175
    esterification 177–8
    methylation 178–80
    oxidative volatilization 174–5
    reductive volatilization 174
    silylation 175–7
  preservation and storage 155–7
  processing 157–68
    semi-volatile compounds 162–8
    volatile compounds 159–62
sampling frequency/rate 76
sampling strategy 155–7
Schrödinger, Erwin 14
Scotch whisky 252, 254
selected-ion monitoring (SIM) 6
semi-permeable membrane devices 161–2
semi-volatile compounds 162–8
  clean-up 162–4
  liquid injection 165–7
  preconcentration 167–8
  validation of preparation 164–5
separation columns
  GC-IRMS 129–33
  LC-IRMS 144–6
shot-noise limit 75–6, 215
silylation 175–7
site-specific natural isotope fractionation-nuclear magnetic resonance spectroscopy (SNIF-NMR) 9, 359–60
SNIF-NMR *see* site-specific natural isotope fractionation-nuclear magnetic resonance spectroscopy
Soddy, Frederick 18
solid-phase extraction 161
solid-phase microextraction (SPME) 253
source apportionment 283–95
  n-alkanes 290–2
  chlorinated hydrocarbons 292–3
  future developments 294–5
  polycyclic aromatic hydrocarbons 283–90
source values 280–3
spirits 249–53
stability 211–18
stable isotopes 19–23
stable isotope analysis
  applications 1
  development of 1–2
  instrumentation 2–3, 50–154
  spectroscopic methods 9
  techniques 3–5
stable isotope ratio *see* isotope ratio
stainless steel capillary 116
standard conversion identity 201
Standard Light Antarctic Precipitation (SLAP) 188
Standard Mean Ocean Bromide (SMOB) 199
Standard Mean Ocean Chloride (SMOC) 189, 199
Standard Mean Ocean Water (SMOW) 188, 190
stationary phases 131–2
stearic acid 259, 264, 265, 267, 268
steroids 273–80
  anabolic 273–4
  endogenous reference 274–5
  isotope analysis 277–8
  synthetic 274
stigmatic focusing 67
storage of samples 155–7
Streitwieser limits 306–7

sulfur
  isotope abundance 24
  isotope ratios 28, 189, 355
  referencing materials 198–9

tea 258–9
tequila 252
tert-amyl methyl ether
  (TAME) 309
testosterone 275
testosterone/epitestosterone
  ratio 276, 277
tetrachloroethene 358
thermal ionization mass
  spectrometry (TIMS) 2
thermospray nebulizer (TSN) 136
Thomson, Joseph J. 18
titanium oxide 116
toluene 309
total organic carbon analysers
  139
transformation reactions, in-situ
  proof 298
transition (Knudsen) flow 54
triacylglycerols 266, 268
trichloroethene 358
trichloroethylene 359
trifluoroacetate/isopropanol
  (TFA/IP) 177, 269
trimethylarsine 174
1,3,5-trinitroperhydro-1,3,5-
  triazine 310
triolein 162
TS-Limestone 194
turbomolecular pumps 54–5
two-point normalization 202–4

uncertainty 185–7

vacuum systems 53–6
  gauges 55–6
  molecular flow 54
  pumps 54–5
  transition (Knudsen) flow 54
vanilla 243, 247, 248–9, 250
vapour-liquid isotope
  fractionation 295
Vienna Canõn Diabolo Troilite
  (VCDT) 189
Vienna Pee Dee Belemnite
  (VPDB) 27, 80, 189
Vienna Standard Mean
  Ocean Water (VSMOW)
  188, 191
vinyl chloride 309, 315
viscous flow 54
volatile compounds 156, 159–62
  extraction and clean-up 159
  preconcentration 159–61
volatilization
  by oxidation 174–5
  by reduction 174

wall coated open tubular column
  (WCOT) 116
wet chemical combustion
  interface 139–41
whole number rule 18
wine 249–53
World Anti-Doping Agency
  (WADA) 273

zirconium oxide 116